HUMAN PERFORMANCE ON THE

Human Performance on the Flight Deck

DON HARRIS
HFI Solutions Ltd, UK

CRC Press
Taylor & Francis Group
Boca Raton London New York

CRC Press is an imprint of the
Taylor & Francis Group, an **informa** business

CRC Press
Taylor & Francis Group
6000 Broken Sound Parkway NW, Suite 300
Boca Raton, FL 33487-2742

© 2011 by Don Harris
CRC Press is an imprint of Taylor & Francis Group, an Informa business

No claim to original U.S. Government works

Printed on acid-free paper
Version Date: 20160226

International Standard Book Number-13: 978-1-4094-2339-3 (Hardback) 978-1-4094-2338-6 (Paperback)

Visit the Taylor & Francis Web site at
http://www.taylorandfrancis.com

and the CRC Press Web site at
http://www.crcpress.com

Contents

PART FOUR
THE MANAGEMENT

List of Figures

List of Tables

Permissions and Acknowledgements

Figure 1.1   From: Harris, D. and Smith, F.J. (1997). What Can Be Done Versus What Should Be Done: A Critical Evaluation of the Transfer of Human Engineering Solutions between Application Domains. In: D. Harris (ed.) *Engineering Psychology and Cognitive Ergonomics* (Volume 1) (pp. 339–46). Aldershot, UK: Ashgate.

Figure 2.1   Adapted from: Wickens, C.D., Lee, J.D., Liu, Y. and Gordon Becker, S.E. (2004). *An Introduction to Human Factors Engineering* (2nd Edition). Upper Saddle River, NJ: Pearson Education; and Baddeley, A.D. (2000). The Episodic Buffer: A New Component of Working Memory? *Trends in Cognitive Sciences*, 4, 417–23.

Figure 3.3   Adapted from: Roscoe, A.H. (1984). Assessing Pilot Workload in Flight. Flight Test Techniques. In, *Proceedings of the Advisory Group for Aerospace Research and Development (Conference Proceedings No. 373): Flight Test Techniques.* Neuilly-sur-Seine: AGARD/NATO.

Figure 3.4   Adapted from: Hart, S.G. and Staveland, L.E. (1988). Development of NASA–TLX (Task Load Index): Results of Empirical and Theoretical Research. In: P.A. Hancock and N. Meshkati (eds) *Human Mental Workload* (pp. 139–83). Amsterdam: North-Holland.

Figure 3.5   From: Young, M.S. and Stanton, N.A. (2001). Size Matters: The Role of Attentional Capacity in Explaining the Effects of Mental Underload on Performance. In: D. Harris (ed.) *Engineering Psychology and Cognitive Ergonomics* (Volume 5) (pp. 357–64). Aldershot, UK: Ashgate.

Figure 4.1   Adapted from: Endsley, M.R. (1995). Toward a Theory of Situation Awareness in Dynamic Systems. *Human Factors*, 37, 32–64.

Figure 5.1   Adapted from: Orasanu, J. (1993). Decision Making in the Cockpit. In, E.L. Wiener, B.G. Kanki and R.L. Helmreich (eds) *Cockpit Resource Management* (pp. 137–72). San Diego CA: Academic Press.

Figure 5.2   From: Jenkins, D.P., Stanton, N.A., Salmon, P.M. and Walker, G.H. (2008). *Decision Making Training for Synthetic Environments: Using the Decision Ladder to Extract Specifications for Synthetic Environments Design and Evaluation.* (HFIDTC/2/WP4.6.2/2). Yeovil: Aerosystems International/HFI-DTC.

Figure 6.1   From: Reason, J.T. (1990). *Human Error.* Cambridge: Cambridge University Press.

Figure 6.2   From: Helmreich, R.L., Klinect, J.R. and Wilhelm, J.A. (1999). Models of Threat, Error, and CRM in Flight Operations. In: R.S. Jensen (ed.) *Proceedings of the Tenth International Symposium on Aviation Psychology* (pp. 677–82). Columbus, OH: The Ohio State University.

Figure 9.1   Photograph courtesy of Peter Jorna.

	Performance and Systems Safety Considerations in Aviation Mishaps. *International Journal of Aviation Psychology*, 1, 97–106.
Figure 17.2	Adapted from: Reason, J.T. (1990). *Human Error*. Cambridge: Cambridge University Press.
Figure 17.3	From: Reason, J.T. (1990). *Human Error*. Cambridge: Cambridge University Press.
Figure 17.4	From: Morley, F.J. and Harris, D. (2006). Ripples in a Pond: An Open System Model of the Evolution of Safety Culture. *International Journal of Occupational Safety and Ergonomics*, 12, 3–15.
Figure 17.5	From: US Department of Defense (1993). *Military Standard: System Safety Program Requirements (MIL-STD-882C)*. Washington, DC: US Department of Defense.
Figure 17.6	From: Ramsey, J.D. (1985). Ergonomic Factors in Task Analysis for Consumer Product Safety. *Journal of Occupational Accidents*, 7, 113–23.
Figure 18.1	From: National Transportation Safety Board (1983). *Accident Investigation of Human Performance Factors*. Washington DC: National Transportation Safety Board Human Performance Group.
Figure 18.3	From: US Department of Defense (n.d.). *Department of Defense Human Factors Analysis and Classification System: A mishap investigation and data analysis tool* (available from http://www.uscg.mil/safety/docs/ergo_hfacs/hfacs.pdf).
Table 5.1	From: Murray, S.R. (1997). Deliberate Decision Making by Aircraft Pilots: A Simple Reminder to Avoid Decision Making Under Panic. *International Journal of Aviation Psychology*, 7, 83–100.
Table 6.1	From: Reason, J.T. (1990). *Human Error*. Cambridge: Cambridge University Press.
Table 6.2	From: Swain, A.D. and Guttmann, H.E. (1983). *Handbook of Human Reliability Analysis with Emphasis on Nuclear Powerplant Operations*. (Sandia National Laboratories, NUREG/CR-1278). Washington DC: US Nuclear Regulatory Commission.
Table 6.3	From: Stanton, N.A., Harris, D., Salmon, P., Demagalski, J.M., Marshall, A., Young, M.S., Dekker, S.W.A. and Waldmann, T. (2006). Predicting Design Induced Pilot Error using HET (Human Error Template) – A New Formal Human Error Identification Method for Flight Decks. *The Aeronautical Journal*, 110, 107–15.
Table 8.1	Adapted from: Arnold, J. (2005). *Work Psychology* (5th edition). Harlow: Prentice Hall.
Table 10.1	From: Holmes, T.H. and Rahe, R.H. (1967). The Social Readjustment Rating Scale. *Journal of Psychosomatic Research*, 11, 213–18.
Table 12.1	From: Harris, D. (2004). Head-Down Flight Deck Display Design. In: D. Harris (ed.) *Human Factors for Civil Flight Deck Design* (pp. 69–102). Aldershot, UK: Ashgate.
Table 14.1	From: Sheridan, T.B. and Verplank, W.L. (1978). *Human and Computer Control of Undersea Teleoperators (Technical Report, Engineering Psychology Program)*. Cambridge, MA: Department of Mechanical Engineering, MIT.
Table 14.2	From: Kelly, B.D., Graeber, R.C. and Fadden, D.M. (1992). Applying Crew-centred Concepts to Flight Deck Technology: The Boeing 777. *Proceedings of the Flight Safety Foundation 45th International Air Safety Seminar*. Long Beach, CA: Flight Safety Foundation.
Table 14.3	From: Bonner, M., Taylor, R., Fletcher, K. and Miller, C. (2000). Adaptive Automation and Decision Aiding in the Military Fast Jet Domain. In: D.B. Kaber and M.R. Endsley (eds) *Human Performance, Situation Awareness and Automation: User-Centred Design for the New Millennium* (pp. 154–9). Madison, WI: Omnipress.
Table 15.1	Adapted from: Shneiderman, B. (1992). *Designing the User Interface* (2nd edition). Reading, MA: Addison-Wesley.

About the Author

Don Harris is Managing Director of HFI Solutions Ltd, a Visiting Professor in the School of Aeronautics and Astronautics at Shanghai Jiao Tong University, China and a Visiting Fellow at the University of Leicester. He has been involved in the design and certification of flight deck interfaces; worked in the safety assessment of helicopter operations for North Sea oil exploration and exploitation and was an accident investigator on call to the British Army Division of Army Aviation. Don is a Fellow of the Institute of Ergonomics and Human Factors and a Chartered Psychologist. He is also a member of the UK Human Factors National Technical Committee for Aerospace and Defence. In 2006 Don received the Royal Aeronautical Society Hodgson Prize and Bronze award for work leading to advances in aerospace and in 2008 was part of the Human Factors Integration Defence Technology Centre team that received the UK Ergonomics Society President's Medal 'for significant contributions to original research, the development of methodology and the application of knowledge within the field of ergonomics'. Don is author or editor of several other Ashgate volumes, including '*Human Factors for Civil Flight Deck Design*', '*Contemporary Issues in Human Factors and Aviation Safety*' (with Helen Muir) and '*Modelling Command and Control*' (with Neville Stanton and Chris Baber).

Preface

When I set out to write this book I deliberately wanted to emphasise the upside of Human Factors in Aviation – what it could contribute to the industry – and not endlessly discuss what happens when the human component in the system fails, or in other words talk about accidents. So, if you are looking for a book packed with descriptions of aircraft crashes and endless analyses picking over the bones of the human failures involved, look elsewhere.

The other thing that I wanted to do was to not concentrate solely on safety (although this does play a big part throughout). Without a doubt, the drive for increased safety has been the primary impetus for the development of the discipline over the last half century, however I believe that this has to change. Human Factors needs to progress to be regarded as a benefit; a positive contribution to the finances and smooth operation of an airline, and not just a cost associated with safety. Taking an integrated, system-wide, through-life approach to the application of Human Factors principles the discipline can now demonstrate that it can be responsible for considerable cost savings.

There are several excellent books on Human Factors in aviation but the vast majority are edited volumes. While the individual chapters in them are excellent no one (in my opinion) has really outlined how it all fits together – no one has provided a systemic overview. This is my attempt to try and explain what it is all about and how it all goes together. I emphasise, though, that this is *my* view and mine alone.

All boundaries and divisions created by man are (to some extent) artificial and subject matter boundaries in the complex world of aviation operations can be both artificial and arbitrary. However, when addressing such a broad topic as Human Factors in aviation in a book, some structure is essential. However, unlike book chapters and university syllabi, the human contribution in the operation of an airline does not fall into neat pigeon holes. Topics like error and training are all pervasive: poor design of flight decks or procedures contributes to error; they also increase workload, which increases the likelihood of error; poor Crew Resource Management (CRM) makes error more likely, and so on. To improve CRM requires training, as does aircraft handling, etc. Appropriate training can also improve decision making (which avoids error). The inter-relationships between chapter topics are complex and manifold, however some structure is required on the chaos. To aid the reader I have provided spider diagrams at the beginning of most chapters to make clear the inter-relationships between chapter content. I have assumed that most readers of the book will not start at the beginning and finish at the end; they will dip into it to find answers to specific questions that they have. Interpreting these spider diagrams is quite straightforward. The closer the boxes containing the other chapter are to the centre of the diagram (where the title of the current chapter resides) the more closely related they are. I hope that these spider diagrams help to guide further reading.

Aviation Human Factors has been largely a requirements-driven, not a theory driven, process. These requirements have been derived from the regulatory demands and from a broader need to operate safely and efficiently. The Human Factors practitioner in the aviation community is most often a user of theory rather than a generator of theory. Theory is often only generated in retrospect after the operational problem has been addressed. Links to further reading where a wider appreciation of the relevant theory are provided at the end of every chapter. However, I emphasise that the selection of any book for further reading is a question of personal taste and not objective choice. This is reflected in my preferences and comments.

My thanks in the production of this book go out once again to Guy Loft at Ashgate (sorry it took so long…) and of course Fiona and Megan who have supported me all the way. Without them this book would not have been possible. Many thanks.

1
A Systems Approach to Human Factors in Aviation

A SHORT (AND SKIMPY) HISTORY OF HUMAN FACTORS IN AVIATION

Human Factors, as a whole, is a relatively new discipline. Its roots lie firmly in the aviation domain with the work undertaken in the UK and North America during and shortly after World War II. It is also a somewhat fragmented and multifaceted subject, for its science base drawing on psychology, sociology, physiology/medicine, engineering and management science (to name but a few disciplines). And, as the Human Factors science base has grown over the last 60 years, with this increasing knowledge has come further specialisation and fragmentation with sub-disciplines in topics such as human-centred design, training and simulation, selection, management aspects (organisational behaviour) and, health and safety.

In a book looking back at the naissance of Aviation Human Factors, Chapanis (1999) reviewed his work at the Aero Medical Laboratory in the early 1940s where, among other things, he was asked to investigate why after landing some pilots occasionally retracted the undercarriage instead of the flaps in certain types of aircraft (particularly the Boeing B-17, North American B-25 and the Republic P-47). In these cases he observed that the actuation mechanisms in the cockpit for the undercarriage and the flaps were identical in shape and located next to each other. The corresponding controls on the Douglas C-47 were not arranged in this way and their methods of actuation were quite different from each other. The pilots flying this aircraft rarely unintentionally retracted the undercarriage after landing. This (what now seems obvious) insight into performance, especially in stressed or fatigued pilots, enabled him to understand how they may have confused the two controls. The remedy to the solution he proposed was simple: physically separate the controls in the cockpit and/or make the shape of the controls analogous to their corresponding component (hence, the flap lever was redesigned to resemble a trailing edge flap and the undercarriage lever to resemble a wheel with a tyre).

US Army Air Corps pilot losses during World War II were roughly equally distributed between three principal causes: about one-third of pilots were lost in training crashes; a further third in operational accidents, and the remaining third were lost in combat (Office of Statistical Control, 1945). This suggested various different deficiencies inherent in the system. In further work investigating cockpit design inadequacies Grether (1949) described the difficulties of reading the early three-pointer altimeter. Previous work and numerous fatal and non-fatal accidents had shown that pilots frequently misread this instrument. The effects of different designs of altimeter on interpretation time and error rate were investigated. Six different variations of altimeter design were evaluated using combinations of one, two or three pointers (both with and without an inset counter also displaying the altitude of the aircraft) as well as altimeters using three different formats of vertically moving scale. The results showed marked differences in the error rates for these different designs. The three-pointer altimeter took slightly over seven seconds to interpret and produced over 11 per cent errors of 1,000 feet or more. Vertical, moving scale designs took less than two seconds to interpret and produced less than 1 per cent of errors in excess of 1,000 feet.

These early examples of control and display design demonstrated that 'pilot error' alone was not a sufficient description for the causes of many accidents. In these cases, pilots fell into a trap left for them by their design (what has become known as 'design-induced' error). Fortunately, blaming the last person in the accident chain (usually the pilot) has lost all credibility in modern accident investigation. The modern approach is to take a systemic view of error and attempt to understand the relationships between all the components in the system, both human and technical. This is partially what this book hopes to achieve.

The principal focus for human performance research in the UK during this time was the Medical Research Council, Applied Psychology Unit in Cambridge. Here fundamental work was being undertaken on issues such as the direction of motion relationships between controls and displays (Craik and Vince, 1945) and Mackworth was developing his famous (if you are a psychology student) 'clock' to investigate the effects of prolonged vigilance in radar operators (Mackworth, 1944, 1948). As a result of the relatively small number of RAF pilots at this time and the large number of sorties being flown during the Battle of Britain, pilot fatigue became an issue. This problem re-surfaced in the latter half of World War II, when longer-range high-performance aircraft began to place increasing demands on their pilots. Losses as a result of fatigue rather than combat began to rise. At the MRC Kenneth Craik developed the 'Fatigue Apparatus' which later universally became known as the 'Cambridge Cockpit'. This was based on the cockpit of a disused Supermarine Spitfire. Even as early as 1940, experiments using this seminal piece of apparatus established that control inputs from a fatigued pilot were different to those from a non-fatigued pilot.

> The tired operator tends to make more movements than the fresh one, and he does not grade them appropriately ... he does not make the movements at the right time. He is unable to react any longer in a smooth and economical way to a complex pattern of signals, such as those provided by the blind-flying instruments. Instead his attention becomes fixed on one of the separate instruments. ... He becomes increasingly distractible. (Drew, 1940)

A further excellent history of the early work in Human Factors can be found in Roscoe (1997).

During these early years of the Human Factors discipline researchers happily turned their skills to many aspects of human performance. However, with increasing knowledge and specialisation, the discipline of Human Factors began to fragment. The sub-disciplines referred to earlier began to develop. Nevertheless, from the late 1950s and early 1960s Human Factors began to make increasingly large contributions, particularly in the three domains of selection, training and flight deck design. However, it was in the 1970s that the

Human Factors discipline really began to take off with the advent of the 'CRM revolution' and the development of the 'glass cockpit'.

The CRM (Cockpit – later Crew – Resource Management) revolution introduced applied social psychology and management science onto the flight deck. CRM evolved as a result of a series of accidents involving perfectly serviceable aircraft. The main cause of these accidents was attributed to a failure to utilise all the crew resources available on the flight deck in the best way possible, for example in the crash involving the Lockheed L-1011 (Tristar) in the Florida Everglades (National Transportation Safety Board, 1973). At the time of the accident the aircraft had a minor technical failure (a blown light bulb on the gear status lights) but actually crashed because nobody was flying it! The crew were 'head down' on the flight deck trying to fix the problem. Other accidents highlighted instances of Captains trying to do all the work while the other flight crew members were almost completely unoccupied; dominant personalities suppressing teamwork and error checking; or simply as a result of a lack of poor crew cooperation, coordination and/or leadership.

The CRM revolution also stimulated further research and changes in practice in the way that flight crew were selected and trained. This coincided with a change in the nature of work of the airline pilot from being a 'hands on throttle and stick' flyer to one of being a flight deck/crew manager of a highly automated machine. This change has been particularly pronounced in the last half of this period.

Throughout the 1950s and 1960s airlines tended to rely heavily on the military for producing already trained pilots. Selection techniques assumed that candidates were trained and competent (e.g. interviews, reference checks and maybe a flight check – see Suirez, Barborek, Nikore and Hunter, 1994). However, in the 1970s and 1980s, particularly in Europe where there was less emphasis on recruiting pilots from a military background and where there was also an increasing demand for commercial pilots, greater emphasis began to be placed upon the selection processes for *ab initio* trainees. In this case it was important to assess the candidate's *potential* to become a successful airline pilot (i.e. not fail the training, which was very costly). Psychometric and psychomotor tests became more commonplace, similar in many ways to the procedures military selection panels had been using for some years. However, as the management role of pilots began to develop (especially with the increasing uptake of CRM) qualities such as judgement and problem solving, communication, social relationships, personality and motivation became as important as the technical skills involved in flying a large jet (Bartram and Baxter, 1996).

Simultaneously, the nature of pilot training began to evolve from the 1970s onwards. Initially, there was increasing use of flight simulators which began to improve in fidelity (and in relative terms) began to decrease in cost. More training was now accomplished on the ground than in the air. The content of the training programmes also began to change (and is still changing). Hitherto, training and licensing concentrated on flight and technical skills (manoeuvring the aircraft, navigation, system management and fault diagnosis, etc.). It addressed issues such as flying the aircraft manually or dealing with emergencies resulting from a technical failure (e.g. engine failure at V_1 or performing flapless approaches). However, with increasing technical reliability, it was evident that the major cause of air accidents had now become human error and that this often resulted from the failure of the flight deck crew to act in a well co-ordinated manner. Non-technical skills training (CRM instruction) was introduced, initially for flight deck crew and subsequently throughout the aircraft to include all the cabin crew. This was partly contingent upon a change in philosophy towards Line Oriented Flight Training (LOFT) where training as a full crew and acting as a team member was increasingly stressed. Performance was evaluated with respect to how the crew handled flying the aircraft, how they dealt with the technical aspects of a problem, and most importantly, how the people on the flight deck (and if necessary elsewhere in the aircraft) were employed to address the issue (see Foushee and Helmreich, 1988).

In terms of the design of the pilot interface, flight deck displays have progressed through three eras: the 'mechanical' era; the 'electro-mechanical' era; and most recently, the 'electro-optical' (E-O) era (Reising, Ligget and Munns, 1998). With the advent of the 'glass cockpit' revolution (where electro-optical display technology began to replace the electro-mechanical flight instrumentation in the early 1980s) opportunities were presented for new formats for the display of information and Human Factors started to play an increasingly important part in their design. However, while the new display technology represented a visible indication of progress (e.g. as implemented on the 'hybrid' Airbus A300/A310 and Boeing 757/767 aircraft) the true revolution resided in the less visible aspects of the flight deck, particularly the increased level of automation available as a result of the advent of the Flight Management System (FMS) or Flight Management Computer (FMC) (see Billings, 1997). The electro-optical display technology was merely the phenotype: the digital computing systems being introduced represented the true change in the genotype of the commercial aircraft. Not only did these higher levels of automation allow opportunities for reducing the number of crew on the flight deck from three to two (eradicating the function of the Flight Engineer – something that had been done in less highly automated, shorter-range commercial aircraft some years earlier) but it also required a change in the skill set required by pilots. Aircraft were now 'managed' and not 'flown'.

With these increasing levels of automation on the flight deck in the 1980s and early 1990s a great deal of research was undertaken initially in the area of workload measurement and prediction, and subsequently in enhancing the pilot's situation awareness. Autoflight systems had certainly reduced the physical workload associated with flying and computerised display systems (coupled with the FMS/FMC) had also reduced the mental workload associated with many routine computations linked to in-flight navigation. However, it was wrong to say that automation reduced workload overall: it had simply changed its nature. Wiener (1989) labelled the automation in these first generation 'glass cockpit' aircraft 'clumsy' automation. It reduced crew workload where it was already low (e.g. in the cruise) but increased it dramatically where it was already high (e.g. in terminal manoeuvring areas).

The new breed of multifunctional electro-optical displays, together with high levels of automation, had both the ability to promote Situation Awareness through new, intuitive display formats and also to simultaneously degrade it by hiding more information than ever before. Automation made many systems 'opaque'. Dekker and Hollnagel (1999) described automated flight decks of the time as being rich in data but poor in information. While the advanced technology aircraft being introduced were much safer and had a far lower accident rate (Boeing Commercial Airplanes, 2009) they introduced a new type of accident (or perhaps they merely exaggerated an already existing underlying problem). Dekker (2001) described these as 'going sour' accidents, where the aircraft was 'managed' into disaster, usually as a result of a common underlying set of circumstances: human errors, miscommunications and mis-assessments of the situation. But as stated previously, it would be very wrong to blame the pilots in these cases. The accidents resulted from a number of factors pertaining to the design of the flight deck, the situation and the crews' knowledge/training which all conspired against their ability to coordinate their activities with the aircraft's automation. It was increasingly apparent that it was almost impossible to separate design, from procedures and from training if it was desired to optimise the whole system of operation, including the human element.

Today the practice of Human Factors in the aerospace industry today is largely an incremental development of the situation described up to this point; however there is now universal acceptance of its importance to safe operations. Human Factors is firmly embedded in the selection, training and design processes and is also a cornerstone of all safety management systems. Good Human Factors practice is effectively mandated through many regulations. For example, an effective safety management system is now an essential part of any Air Operator's Certificate (AoC) – for instance see EU-OPS (2009): since September 2007 Certification Specification (CS) 25.1302 and AMC (Acceptable Means of Compliance) 25.1302 have been adopted by EASA to mandate for 'good' Human Factors design on the flight deck: training in Human Performance and Limitations and CRM is now mandatory for all pilots, and recurrent training in these topics is needed for all professional pilots (see Civil Aviation Publication 737: CAA 2006). In some cases, though, more control has been delegated back to the airlines. Training curriculum requirements may now be delegated back to suitably experienced airlines. The approach of developing training needs directly from line operational requirements reflects the training philosophy outlined by the Federal Aviation Administration (FAA) in the Advanced Qualification Program (AQP) and also being adopted elsewhere (e.g. by the European Aviation Safety Agency – EASA). The emphasis in the AQP is away from time-based training requirements to fleet-specific, proficiency-based requirements (see AC 120-54; FAA, 1991). In the AQP process, the airline develops a set of proficiency objectives based upon their requirements for a specific type of operation (e.g. based upon a threat and error management process developed from a Line Operations Safety Audit – LOSA – part of the Safety Management programme). The AQP is based upon a rigorous task analysis of operations but with emphasis placed firmly upon the behavioural aspects of the flight task, such as decision making or the management of the aircraft's automation.

For 50 years there has been a general decline in the commercial aircraft accident rate, however during the last decade (or so) the serious accident rate has plateaued at approximately one per million departures (Boeing Commercial Airplanes, 2009). With the projected increase in the demand for air travel, if this rate remains unchanged by 2015 there will be one major hull loss per week. As the reliability and structural integrity of aircraft has improved, the number of accidents directly resulting from engineering failures has reduced dramatically, hence so has the associated rate. However, human reliability has not improved to the same degree. Figures vary, but it is estimated that up to 75 per cent of all aircraft accidents now have a major Human Factors component. Human error is now the primary risk to flight safety (CAA, 2008).

TOWARDS AN INTEGRATED VIEW OF HUMAN FACTORS

Human Factors in aviation as a discipline has come of age. It is all pervasive and is now completely integrated into all aspects of aircraft design and flight operations. To date, though, almost the sole impetuous for Human Factors has been safety-oriented.

From an overall perspective, though, three generic, quasi-antagonistic parameters can be applied to measure system functioning: safety, performance and cost. Airworthiness authorities are concerned solely with safety aspects of aircraft design, pilot training and airline operations. However, airlines are required to balance the requirement for safety against both cost and performance considerations, but the Human Factors discipline has until recently concentrated almost exclusively on the safety aspects of this organisational performance troika.

In the domain of commercial aviation Human Factors has become regarded almost as a 'hygiene factor'. From a flight deck design perspective it goes without saying that a failure to fully consider the human requirements of the pilot–aircraft interface will result in a product which is difficult to use and promotes error. However, playing the Devil's advocate, from a manufacturer's perspective, providing a 'better' human-system interface does not 'add value': a failure to provide a pilot-friendly interface merely detracts from its value (see Harris, 2008). As a result, it is often difficult to make a convincing cost-based argument for a manufacturer to invest heavily in Human Factors research. All modern flight deck interfaces are more than adequate for their job. They do not unduly promote error. Minor deficiencies become a training issue to be dealt with by the airline.

The military tend to put a high premium on Human Factors. Military personnel *must* be able to use the equipment they are provided with in a range of stressful, high-pressure situations. However, the customer for a military aircraft has the further advantage in that they can provide dedicated, comprehensive training for their crew. Indeed anything that is not operational flying is training. This is not a luxury that commercial airlines possess. Nevertheless, a great deal of the time the emphasis still tends to err towards the role of Human Factors as providing a means to avoid poor performance. With 'good' Human Factors the pilot will be able to exploit the full capability of their aircraft.

However, Human Factors in aviation is now entering a new era with the potential to take a new, more integrated approach. During the late 1990s the notion of Human Factors Integration (HFI) began to appear, which many regard as a sub-discipline of Systems Engineering. This approach begins to provide an integrative framework. HFI originally encompassed six domains regarded as essential for the integration of the human element into a system (UK Ministry of Defence, 2001). These were *Staffing* (how many people are required to operate and maintain the system); *Personnel* (what are the aptitudes, experience and other human characteristics required to operate the system); *Training* (how can the requisite knowledge, skills and abilities to operate and maintain the system be developed and maintained); *Human Factors Engineering* (how can human characteristics be integrated into system design to optimise performance within the human/machine system); *Health Hazards* (what are the short- or long-term hazards to health resulting from normal operation of the system) and *System Safety* (how can the safety risks which humans might cause when operating or maintaining the system be identified and eliminated, trapped or managed)? Subsequently a seventh domain was added, the *Organisational and Social* domain, which encompasses issues such as information sharing and interoperability.

When put together, taking a system-wide HFI approach means that Human Factors is no longer necessarily merely a 'hygiene factor'. 'Good' Human Factors can now make positive benefits to enhancing performance and reducing both operational and through life costs. Human Factors *can* now 'add value'. Examples of this are already appearing in the military domain (Human Factors Integration Defence Technology Centre, 2006a,b). Broadly speaking, taking an end-to-end system perspective, good flight deck interface design simplifies operating (and hence training) requirements, making training faster and cheaper (less time is spent by pilots in unproductive – not revenue producing – work). Training better targeted to the airline's requirements is more efficient. Simultaneously, the flight deck interfaces and better specified training produce superior, more error-free (safer) performance. Careful selection processes may be more expensive initially but they subsequently reduce the drop out and failure rate in pilot training (which is very expensive). Analysis and modification of crew rostering practices can produce rotas which produce more efficient utilisation of flight crew, reduce crew fatigue, increase well-being and simultaneously enhance safety. Such efforts can also reduce stress and decrease employee turnover. At the same time a well-considered Human Factors aspect in a company's safety management system makes it cheaper to run and produces the information required to promote safer operations. The key to demonstrating the utility of Human Factors in aviation is not to count the cost of investing in it, but to calculate the savings that it makes on a through-life basis. However, these efficiency gains are best accrued by taking a systemic approach and avoiding the urge to promote uncoordinated, local solutions to problems. Half of the time these tend to address the symptoms and not the underlying disease.

To make these gains, though, Human Factors as a discipline must avoid its natural inclination to rush to claim the moral high ground by marking its territory solely within the realm of aviation safety. Furthermore, the discipline must also coalesce once again in order that the maximum benefit from an integrated, through-life, systemic approach can be realised. While increasing levels of specialisation have served to develop the science base it has also simultaneously mitigated against its coherent application in commercial aviation. However, the opportunity now exists to begin to capitalise on the developments made in the last half century.

It has to be stated right from the start that this book *is not* about Human Factors in aircraft accidents. There is a great temptation to illustrate many Human Factors principles with an accident illustrating the consequences of *not* applying said principle. This is superficially an attractive thing to do: it has shock and awe value and every accident is usually in itself a good story (and everyone likes a good story). To be honest, this will happen from time to time in this book, but it should always be remembered that in such cases what is being illustrated is a *failure* of Human Factors. This book is about getting Human Factors right. Unfortunately, when Human Factors is applied properly and systematically no one notices: everything goes right with minimum effort and there are no unforeseen consequences … but there is still a long way to go.

A SYSTEMS PERSPECTIVE

Commercial aviation is a socio-technical system. Socio-technical systems regard organisations as consisting of complex interactions between personnel and the technology in the workplace. This approach can also encompass the wider context to include the societal infrastructures and behaviour in the organisation. Socio-technical systems contain people, equipment and organisational structures. These are linked by functional processes (which are essential for transforming inputs into outputs) and social processes which are informal but which may serve to either facilitate or hinder the functional processes (McDonald, 2008).

Flying a commercial airliner is not just about the integration of the pilots (hu*Mans*) and aircraft (*Machine*) to undertake a particular flight (or *Mission*) within the constraints imposed by the physical environment (*Medium*). This approach, as first expounded by Edwards (1972) in the SHEL model of Ergonomics (Software, Hardware, Environment and Liveware) needs extending further to take into account the societal environment (a further aspect of the *Medium*). In commercial aviation, the role of *Management* is crucial. This five Ms system approach (Harris and Smith, 1997; Harris and Harris, 2004; Harris and Thomas, 2005) is extended and adapted from the system safety model of Miller (1988).

The hu*Man* aspect of the five 'M's approach encompasses such issues as the size, personality, capabilities and training of the end user, in this case a pilot. Taking a user-centred design approach, the human component is the ultimate design forcing function. The Human Factors specialist involved in flight deck design has to operate within the core abilities of the pilot(s). The hu*Man* and the *Machine* (aircraft) components come together to perform a *Mission*. It is usually the *Machine* and *Mission* components on which the engineering design processes focus. However, Human Factors Engineers must not only work within the parameters imposed by the pilots, the technology and the physical aspects of the *Medium* in which the task (flight) is undertaken, they are also bound by the airworthiness rules and the norms of society (both societal aspects of the *Medium*). It is always worth noting that performance standards for human–machine systems of any kind are determined by societal norms (regulations); for example, the level of safety required or the minimum standard of user competence. Sometimes these may vary from culture to culture, even though aviation is an international activity. Airline *Management* must work within these rules. The *Management* aspect also prescribes secondary performance standards through the selection and training of crew or the required technical performance of the aircraft. The *Management* is the key link between the hu*Man*, *Machine*, *Mission* and *Medium*. It plays the integrating role that ensures compliance with the regulations and promotes safe and efficient operations.

The inter-relationships between the five Ms can be demonstrated in a simple diagram (Figure 1.1). In the case of an airliner, the pilots fly the aircraft to achieve a well-defined goal. This exemplifies the union of hu*Man* and *Machine* to perform a *Mission*. The *Management* tasks this *Mission* and is responsible for ensuring that the crew and aircraft conform to the regulatory requirements and that the aircraft is capable of enduring the physical demands placed upon it by the aeronautical *Medium*.

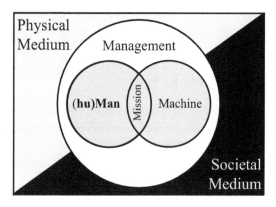

Figure 1.1   The Five Ms Model (Harris and Smith, 1997; Harris and Harris, 2004; Harris and Thomas, 2005)

Socio-technical system theory evolved alongside the development of open systems theory, which itself was derived from General Systems Theory (von Berthalanfry, 1956). Katz and Kahn (1978) however, assert that open systems theory is better characterised merely as a framework within which the workings of a system may be understood. All industries are open systems (i.e. they must interact with their environment). As Schein (1992) stated:

> The environment places demands and constraints on the organization in many ways. The total functioning of an organization cannot be understood, therefore, without explicit consideration of these environmental demands and constraints. (p. 101)

Katz and Kahn (1978) on the other hand argued that organisations are only selectively open, in that although they interact with their environment they also need boundaries in order to exist. Some aviation organisations, however, are a great deal more open than others (Harris, 2006b). Military aviation is a much less open system than is commercial civil aviation. The armed forces exert far more control over flight operations and, at least in peacetime, the operating environment. Not only do the military own the aircraft that they operate, they also have considerable influence over their design and development. The military is also responsible for a great deal of the maintenance of their aircraft. Air forces operate their own airfields and provide their own Air Traffic services. They train their own pilots and engineers and personnel are indoctrinated into the military culture. In contrast, civil airlines operate into a wide range of airports (none of which they own); maintenance is often provided by third parties; ramp servicing is provided by a range of external suppliers and Air Traffic services are provided by the countries into which they either operate or overfly. In the case of the new generation of low-cost carriers these organisations may not even own their own aircraft, employ their own ground and check-in personnel, and in extreme cases, they may not even employ their own pilots. There are a great many more organisational boundaries that information and resources must cross in the operation of commercial aircraft compared to military operations. Airlines can be considered to be considerably more 'open' systems than the military.

In fact commercial aviation is more than just a socio-technical system. It is a socio-technical 'system of systems' (Harris and Stanton, 2010). Maier (1998) characterised a 'system of systems' as possessing five basic traits: operational independence of elements; managerial independence of elements; evolutionary development; possessing emergent behaviour and having a geographical distribution of elements. In the context of commercial aviation, these systems have distinct operational independence (aircraft operations; maintenance; Air Traffic management/control) and each of these aspects has managerial independence (they are offered by independent companies or national providers) however they are bound by a set of common operating principles and international regulations for design and operation. All aspects of aviation encompass technical, human and organisational components encompassing critical Human Factors considerations such as usability, training, design, maintenance, safety, procedures, communications, workload and automation. In the operation of civil aircraft there are a great number of inter- and intra-organisational boundaries that information and resources must cross in this 'system of systems'. It is fair to say, though, that the aviation 'system of systems' was never designed: it is a legacy system that has evolved over the past century.

The pilot is usually at the day-to-day operational centre of this socio-technical 'system of systems'. When sitting on the airport gate the flight deck crew are the nexus for information stemming from and being passed on to other systems, such as Air Traffic services, ramp personnel, dispatch, fuellers, maintenance, etc. Data are constantly being passed, combined with other data, stored and processed to a series of disparate, yet related, ends. The old conception of the pilot's task was simply expressed in terms of a hierarchy of functions: 'aviate' (fly the aeroplane); 'navigate' (know where you are and where you are going) and 'communicate' (with other crew, ATC and the ground). Several years ago this was recognised as no longer being enough, so a fourth generic requirement was added: 'manage'. This was originally conceptualised as managing the human resources in the aircraft (CRM) however it began to be expanded to further encompass such things as the management of the automation in the aircraft. It now has to be expanded outside the aircraft in two further respects. The 'manage' aspect now embraces the human resources on the ground in related support systems, and it must also take on the management of information. The pilot's task has changed. Being a 'flyer' is no longer enough.

De-composing the five Ms model a little further, some basic, high level representations of the manner in which Human Factors operate in the airline industry can be proposed which incorporates many of the sub-disciplines alluded to in the opening paragraph of this chapter (see Figures 1.2–1.5). These figures should by no means be taken to represent a comprehensive, unifying model of Human Factors in aviation: they are just a start to provide a framework for what follows. They are merely an attempt to make explicit the relationships between the various Human Factors disciplines and how they contribute to the overall function of an airline.

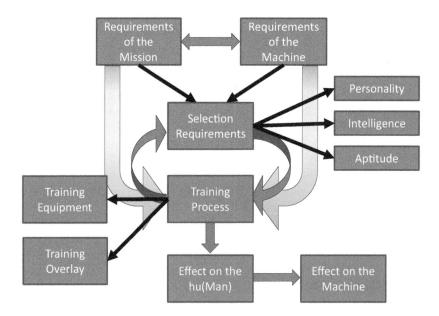

Figure 1.2   A system-wide depiction of the interactions between the various Human Factors sub-disciplines producing a positive effect on the pilot

At the top of Figure 1.2 it can be seen that the requirements of the flight task (*Mission*) drive the design of the *Machine* (and hence the design of the flight deck). The role of the flight deck is to support the four basic tasks of the flight crew (Aviate; Navigate, Communicate; Manage) and to protect the crew from the physiological stressors imposed by the *Medium* (see Figure 1.3). These basic functions of the flight deck are pivotal to the design of the training (essentially a process of modifying the hu*Man*). Training is all about teaching someone how to do something, with something. To train a pilot you need training devices (a full flight simulator approved by the regulatory authorities for training and licensing purposes, equipped with a daylight visual system and six-axis motion platform; cockpit procedures trainers; other part-task trainers; computer-based training facilities for aircraft systems and procedures; and also 'regular' classroom facilities, etc.). The training overlay is more than just a curriculum: it also dictates the training devices or media that are best suited for delivering the various aspects of pilot training. Simulators do not provide training – they are just a useful tool to aid in the delivery of training if used properly.

The nature of the task also dictates the type of person required: what are the basic characteristics, aptitudes and abilities (individual differences) a candidate must possess that will make it likely they will successfully complete the training and become a safe and efficient pilot? Basic 'stick and rudder' skills are now only a very small part of the make-up of the modern pilot. Cognitive and team working (flight deck *Management* skills) are now essential. There is an intimate relationship between all these components, all of which help to inform the design of each other. Hopefully, as a result of this process there will be a positive, beneficial effect on the pilot undertaking the training who, through the use of a well-designed flight deck, will have the desired effect on the aircraft (i.e. they will use it effectively to complete the *Mission* tasked by the *Management*).

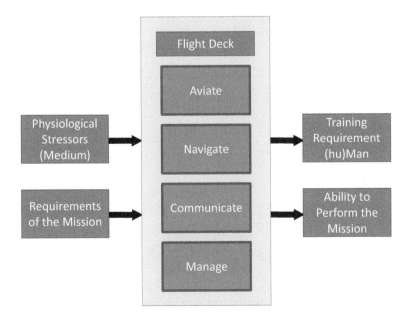

Figure 1.3   **A system-wide depiction of the human-related functions of the flight deck**

There are, however, a great many negative factors that can be detrimental to flight crew's health and performance if they are not controlled properly. Some of these are outlined in Figure 1.4. These stressors fall into two major categories: physiological and psychological stressors. The physiological stressors are in many ways much easier to deal with. These are mostly a product of the physical *Medium* (noise, vibration, temperature, etc.) and it is one of the roles of the *Machine* to protect the pilots from these stressors. Mostly these are aeromedical issues but they should not be thought of exclusively in this manner. Minor issues in these areas, even if they are not detrimental to health, can accumulate and erode performance.

The psychological stressors are more complex and difficult to control. Before they can be controlled, though, their source needs to be identified (the middle column in Figure 1.4). For example, the flight deck and the flight task can never eradicate workload: these factors can only be designed to minimise the workload imposed on the pilots. Similarly, factors such as jet lag, the requirement to work anti-social hours and separation from family and friends are things that can only be controlled by sympathetic *Management* (although stress from these factors can also be imposed by unsympathetic *Management*). These are analogous to Herzberg's Hygiene Factors (Herzberg, 1964): they can never contribute positively to the psychological health and motivation of flight crew, however a failure to control them properly will be detrimental.

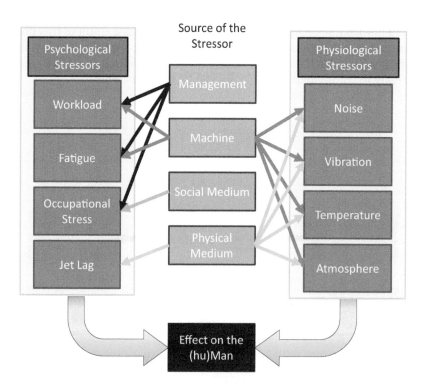

Figure 1.4   A system-wide depiction of the sources of the various aspects of Human Factors producing a detrimental effect on the pilot

Again, it can be seen how the role of *Management* is central to just about all aspects of Human Factors in aviation. Its role can have a positive effect in producing efficient, safe operations or it has the potential to be equally damaging. As noted previously, the role of airline *Management*, in addition to making money, is to ensure that all operations fall within the legislative requirements (e.g. for flight crew licensing and aircraft maintenance) required by society – Figure 1.5. This is essentially a safety management role which itself contains lots of Human Factors issues associated with confidential reporting schemes, safety culture and the analysis of accident and incident data. However, there is also a crucial *Management* role within the aircraft itself: CRM. The Captain of a modern airliner is as much a manager as a flyer and in many ways is as responsible as the airline's management for ensuring that a flight remains safe and legal. In fact, the Captain is the hands-on, minute-to-minute manager with the majority of operational responsibility. It is also the Captain (and the rest of the crew) that has to deal with the unique demands imposed by operating across a diverse set of national cultures. Most airlines may be national carriers but aviation is an international activity.

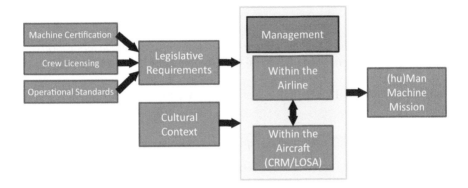

Figure 1.5   **A system-wide overview of the inputs and effects of Management, both within the airline and on the flight deck**

CONCLUSIONS

This is merely the end of the beginning (apologies to Winston Churchill). This chapter has provided (I hope) the foundations for a framework within which to place everything that follows. It also illustrates the inter-connectedness of the various Human Factors activities in and around the flight deck. By taking an integrated approach to the application of Human Factors the discipline can have a much greater impact in reducing costs, increasing efficiency, enhancing performance and improving safety, especially if a systemic approach is adopted. Human Factors is not all about safety and if applied judiciously, can actually produce a return on investment.

The following chapters attempt not only to flesh out the contents of the boxes in the previous diagrams but also to provide a slightly more in-depth appreciation of some of the theory underpinning this applied discipline. It is not enough simply to know *that* something works; it is also necessary to how *why* something works. The first part of the book examines the basic underpinning psychological science of Human Factors in aviation. The following three parts then look at the application of this science base to the Hu*Man* in the system; the Hu*Man* interfaces with the *Machine*; and the *Management*, which is essentially concerned with ensuring the optimum 'fit' between the Hu*Man* and the *Machine* to undertake the *Mission* safely and expediently within the demands imposed by society. In this case, though, the emphasis is firmly on safety management. The book concludes with a few thoughts on the future direction for Human Factors in the civil aviation context.

FURTHER READING

For comprehensive (and excellent) overviews of the early years of Human Factors, that emphasise the discipline's roots in aviation see:

Roscoe, S.N. (1997). *The Adolescence of Engineering Psychology.* Santa Monica, CA: Human Factors and Ergonomics Society.

Or:

Chapanis, A. (1999). *The Chapanis Chronicles: 50 years of Human Factors Research, Education, and Design.* Santa Barbara, CA: Aegean Publishing Company.

An interesting historical view of the design and development of the aircraft flight deck can be found in:

Coombs, L.F.E. (1990). *The Aircraft Cockpit.* Wellingborough, UK: Patrick Stephens Ltd.

For more on sociotechnical systems (but not from an aviation perspective) try:

Mumford, E. (2003). *Redesigning Human Systems.* London: Information Science Publishing.

If you don't believe that the application of 'good' Human Factors can save you money, there are two short (and freely available) UK Ministry of Defence information booklets available which contain outline principles and case studies:

Human Factors Integration Defence Technology Centre (2006a). *Cost Arguments and Evidence for Human Factors Integration.* London: UK Ministry of Defence.
Human Factors Integration Defence Technology Centre (2006b). *Cost-Benefit Analysis for Human Factors Integration: A Practical Guide.* London: UK Ministry of Defence.

PART ONE

The Science Base

Human Factors (or Ergonomics – the two terms tend to be used interchangeably) as defined by the UK Institute of Ergonomics and Human Factors, is *'the application of scientific information concerning humans to the design of objects, systems and environment for human use'*. Parts Two, Three and Four of this book address these applied issues, however this first part examines its underpinning science base.

There is a very broad dichotomy in the Human Factors discipline: Cognitive Ergonomics and Physical Ergonomics. The two are not mutually exclusive, though: both are necessary and must work in harmony.

Physical Ergonomics is concerned with issues such as the size, shape, strength and the environmental tolerances of the human being, and the design of the workplace (in this case the flight deck) that accommodates them. Unlike engineering components it has to be recognised that people come in a range of different shapes and sizes and the workspace has to be able to adjust to this fact. The operator must not get too hot or too cold; they need food and oxygen; they need protecting from certain frequencies and amplitudes of vibration and they get fatigued, both mentally and physically depending upon the nature of the task. All of these things can be detrimental to health but these physical aspects of the workplace also have a direct effect on the cognitive function of the person in it. A failure to recognise these physical limitations of the human will also result in an increase in error proneness.

This book touches on several of these topics but it approaches Human Factors in Commercial Aviation largely from the perspective of Cognitive Ergonomics. As a gross generalisation, Cognitive Ergonomics regards the human operator as an information processing component within a wider system. Within this system there are two major types of information processing bottleneck that can occur; resource limited processing and data limited processing. Resource limited processing occurs when the rate of tasks that the pilot is required to perform is outside his/her capability or that the information integration aspects of a task are too great for human cognitive capacity. Data limited processing occurs when there is not enough information flowing into the pilot to enable them to perform their job or when the data transmission rate is too slow. If the data arrives in an inappropriate format then the pilot also has problems (see Part Three – The Machine). They need to re-code the data into a usable format and hence the task may subsequently become resource limited (and prone to error). There is also a third, slightly different category of information processing bottleneck, where two pieces of information arrive simultaneously but require processing by the same cognitive resource (resource competition).

Three psychological disciplines provide the underlying scientific basis for many aspects of Human Factors in aviation: Human Information Processing, Individual Differences and Social Psychology. In the first chapter in this part there is a brief overview of the Human Information Processing science base. Individual differences are discussed in the last chapter in this opening part. The relevant Social Psychology concepts are briefly described in Chapter 16 on Crew Resource Management and Line Operations Safety Audits (LOSA) in Part Four (The Management).

Human Information Processing begins to provide an explanation of the nature of some of those bottlenecks just described. When the cognitive system becomes highly loaded this is experienced as Mental Workload. Providing information in the right format at the right time to the pilot enhances their Situation Awareness. While Mental Workload and Situation Awareness are two topics with their foundations firmly in the aviation industry (where their science base was developed) these issues are now pervasive throughout all aspects of practice of Human Factors. There is an intimate relationship between Situation Awareness and Decision Making: poor Situation Awareness leads to poor decisions. You can't make good decisions if the information is not there. High workload also leads to error. When workload becomes excessive there is a tendency to react to the situation rather than analysing it and responding to it appropriately. This is undesirable on the flight deck.

High workload, poor Situation Awareness, poor decisions and error are all the symptoms of a poorly designed system that doesn't take into account the nature and limitations of the Human Information Processing system. However, this is the negative side. With good design workload can be managed and Situation Awareness can be enhanced, enabling the inherent adaptability and flexibility of human cognition to be exploited to best effect. Selecting the finest candidates for training as pilots, a process with its foundations in the discipline of Individual Differences, forms the basis for enhancing decision making and minimising error.

The human component is central to the safe and efficient operation of the modern commercial airliner but it is not separate from it. There is a philosophically different approach employed by the Human Factors Engineer and traditional engineers. Traditional engineers expect a causal effect between elements in the system: the same input will always result in the same output. The Human Factors Engineer (or Ergonomist) merely expects a similar output given the same input. In Human Factors terms, the relationship between inputs and outputs, if directed through a human, is probabilistic rather than causative. When faced with a bottleneck in a system the traditional engineer can redesign the system to operate faster or increase its capacity. This option is not available to the Human Factors specialist. This is why when designing systems which require a human operator (such as an airliner) the starting point is that operator. The system hardware and software must be designed around them. Pilots should not be expected simply to cope with the demands imposed on him/her by the system. Finally it has to be recognised that human beings make errors and pilots are no different. They do not intend to make errors but they occur nonetheless: making errors is simply part of the human condition. In fact, making errors can even be regarded as an essential part of the human condition. Without making errors we would not learn. However, there is a time and a place to make errors and an aircraft flight deck is not one of these. Thus Human Centred Design should attempt to minimise the opportunities for error to occur or if an error does perchance happen, its consequences should be minimised. The system should be made 'error tolerant'. To do this, though, requires an understanding of the nature of human error. Errors on the flight deck are no different from those occurring in other aspects of life to people in other professions. It is only in their context and potential consequences that they differ.

All these issues are addressed in this book. All the parts, chapters and concepts in this volume are interdependent but you have to start somewhere, so let's start with an overview of some essential aspects from the psychological discipline.

FURTHER READING

This book covers only superficially the topic of Aviation Medicine. For a comprehensive treatment of the subject see the most comprehensive and exhaustive tome on the topic:

Rainford, D.J. and Gradwell, D.P. (2006). *Ernsting's Aviation Medicine* (4th edition). London: Hodder Arnold.

Although not specifically concerned with aviation, for coverage of the physical aspects of Human Factors try:

Salvendy, G. and Karwowski, W. (eds) (2010). *Advances in Physical Ergonomics.* Boca Raton, FL: CRC Press.

2
Human Information Processing

Human Information Processing provides a system model describing how people receive, store, integrate, retrieve and use information. It deals with everything from stimulus to response while also providing for complex problem solving and decision making. Human Information Processing theory forms the basis of other key concepts such as Workload and Situation Awareness, both covered in more detail in the following chapters. It also provides the foundation for many aspects of the design of the flight deck, particularly the information systems (displays and computer interfaces) and automation (see Part Three of this volume: The Machine).

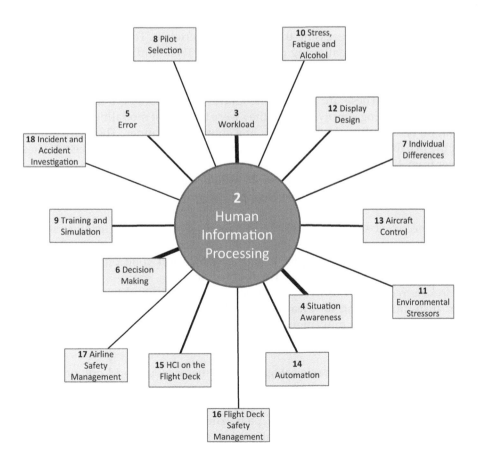

Human Information Processing has an intimate relationship with both workload and Situation Awareness. Both of these are products of load on Working Memory (WM). All of these elements are critical to good decision making. Error is a second order product of Human Information Processing. The application of the principles of Human Information processing can be most closely seen in the design of flight deck displays, automation, HCI on the flight deck and the manual control of the aircraft.

Huitt (2003) suggests that there are several basic principles that most cognitive theorists agree on. The Human Information Processing system has:

- limited capacity to process information;
- a central executive controlling the encoding, transformation, processing, storage, retrieval and utilisation of information; and
- the ability to combine sensory input with stored information to produce novel solutions.

There also tends to be consensus that information processing proceeds through a number of stages (Atkinson and Shiffrin, 1968) involving various memory structures (Sensory Memory; Working Memory; Long Term Memory; Response Formulation) before an overt response is effected. The nature of these components, their organisation and their control mechanisms is, however, debated among theorists. Before proceeding further the reader should also be aware of the context within which a great deal of the research that developed information processing theories was undertaken. The work essentially started in the 1950s before really gathering pace during the 1960s and 1970s. Should you choose to follow up any of the readings at the end of this chapter you may well end up coming to the conclusion that the human being only possesses two senses: the visual sense and the auditory sense. These are the two modalities which dominate the study of Human Information Processing simply because they were the easiest modalities to manipulate the presentation of stimuli in the laboratory in a controlled manner. There are other senses, but not the five that are normally touted (sight, sound, touch, smell and taste). These are a hangover from Aristotle with little or no psychological or physiological basis. Somewhere between nine and 21 senses have been identified, depending upon what classification scheme is used. One commonly used basis for categorisation of the senses is:

- *Chemoreception* – this includes the olfactory sense (smell) and gustatory sense (taste).
- *Photoreception* – this involves sight in the eye using both rod (monochromatic, low light receptors in the periphery on the retina) and cone photoreceptors (colour sensitive receptors in the central, foveal region).
- *Mechanoreception* – this involves receptors which respond to mechanical pressure or distortion, giving an awareness of the relative position of parts of the body (proprioception) or movement and acceleration of the limbs (kinesthesia).
- *Thermoception* – this is the sense by which temperature is perceived.

In addition to the above basic categorisation, many would add one or more further sensory categories:

- *Nociception* – this is the sense of pain. Three types of pain receptors have been identified: cutaneous (skin); somatic (joints and bones) and visceral (body organs).

- *Equilibrioception* – this is the sense of balance, mainly signalled by the detection of accelerations in the inner ear (although some argue this is another aspect of mechanoreception).

There are also other non-human senses such as magnetoreception (the ability to sense the Earth's magnetic fields, as used by homing pigeons) and electroreception (the ability to detect electric fields, found in several species of fish such as sharks and rays). One of these non-human senses would potentially be of some use to pilots.

The remainder of this chapter will tend to assume that there are just the two senses, however there is beginning to be more work done in laboratories on haptic interfaces (which involve mechanoreception – or the sense of touch) which also has implications for current functional models of the Human Information Processing system. And it has to be emphasised, what is presented in the following pages is merely a schematic representation of human cognition. If you were to take off the top of a pilot's head you would not be able to recognise any of the structures discussed on the following pages. Nevertheless, such models do provide a useful representation of how Human Information Processing works and can be used to inform the design of flight decks, procedures and training.

Figure 2.1 contains such a schematic model of Human Information Processing (adapted from Wickens, Lee, Liu and Gordon Becker, 2004). This is typical of many models of this type. It can be seen that human response is broken down into three basic stages: perceptual encoding (what is it?, where is it?, what does it do?, etc.); central processing (what shall I

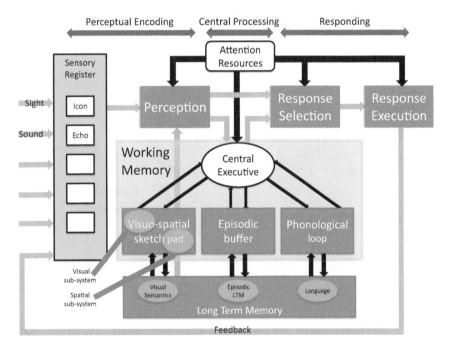

Figure 2.1   Adapted form of Wickens's schematic model of Human Information Processing incorporating Baddeley's proposed structure of Working Memory

do about it?); and responding (doing something about it, if required). In the companion figure (Figure 2.2) it can be seen that this is a Multiple Resource Theory (MRT) of human cognitive resources. As will be discussed later, there are also single resource theories of cognitive resources (e.g. Kahneman, 1973, or Norman and Bobrow, 1975).

Sensory data arrives via the senses into the Sensory Register, which is a very short term, very high capacity, memory store. It is assumed that there is one of these stores for each sense (for the visual sense the Sensory Register component is sometimes referred to as the *Icon*; for the auditory senses it is called the *Echo*). Somewhere around this stage perception occurs, where these sensory data begin to have properties attached to them to give them meaning. Perception is an active process (note the link to Long Term Memory). This is a process by which people begin to interpret and organise the sensory data to produce a meaningful experience of the world. It is essential to distinguish sensation from perception. Sensation refers to the immediate unprocessed stimulation of the sensory receptors and it must precede perception. Perception involves further processing of these inputs to begin to extract their meaning.

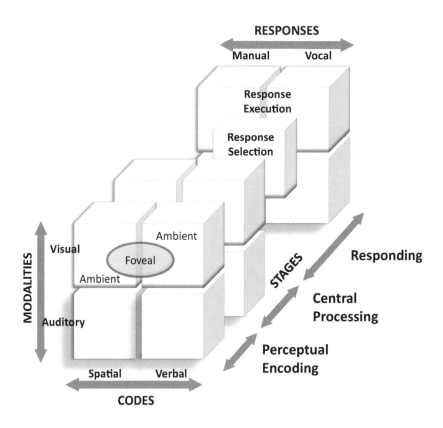

Figure 2.2   Adapted form of Wickens's (1984, 2002) Multiple Resource Theory (MRT)

When considering the processing of information it is sometimes useful to bear in mind Ackoff's (1989) categorisation of what may broadly be termed 'information':

- *Data* – Basic building blocks/symbols;
- *Information* – Data that have been combined and processed concerning questions such as 'who?', 'what?', 'where?' and 'when?'.
- *Knowledge* – Applies information to questions concerning 'how?'
- *Understanding* – Begins to provide an appreciation concerning 'why?' questions; and
- *Wisdom* – Provides an evaluated understanding of the situation.

This categorisation will be re-visited in the discussion of Situation Awareness in Chapter 4 and Chapter 12 looking at the cognitive aspects of flight deck design in Part Three. However, one way of characterising Human Information Processing is as a data transformation process.

In the case of a reflex action, perception may immediately trigger the execution of a response with no conscious thought. However, of more interest in this case is the process which involves thought and decision making on the data/information selected for further processing (we will come back to the hypothesised mechanisms by which this occurs). For now it is enough to know that only a relatively little amount of the input to the senses is selected, based upon its pertinence to the situation. This further processing occurs in WM.

A crude analogy would be to equate WM with RAM (Random Access Memory) in a computer in as much as WM is the place where all the major processing happens but it also has only a limited capacity and only so much information can be stored and manipulated in it, irrespective of the source of the information (i.e. the senses or Long Term Memory – LTM). However, unlike RAM in a computer, WM is not of a set size. The effective size of WM (and the Human Information Processing system as a whole) is dictated by the level of arousal. It is also hard to define how large the system is as it is difficult to establish how large a 'unit' of information is, in psychological terms!

LTM, in a similar way, can almost be characterised as the computer's hard-disk, where data/information and programs for manipulating this material are stored. Again, this is about as far as any analogy with a computer should be pushed. Information in LTM is stored and organised in a very different manner. However, it can be noted from Figure 2.1 that there is a dynamic link between the Sensory Register/perceptual processes and LTM and also between WM and LTM. Information processing is not a passive process: it is an active, dynamic process.

Attention is intimately related to the concept of cognitive resources. It is the process that allocates cognitive resources to salient features to support (or re-direct) behaviour. It is a basic function that is a precursor to all other cognitive functions. The attentional process has to both sustain focus on relevant information while filtering out extraneous information. However, the same processes that support cognitive focus also need to be able to switch this focus rapidly to other issues in the environment if these are more important than the task at hand. These attentional processes are under the control of a higher-order executive function.

Finally, the capacity of the Human Information Processing system seems to be dictated by the level of arousal. Arousal is a very general term used to refer to the level of cognitive activity of a person or their readiness to engage in activity. Some of the earliest psychological experiments conducted (in the first decades of the twentieth century) investigated the concept of arousal and its relationship to performance. These experiments led to the famous (within psychology) inverted 'U' hypothesis postulated by Yerkes and Dodson (1908). High levels of arousal are thought to reduce overall cognitive capacity. Low levels of arousal cause attention to become labile and less focused, decreasing performance. Hence, there is an optimum level of arousal, which maximises cognitive capacity, enhances attentional focus and therefore underpins optimal performance.

The following sections describe these components of the Human Information Processing system in a little more detail. This discussion commences with an overview of the concept of attention, which can be considered as part of the controlling executive that directs cognition.

MODELS OF ATTENTION

Attention is difficult to define rigidly, however it can be characterised as the mental phenomenon of concentration, which is selectable (you can choose what you want to pay attention to); shiftable (attention may shift to a more pertinent element in the environment); and divisible (you can pay attention to more than one thing at a time, within reasonable bounds). The 'amount' of attention (in effect the amount of cognitive capacity, or 'thinking power' available) is a direct product of the level of arousal of the person. Attention is fundamental to the coordination and control of cognitive resources. It is also generally acknowledged that these cognitive resources (either singly or in combination) are of limited resource capacity. The concept of attention has been metaphorically likened to a 'searchlight' that directs consciousness.

There are four distinct areas of attention:

- *Selective Attention* – the selection for further processing, of sensory inputs and central processing events.
- *Focused Attention* – ignoring of certain inputs to focus on inputs and events associated with the chosen task.
- *Divided Attention* – the concurrent monitoring and processing of inputs and events associated with two or more tasks.
- *Sustained Attention* – the process of attending to relevant inputs and events over a sustained period.

A dual-task performance paradigm has often been used in the laboratory to explain information processing limitations (single versus multiple resources). The capacity of these cognitive resources also serve to define mental workload limitations, in terms of what can and cannot be done within a given period of time. Theories of dual-task performance focus on the notion of competition for limited information processing resources which, depending upon the nature of the tasks performed, may result in interference on one or both of the tasks.

As noted earlier, there are two general categories of theories of attention: single resource theories and multiple resource theories. Early models of attention were categorised as 'channel' or 'filter' theories. These were based on the idea that information processing was restricted by channel capacity. These channels were thought to correspond to inputs (left ear, right ear, eyes, etc.). Attention was thought to control the input channel attended to. In experiments it was noted that information on non-attended channels did not reach consciousness, hence it was thought that they were effectively filtered out. Filter theory, proposed by Broadbent (1958) was one of these models of 'early' selective attention (in terms of the stages of cognitive processing). Filtering of inputs occurred before any low level cognitive processing.

Triesman (1960) proposed a modification to Broadbent's model which suggested that inputs from the senses were attenuated, rather than filtered out entirely. This can be demonstrated by the 'cocktail party' phenomenon. You may be in deep conversation with someone in a noisy cocktail party, essentially filtering out all the background noise. Suddenly you orientate to another group in the room from where you think you have just heard your name spoken. The noise from the party has not been completely filtered out, it has merely been attenuated. When something highly semantically meaningful is present in the 'noise', it produces a stimulus strong enough for you to process it and divert the focus of your attention. The same phenomenon may also cause a pilot to immediately respond to their call-sign in a dense air traffic environment while largely ignoring those of other aircraft. This model of attention also implies that the content of the sensory channels is subject to a large degree of pre-attentive semantic analysis. Deutch and Deutch (1963) developed these early channel-based theories of attention further into a 'late' selection theory in which all inputs are fully processed but only a single input is responded to on the basis of its importance in the situation. This allows memory to better control and direct actions, rather than simply responding to environmental stimuli.

Kahneman (1973) proposed a different approach to these channel/filter theories. He proposed a single, undifferentiated pool of cognitive resources (but one which had an upper capacity limit). The ability to perform two separate tasks concurrently was hypothesised to depend upon the allocation of attention to each task. Interference between tasks was a product of the cognitive capacity required by each task. When the combined requirements imposed exceeded the cognitive capacity of the person, performance began to decline on one or both of the tasks (peak mental workload capacity was exceeded). Norman and Bobrow (1975) extended this model, demonstrating that performance could be adversely affected in two distinct ways: by either the quality of the input data (data-limited – e.g. poor quality flight deck displays) or by limitations on cognitive processing resources (resource-limited). Only in the latter case was the participant's cognitive capacity the performance limiting factor. As an addition to this observation, it was noted by Shiffrin and Schneider (1977) that highly skilled processes (e.g. the psychomotor aspects of driving in an experienced driver) were not really available to consciousness and also took up very few cognitive resources (very little cognitive capacity). This led them to distinguish between 'controlled' and 'automatic' processes: practice reduces the attentional capacity requirements of a task, which has implications for the 'divided attention' experimental paradigm. Performance was deemed to be a product of experience as well as cognitive capacity. From a Human Information Processing perspective, one of the fundamental characteristics of expert performance is that it requires fewer cognitive resources than the same level of performance in a novice. Hence this also has implications for the measurement of mental workload, if it is assumed that workload is directly related to cognitive capacity (see later).

Multiple cognitive resource theories, for example the MRT model (Wickens, 1984; Wickens, 2008; Wickens and Hollands, 2000) posit that there are separate pools of information processing resources. Wickens (1984) noted that in certain circumstances (e.g. when both tasks were verbal or spatial in nature) increasing the difficulty of one task reduced performance on the other task. However, when one task was spatial and the other verbal or vice versa, increasing the difficulty of one task did not necessarily reduce performance on the other task. In the case of Wickens's MRT three dimensions are hypothesised: processing stages, input modalities and input codes (see Figure 2.2). Two manners of responding are defined: manual and vocal.

Perceptual encoding, central processing and responding represent discrete stages in information processing. The perceptual encoding and central processing stages compete for the same cognitive resources but the resources used for the later responding stages (response selection and response execution) utilise a different cognitive resource. Pashler (1998) proposed that the primary information processing bottleneck is at the stage of response selection: this is a single, undifferentiated resource allocated on an 'all or nothing' basis. However, it is independent of all the other cognitive resources required at the other stages of information processing. It does not affect response execution which can proceed concurrently with perceptual encoding, central processing and response execution involving other stimuli. Changing the required difficulty in the responses required in one task has no effect on performance of a concurrent perceptual encoding/central processing task. Note that in Figure 2.2, which is an adapted version of Wickens's MRT theory, the response selection phase is represented as an undifferentiated resource (i.e. it is not tied to any of the modalities or codes at the input or central processing stages). This is where Pashler (1998) proposed that the primary information processing bottleneck occurred.

A further MRT dimension is input modality: visual or auditory. This dimension is crossed with the dimension of input code: spatial or verbal. Thus spoken language can be characterised as a verbal, auditory input; written language as a verbal, visual input; map reading as a spatial, visual input, and orienting to a sound in the environment as an auditory, spatial input. Wickens (2002) added a fourth dimension to the MRT model in 2002. He separated the visual channels into focal and ambient resources to account for situations where it was possible to perform two visual tasks at the same time (i.e. a focal and a peripheral/ambient visual task). Other research is now also beginning to indicate that the tactile senses (also referred to as haptic senses) may be a further independent information processing resource that can be used in parallel to the auditory and visual channels (e.g. Sklar and Sarter, 1999; Ho, Tan and Spence, 2005).

MRT makes certain predictions about human performance. For example, performance of two simultaneous tasks will be better if one task is presented visually and the other is presented via the auditory channel, rather than using the same modality for both. Similarly, don't ask a pilot to perform two spatial or two verbal tasks at the same time. Although somewhat simplistic in as much as it only incorporates two of the senses, MRT is useful for the scheduling of flight deck tasks and making some basic, predictions to avoid potential workload bottlenecks. Wickens has developed MRT to produce a prospective measure of workload (see Wickens, 2005, discussed further in Chapter 3 on Workload).

SENSORY REGISTER

The Sensory Register is a *very* high capacity *very* short term store that is modality specific (i.e. there are separate Sensory Registers for each of the senses). It is also a very high fidelity store of sensory information (hence the requirement for high capacity). The most highly researched aspect of the Sensory Register is the visual component, followed by the auditory component. As noted earlier, this is probably merely because these are the most amenable senses for study.

The operation of the Sensory Registers does not depend upon the allocation of cognitive resources – there is no requirement for a person to have to consciously attend to material in order for it to lodge there. The duration of material in the Sensory Register is approximately 250–300 msec for the icon (the visual Sensory Register) and about 2.5–3 seconds for its auditory counterpart, the echo. The difference in storage times seems to reflect the dissimilar nature of the manner in which the raw data impinges on these senses. Data concerning the visual environment can be thought of as being stored in an array (it is all available simultaneously) whereas auditory data arrives in a sequence and, as a result, meaning can only be determined once a reasonable amount of the data stream has been stored and undergone initial processing (especially in the case of language). Material in the Sensory Register is lost by decay, which is simply the loss of data as a result of the passage of time.

There seem to be three basic functions of the Sensory Register:

- *Limits sensory input* – it stops the information processing system being overwhelmed by many inputs as material is lost relatively quickly, simply by decay.
- *Acts as a buffer* – it allows the Information Processing system the time necessary to determine if an input needs to be processed further.
- *Provides stability* – of visual images, making visual perception appear smooth and continuous despite interruptions (e.g. blinking or moving your eyes).

It is in the Sensory Register that the initial processes of feature detection and pattern recognition occur. Salient data are then transferred to the perceptual encoding process and the higher parts of the Human Information Processing system before it decays (see Figure 2.1).

PERCEPTION

Sensation and perception are not synonymous terms. Sensation refers to the abilities of the human sensory system to detect inputs which impinge on the sensory channels or to detect differences in the magnitude of these inputs. Perception refers to the manner in which the Human Information Processing system interprets these sensory inputs and converts them into something meaningful. There is often a mismatch between what is sensed and what is perceived (this is the basis for visual illusions).

A large amount of data is being sensed at any one time and is transferred to the Sensory Register, however only a small amount of it is selected for further processing. Necessarily, sensation must precede perception. Sensation is a passive process: perception is an active process involved in selecting, organising and interpreting the data from the senses selected for higher processing. This can be demonstrated quite simply. The eye is actually a very simple camera. As a result, the size of an image falling on the retina not only corresponds to the size of the object but also to its distance from the viewer. Thus two objects of the same size at different distances will produce different size images and two objects of different sizes at different distances may produce the same size image on the retina, yet humans are (usually) able to distinguish between these two conditions. This is because people 'learn' the skill of depth cueing as they develop. Memory plays an essential part in this scheme. People intuitively 'learn' how large an object is; they 'learn' what shapes objects are and 'learn' to distinguish between figure and ground. As a further example, a door is always perceived as being rectangular even when the image on the retina is anything but rectangular.

This 'learning' is the basis for visual illusions, where what is learned to be true (so is perceived to be 'true') is in fact false. As a result of training and experience, pilots develop a mental image of the expected relationship between the length, width and shape of a typical runway during final approach. However, a final approach over flat terrain with an up-sloping runway can produce the illusion being high on final approach, encouraging the pilot to descend. Conversely, an approach over a flat terrain with a down-sloping runway can produce the illusion of being low, causing the pilot to arrest the rate of descent, potentially resulting in a go-around. The 'black-hole' approach illusion can happen during final approach over water or unlit terrain to a brightly lit runway on a dark night with no stars or moonlight and with no horizon. As a result of this lack of cues, particularly in the periphery of the eye, there may be the illusion of flying level and of the runway sloping. With bright lights beyond the runway there may be the illusion of being too high on approach encouraging the pilot to pitch down. Pilots may also orientate themselves to false horizons (which are not horizontal) for example when flying over cloud or featureless terrain.

There are two major approaches to the theory of perception. 'Bottom-up' theories of perception suggest that we construct our perception of reality from a set of primitive 'features' to build up our internal representations. Conversely, 'top-down' approaches posit that perception starts with a set of primitive features but our experience is also influenced by higher-level factors from memory, such as knowledge and context.

There are various examples of different 'bottom-up' approaches to perception:

- *Exemplar theory* puts forward the premise that examples of all objects seen are stored as exemplars. Perception involves comparing the sensory pattern received with the bank of exemplars until a match is found.
- *Prototype theories* of perception hypothesise that instead of storing many exemplars a more generic prototype of an object is formed and stored. Perception is therefore the process of comparing the sensory input with the closest match to stored prototypes.
- *Feature theory* puts forward that the feature primitives stored are much less complex than exemplars or prototypes. These primitives are simply aspects such as lines and corners. Perception starts with the identification of these basic features which are then assembled hierarchically into more and more complex objects until the object is identified.
- *Structural description theory* can be conceptualised as a 3-D version of feature theory. In this case the primitive features stored are simple geometric shapes (geons) rather than lines and corners, etc.

'Bottom-up' theories of perception alone are not enough. 'Top-down' theories of perception hypothesise that higher cognitive processes help to determine what is perceived. 'Top-down' perception requires some 'bottom-up' processing of the image to occur, the results of which are then analysed by higher level processes to suggest objects or to provide further information, such as determining the object's size or distance from its context. 'Top-down' perception relies on the fact that there is a great deal of semantic redundancy in most visual scenes: you see what you expect to see. For example, Biederman

(1981) observed an increase in errors in identifying objects such as fire hydrants or sofas when they were presented out of their normal context (e.g. in a kitchen or floating over a city street). The same processes apply to the illusions encountered when making a final approach discussed earlier.

The perception of depth and motion is important in piloting an aeroplane and both these aspects demonstrate how perception is a learned, active skill rather than a passive process. Depth perception results from a number of environmental cues (all of which demonstrate the inadequacy of 'bottom-up' driven approaches to perception when taken alone, as these cues need to be learned from experience). For example:

- *Motion parallax* – when moving, nearby things appear to pass more quickly than objects further away, which can almost appear to be stationary.
- *Kinetic depth perception* – as moving objects become smaller, they appear to move farther away, and vice versa.
- *Perspective* – parallel lines appear to converge at infinity (however, you need to 'know' that the lines are actually parallel).
- *Relative size* – if you have a prototype of the typical size of an object (e.g. a car) the visual system uses the relative size of familiar objects to help perceive distance.
- *Luminance contrast* – objects at a greater distance away have lower luminance contrast and colour saturation.
- *Occlusion* – objects behind other objects are further away.
- *Texture gradient* – texture becomes less easy to determine as distance increases.

There are also other cues to depth and motion that originate from the architecture of the human visual system itself:

- *Stereopsis* works as a result of humans having two eyes set slightly apart on their head which produce two images of the same scene from slightly different angles. If an object is far away, the disparity between the images falling on both retinas is small: when the object is nearer the disparity is larger.
- *Convergence* is a further product of stereopsis. As an object gets closer the two eyes focus on it and in doing so converge, which stretches the extraocular muscles. Furthermore, the angle of convergence of the eyes is smaller when fixating on objects further away.
- *Accommodation* is another muscular cue to depth perception. When focusing on distant objects the ciliary muscles stretch the eye's lens to make it thinner. This kinaesthetic sensation is used as a cue for interpreting depth.

Again, all of this illustrates that perception is an active, learned process requiring access to longer-term stores of information in the cognitive system.

WORKING MEMORY

After perception has taken place, the next destination for information is WM. Transfer of information between is controlled by the attentional processes. The role of WM is a longer term store of information (in the region of 30 seconds or so) although the length of information storage can be increased by conscious control. WM has a much lower storage capacity than the Sensory Register.

The concept of 'Working Memory' has replaced the older concept of 'Short Term Memory' (STM). STM was conceptualised as a much more passive store of information, in contrast to the functioning of WM, which is more as a central information processor (or RAM to use the computer analogy). The characteristics of STM were held to be that:

- It was of a very limited capacity, typically 7±2 'chunks' (from the famous work by Miller, 1956; 'The Magical Number Seven, Plus or Minus Two').
- If rehearsal was prevented, information was lost through decay and overwriting by subsequent information.
- Information was held in an Auditory, Verbal, Linguistic (AVL) code and as a result the contents of STM were liable to acoustic confusion.
- Entry of information to STM was via limited capacity channels from the senses.

The concept of a memory 'chunk' to describe the capacity of STM (or WM) is interesting. In a classic series of experiments in the mid 1950s, it was established that the size of STM was 7±2 'chunks' of information. However, the size of a 'chunk' in Human Information Processing terms does not correspond in any way to the concept of a 'bit' of information, as used in information theory. This is easily illustrated. Consider the following string of numbers:

1   0   6   6   1   9   3   9   1   8   1   5

It would be expected that in a simple recall task participants should be able to retrieve between five to nine of these numbers (approximately two-thirds of them from the beginning of the sequence as a result of storage in the phonological loop and one-third from the end of the list from the acoustic buffer – the so called 'primacy' and 'recency' effect). In this case one number corresponds to one 'chunk' of information. However, the more astute of you may be able to recall all of these numbers if you recognise something else about them:

1066   1939   1815

If the later strategy is employed then far more *numbers* can be remembered but only about the same number of *dates* would be remembered. Hence, it is difficult to establish how big a 'chunk' is and again, the interactive nature of STM (or WM) and LTM is emphasised. It should also be noted that these dates are far more meaningful for the British readers than for readers from other countries (Battle of Hastings; Britain's entry to World War II; Battle of Waterloo). STM/WM does not exist alone as the memory for these dates implies that they are recognised as being important dates (which are culturally dependent) already encoded in LTM.

The fact that information is held in an AVL code and is subject to acoustical confusions (e.g. 'K' instead of 'A'; 'V' instead of 'B') has implications for task and flight deck design. Consider the way in which most information is currently passed to the pilot (via radio). If it is a long stream of information (exceeding 7±2 'chunks') some of it is likely to be lost, or even if parts of it are not lost and the entire list is perceived correctly, some parts of communications such as an ATC clearance may be susceptible to acoustical confusions.

Computer data link provides a much better option when these limitations to the Human Information Processing system are considered (it is not transient – the clearance can be referred to again). However, the Human Information Processing system has another trick up its sleeve to be found in LTM ...

The concept of STM could be described as a 'necessary but not sufficient' description of the role and function of this aspect of the Human Information Processing system. In effect, the features of STM only describe one aspect of WM.

There is a counterpart to STM in the information processing system that deals with longer-term storage of visual information – VSTM (Visual STM). It was demonstrated that visually-presented objects, deliberately designed so that it was not possible to label them using a linguistic code, could still be stored by people for a short period of time. The capacity of VSTM was difficult to determine, though, as objects (unlike words, letters or numbers) can have many different properties associated with them (e.g. size, shape, colour, position) and these factors influenced performance. Nevertheless, it was established that there was a component of STM that did not hold information in an AVL code.

To account for these findings Baddeley and Hitch (1974) proposed that WM had two main storage components: the Phonological Loop (cf. STM) and the Visuo-Spatial Sketch-Pad (cf. VSTM). The phonological loop deals with verbal information. The Visuo-Spatial Sketch-Pad serves a similar purpose but for non-verbal information. Logie (1995) proposed that the Visuo-Spatial Sketch-Pad could be further broken down into two sub-systems, a visual subsystem (dealing with shape, colour or texture) and a spatial sub-system (dealing with the location of objects). The two-component model was subsequently extended by Baddeley (2000) to encompass a third component, the Episodic Buffer. These components were all under the control of a Central Executive (see Figure 2.1, which incorporates Baddeley's revised model of WM).

Baddeley (2000) described the Episodic Buffer as being:

> ... a limited capacity system that provides temporary storage of information held in a multimodal code, which is capable of binding information from the subsidiary systems, and from long-term memory, into a unitary episodic representation ... focussing attention on the processes of integrating information, rather than on the isolation of the subsystems. (p. 417)

Baddeley argued that the Episodic Buffer was required as neither the Phonological Loop nor the Visuo-Spatial Sketch Pad could be regarded as a general memory store in which several kinds of information can be combined. He characterised the Episodic Store as being:

• Capable of integrating information from a variety of sources (the senses and LTM).
• Controlled by the Central Executive through conscious awareness.
• Holding episodes (memory events) across space and over time.
• Having a central role in the storage and retrieval of items in episodic LTM.

The three components of WM were posited to be under the control of a Central Executive function which was responsible for coordinating performance on two (or more) tasks, controlling memory retrieval strategies and endowing the information processing system with the ability to attend selectively (Baddeley, 1996).

It will be noted that all aspects of WM and perception have an implicit link to a much longer term store of information that is required for them to operate. LTM, however, is structured very differently to these other aspects of the information processing system and the manner in which data/information is stored is also unlike that used previously. Not all information held in WM is transferred to this longer-term store. The primary method by which data (or maybe information by now) is transferred (encoded) from the various components of WM into LTM is by elaboration (mental rehearsal).

LONG TERM MEMORY (LTM)

The structure of LTM has proven to be very difficult to determine. It is unlikely that information is held in either an AVL code or as a visual representation. The memory code in LTM is probably semantic in nature (represented in terms of meaning).

LTM holds two basic types of memory: procedural and declarative. Declarative memory has two further sub-components: semantic and episodic memory:

- *Procedural Memory* contains knowledge of skills, strategies and procedures concerning how to undertake tasks. Procedures are stored as a series of steps (stimulus-response pairings) which when activated trigger subsequent steps. Procedural memories are not usually available to consciousness.
- *Declarative Memory* concerns memories that are consciously available:
  - *Episodic Memory* is the ability to recall past experiences and events, including information related to those events. Episodic memory is stored as images.
  - *Semantic Memory* is the memory for facts, rules, concepts, principles and problem-solving skills. Memories are stored as networks or schemata.

Schemata are memory structures used to organise knowledge in LTM. There are many different types of schemata, for example social schemata and role schemata, concerning how to behave in certain social situations, and person schemata containing information about individuals. Event schemata involve what happens in specific situations and are often referred to as 'scripts'. Schemata are often highly related to each other and can be arranged hierarchically (e.g. animal: mammal; human: pilot).

Schemata provide a very effective strategy for understanding the world and for organising new information into LTM. Information is more easily transferred and stored in LTM when it is meaningful, which in this case can be defined as related to well-established, existing schemata. This is why good text book chapters should start with a generic framework into which increasingly detailed information can be introduced. It begins to provide an overall framework and describes the relationship between elements (see Figure 2.1!). Various schemata are developed over time and become very robust (even if they are incorrect). As such, they can both influence and hinder the uptake of new information. Schemata in the form of scripts help to explain why most ATC clearances are usually acted upon correctly. Most clearances come in the form of a standard pattern. Thus, the pilot just inserts the appropriate information into the schemata template she/he already has for that situation. If ATC clearances are not given in the prescribed format, then they are likely to be recalled poorly. However, schemata can also act against the accurate storage (or recall) of information. When interviewing accident-involved pilots, the description of events can be distorted by well-developed schemata for the *assumed* sequence of events rather than the actual sequence of events. This is not a deliberate distortion of the truth. It is simply a product of the way in which the brain stores and manages information.

It is debatable if things are actually forgotten from LTM (cf. WM, where items are lost through decay, interference or over-writing). What is commonly referred to as forgetting is actually a failure to retrieve. Most people are familiar with the 'tip of the tongue' phenomenon where something can almost be retrieved, or what could not be retrieved on one day is suddenly remembered the next day. This can be demonstrated using two different experimental paradigms to investigate the properties of LTM: recall versus recognition recall strategies. In recall memory experiments the participant is required to simply reproduce information from memory. They may be given a list of words to learn and then asked to recall them several days later. In doing so they will often make semantically related substitutions (e.g. 'truck' instead of 'lorry' or 'aircraft' instead of 'aeroplane') and/or there will be quite a high proportion of failures to retrieve. The recognition experimental paradigm cues memory recall, so it can be thought of as a less complex task. In this case the experiment starts off in the same manner (learning a list of words), however, when asked to undertake the remembering part of the task the participant is presented with a word and asked whether it was present in the list or not (a *recognition* task rather than a pure *recall* task). In this case memory performance is considerably better. It would appear that a memory trace needs 'activating' to help restore it. One of the goals of software design is to minimise the amount of semantic knowledge that the user needs to acquire or retain to use a computer, or to ease the acquisition of that knowledge. This is why a Windows™ graphical user interface is easier to use than an old-style, command language DOS-based interface. DOS operating systems were based entirely on recall memory (e.g. you had to know and retrieve accurately the format of the commands to copy a file from one location to another). Windows eases LTM retrieval using recognition principles cued from icons and menus. One of the simple principles of the working of LTM is that recognition-prompted recall is far better than pure recall.

The Spreading Activation Theory of LTM storage posits that the more often a memory trace in LTM is activated, the stronger it becomes and hence it also becomes more available to the person without having to activate it using a priming stimulus (Anderson, 1983). LTM storage consists of a set of cognitive units, which are comprised of a node and a set of related elements. Material from WM is more likely to be encoded into LTM (as a memory trace or cognitive unit) as a result of rehearsal. The more often a cognitive unit is retrieved from memory the stronger the memory trace becomes and the easier retrieval becomes. Cognitive units also become associated with each other as a product of the number of times that they are retrieved together, basically forming the basis for scripts and schemata (see Figure 2.3).

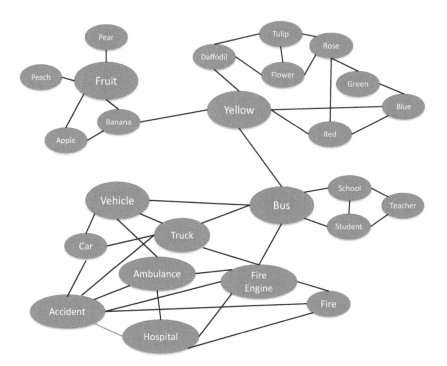

Figure 2.3   Diagrammatic representation of the storage of knowledge in LTM according to Spreading Activation Theory (adapted from Anderson, 1983)

AROUSAL

Arousal is a very general term used in psychology to refer to the cognitive activity of a person or their readiness to engage in activity. The concept is thought to have its roots in the primitive 'fight or flight' response. Arousal increases in a stressful situation in an attempt to maximise the body's ability to respond (principally physically). Perception of a potentially threatening situation causes the reticular activating system to increase general cortical levels of arousal, while simultaneously the autonomic nervous system prepares the body to use a large amount of energy, for example by an increase in heart rate, dilation of the bronchioles in the lungs and liberation of glucose, in preparation for either fight or flight. This is also the basis for certain types of stress (see Part Two).

Any discussion of this topic will invariably lead at some point to the inverted 'U' hypothesis postulated by Yerkes and Dodson (1908). Go and read the paper. The inverted 'U' hypothesis comes from one page (or so) towards the end of it and the data are derived from administering electric shocks to mice. Now remember that you are probably dealing with pilots!

The Yerkes–Dodson 'law' suggests that there is a relationship between arousal and performance. As arousal increases, performance increases, but only to a certain point (Figure 2.4). After this point arousal levels start to become too high and performance declines. The ascending component is associated with the energising effect of arousal whereas the descending leg is a result of the negative effects (stress) on cognitive processes.

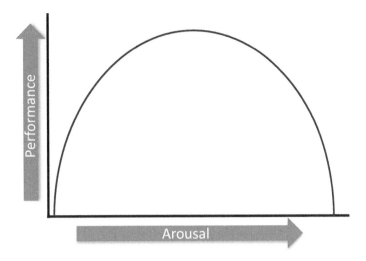

Figure 2.4   **Inverted 'U' hypothesis of arousal and performance (with its origins from Yerkes and Dodson, 1908)**

Easterbrook (1959) proposed that as arousal increased, so did the capacity of (what is now referred to as) WM, which explains the increase in performance on the ascending leg of the curve. Kahneman (1973) suggested that the level of arousal affected the capacity of the whole of the information processing system (a 'reservoir of mental energy'). The amount of cognitive capacity allocated to a task was a product of:

- involuntary processes;
- momentary intentions;
- evaluation of task demands; and
- the effects of arousal produced by external stressors.

However, after a certain point heightened arousal serves to decrease performance as cognitive capacity begins to shrink. This subsequently results in the progressive reduction in the range of environmental cues utilised. Irrelevant cues are filtered out in an attempt to maintain performance with reduced cognitive capacity. Further increases in arousal, for example as a result of the acutely stressful nature of a situation (e.g. an emergency) may result in a further reduction in cues being used (tunnel vision) and a further reduction in performance as a product of relevant cues being missed. It is assumed that a person is constantly attempting to keep their arousal at the optimum level, i.e. on the peak of the curve, to maximise the amount of cognitive resources available. In a dull, boring situation attention 'wanders' (it becomes more 'labile') in an attempt to find something more interesting and stimulating.

There would seem to be a curvilinear relationship between arousal and performance but it is very doubtful if this performance is as perfect as that depicted in Figure 2.4. This may be attributable to variations in attentional lability or cognitive capacity. Other researchers have suggested alternative (or complementary) explanations for the relationship between arousal and performance. The personality theorist Eysenck has suggested that people with an extrovert personality are naturally low on arousal, so seek out arousing situations to optimise their arousal levels. Conversely, introverts are naturally high in arousal, so avoid highly arousing situations in order to attain optimal arousal levels (Eysenck, 1967). As a result, personality will have an effect on cognitive performance when undertaking certain types of task and will also determine the suitability of certain types of people for certain types of job. As an example, people with an introverted personality are highly diligent and probably best suited to tasks requiring prolonged periods of monitoring, whereas those with an extroverted personality are most likely to be socially bold and enjoy the company of other people, as when working in a team. This directly leads to the other topics of individual differences (see later in this part) and Chapter 8 in Part Two concerned with pilot selection.

DISTRIBUTED COGNITION

Despite everything written in the preceding sections, Human Information Processing theories alone are not enough to explain human cognition. We need to look further than just the human being. Hutchins (1995a) developed the concept of distributed cognition. He argued that the approach taken by cognitive psychology is not wrong, but is limited. Its frame of reference was only the processes that occurred within an individual. Basically, it is argued that you cannot study human cognition without the context in which it takes place, as human beings use 'Things That Make Us Smart' (Norman, 1993). Everyone, either knowingly or unknowingly uses artefacts around them to enhance their cognitive abilities. Distributed cognition is the co-ordination between individuals, artefacts and the environment. At the simplest level possible, writing things down on a piece of paper using a pencil improves human memory, either in the long term (e.g. a diary) or in the short term (when doing long division). A pencil and paper makes us smarter. You don't forget soap powder at the supermarket as a result of overwriting of the memory trace. You can do long division with the aid of a pencil and paper, as the main information processing limitation is now not the size of WM but the accuracy with which you can recall and execute the procedure for doing such a calculation. The role and function of memory is now distributed between a human and a non-human component.

Rogers (1997) describes four generic properties of distributed cognition:

- Cognitive systems comprising more than one person have properties over and above those individuals making up the system (e.g. a flight deck).
- The knowledge possessed by members of such a system is highly variable and redundant: teams working together on a collaborative task will possess different kinds of knowledge and so will engage in interactions that allow them to pool their cognitive resources.
- Knowledge is shared by the individuals through communication but also by implicit communication and prior knowledge of each other, enabling them to engage in heedful interrelating during tasks.
- Distribution of access to information and sharing access and knowledge promotes coordinated action.

Hutchins (1995b) in a classic paper 'How A Cockpit Remembers Its Speeds' illustrates the manner in which the cognitive representation of speed and the processes for calculating target speeds are distributed across human and machine agents on a flight deck. The agents used include pilots' LTM, speed bugs on altimeters, speed reference cards and flight deck/ATM procedures. Speed calculation and speed awareness *is not* a problem in WM. It is a problem in distributed cognition across the flight deck which ultimately results in smarter flight crew and a more resilient system. Speed representation and calculation is best understood on a system-wide basis.

FURTHER READING

For further, more in depth treatment of Human Information Processing try:

Eysenck, M.W. and Keane, M.T. (2005). *Cognitive Psychology: A Student's Handbook (5th edition)*. Hove, UK: Psychology Press.

For a *slightly* more applied approach to the topic but with a greater aviation flavour to it, I would thoroughly recommend either:

Wickens, C.D. and Hollands, J.G. (1999). *Engineering Psychology and Human Performance (3rd edition)*. Upper Saddle River, NJ: Prentice Hall.

Or:

Wickens, C.D., Lee, J.D., Liu, Y. and Gordon Becker, S.E. (2004). *An Introduction to Human Factors Engineering (2nd edition)*. Upper Saddle River, NJ: Pearson Education.

The reader interested in Distributed Cognition is referred to the classic:

Hutchins, E. (1995a). *Cognition in the Wild*. Cambridge, MA: MIT Press.

3
Workload

Workload is an enduring concern for all personnel in the aerospace industry, whether they are involved directly with flight operations or indirectly through research and development. High workload is associated with increased error rates (and hence a concomitant decrease in safety margins) as well as having the effect of reducing overall productivity and increasing occupational stress (e.g. Moray, 1988; Sharit and Salvendy, 1982).

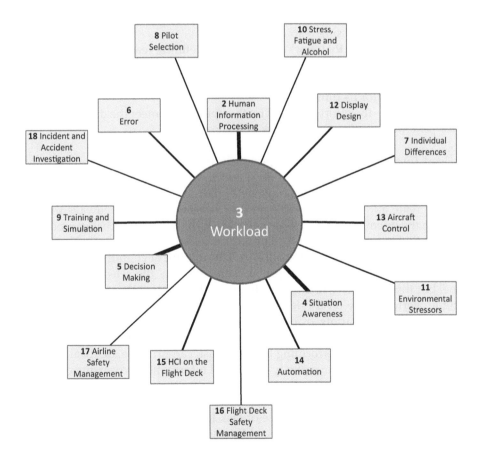

Mental Workload is a product of information load in WM, a characteristic that it shares with Situation Awareness. Mental workload is also highly related to the quality of pilot decision making. Workload is imposed by aircraft characteristics, such as the design of the controls, displays and automation, and the quality of pilot training.

This chapter commences with a description of the concept of mental workload followed by a brief overview of the four major approaches to its assessment. The chapter concludes with some application guidelines for the appropriate use of these measures.

WHAT IS MENTAL WORKLOAD?

There is no universally accepted definition of mental workload. There are almost as many definitions of workload as there are researchers in the area but in general it tends to be defined as the 'cost' (in information processing terms) of performing a given flight task. It is intimately related to Human Information Processing theory and specifically the finite capacity of cognitive resources. Theories of dual-task performance attempt to explain the information processing limitations that ultimately define a pilot's workload limits in terms of the number and nature of tasks that can (or cannot) be performed within a unit of time. The underlying concept is that cognitive workload is a product of competition for limited information processing resources.

Jahns (1973) suggested that workload has three defining features:

- *Input load* (also referred to as task load) – which is a product of the amount of 'work' to be done and/or its difficulty.
- *Operator effort* – the effort supplied by the operator to meet the demands of the task.
- *Task performance* – the actual results of their efforts.

Task load (input load) is imposed primarily by the *Mission* requirements and how these are translated into the pilot's control and monitoring actions by the *Machine* (i.e. the aircraft). Ideally the *Machine* should help alleviate workload imposed by the task rather than create additional demands on the crew but this is not always the case (see comments in the opening chapter). Some aspects of the *Mission* imposing input task load can be controlled to some degree, for example, some elements of task sequencing and the duration of the *Mission* (see Johannsen, 1979). The amount of cognitive capacity available to meet the task demands can be improved somewhat by the selection and training of personnel but this capacity will also be adversely affected by other factors such as stress and fatigue. Performance on the task is strongly influenced by the demands required by the task versus the supply of cognitive capacity available to meet these demands. However, workload *is not* performance (as we will see shortly when considering primary task measures). Two pilots performing the same task can produce identical performance; however, the first pilot may have plenty of residual information processing capacity left to allocate to other concurrent tasks whereas the second pilot may not be so lucky (Vidulich and Wickens, 1986; Yeh and Wickens, 1988).

Workload is the subjective experience of task load and, as alluded to in the previous paragraph, is moderated by individual differences. Nevertheless, regardless of these individual differences, the level of effort required to perform a task is a product of task requirements and flight deck design and is moderated by the assistance provided to the pilot. However, increasing the levels of flight deck automation does not necessarily decrease the level of workload: it often does quite the opposite (e.g. Wiener and Curry, 1980). Automation merely alters the nature of the task requirements hence the nature of the workload experienced by the crew also changes. Moray (1988) distinguishes between *cognitive workload* ('thinking') and *perceptual motor load* ('doing'). Moray suggested that *cognitive workload* is most likely to be increased by automation as a result of the aircraft systems primarily relieving the pilot of the perceptual motor tasks (flying the aircraft). This results in the pilot becoming a '*goal-orientated planner and decision maker*' which places greater emphasis on their cognitive skills. More importance is now being placed on managing the cognitive elements of workload through appropriate flight deck design and supplying more help and guidance to the pilot for flight planning purposes. However, workload can now be reduced by more appropriate implementation of automation but possibly at the expense of Situation Awareness, leaving the designer with such abstruse decisions to make as 'how many "units" of workload would one trade for one "unit" of situation awareness?' (Kelly, 2004).

Mental workload assessment has been a key aspect of the flight deck certification process since 1993. Indeed until the implementation of the recent flight deck certification requirement (discussed in Part Three of this book) the assessment of pilot workload was the primary rule associated with Human Factors on the flight deck. FAR/CS 25.1523 (Minimum Flight Crew) requires that:

> The minimum flight crew must be established (see AC 25.1523-1) so that it is sufficient for safe operation, considering –
>
> (a)   The workload on individual crew members;
>
> (b)   The accessibility and ease of operation of necessary controls by the appropriate crew member; and ...

> (Code of Federal Regulations, 2003)

Appendix D to FAR/CS 25.1523 and Advisory Circular AC 25-1523-1 (FAA, 1993) define six basic workload functions and ten workload factors. Workload functions can be conceptualised as the basic tasks required to fly the aircraft. These functions impose workload on the pilots and they are:

- Flight path control;
- Collision avoidance;
- Navigation;
- Communications;
- Operation and monitoring of aircraft engines and systems; and
- Command decisions.

The workload imposed by the basic flight functions can be either ameliorated or exacerbated by the workload factors. These are things relating directly to the design of the aircraft and/or its operating procedures, and/or may be the product of other issues, including environmental factors and emergency/abnormal situations that occur during operations. The ten workload factors defined in Appendix D to FAR/CS 25.1523 and Advisory Circular AC 25-1523-1 are:

- The accessibility, ease, and simplicity of operation of all necessary flight, power, and equipment controls, including emergency fuel shut-off valves, electrical controls, electronic controls, pressurisation system controls, and engine controls.
- The accessibility and conspicuity of all necessary instruments and failure warning devices such as fire warning, electrical system malfunction, and other failure or caution indicators. The extent to which such instruments or devices direct the proper corrective action is also considered.
- The number, urgency, and complexity of operating procedures with particular consideration given to the specific fuel management schedule imposed by centre of gravity, structural or other considerations of an airworthiness nature, and to the ability of each engine to operate at all times from a single tank or source which is automatically replenished if fuel is also stored in other tanks.
- The degree and duration of concentrated mental and physical effort involved in normal operation and in diagnosing and coping with malfunctions and emergencies.
- The extent of required monitoring of the fuel, hydraulic, pressurisation, electrical, electronic, de-icing, and other systems while en route.
- The actions requiring a crew member to be unavailable at his assigned duty station, including: observation of systems, emergency operation of any control, and emergencies in any compartment.
- The degree of automation provided in the aircraft systems to afford (after failures or malfunctions) automatic crossover or isolation of difficulties to minimise the need for flight crew action to guard against loss of hydraulic or electric power to flight controls or to other essential systems.
- The communications and navigation workload.
- The possibility of increased workload associated with any emergency that may lead to other emergencies.
- Incapacitation of a flight crew member whenever the applicable operating rule requires a minimum flight crew of at least two pilots.

Workload doesn't just 'happen'. It is a stressor that needs to be controlled. These workload factors can all be managed by the good design of procedures and systems (see Figure 1.4 in Chapter 1). Certification is all about the measurement of the workload imposed by the aircraft and its operation to demonstrate that it is within acceptable bounds for safe fight, even in non-normal and emergency situations.

APPROACHES TO THE MEASUREMENT OF MENTAL WORKLOAD

Moray (1988) suggested three basic approaches to the measurement of workload: behavioural, physiological and subjective. To this list a fourth category of 'analytical approaches' can be added. All of these general approaches have relative strengths and weaknesses. O'Donnell and Eggemeier (1986) suggested the following criteria for the evaluation of a workload assessment technique:

- *Sensitivity* – the capability to discriminate significant variations in workload imposed by a task.
- *Diagnosticity* – the capability to discriminate between the amount of workload imposed on different cognitive resources (e.g. perceptual load versus central processing versus psychomotor control).
- *Intrusiveness* – the degree to which a measurement technique causes degradations in ongoing primary task performance.
- *Implementation* – the ease of implementing a workload technique (e.g. instrumentation requirements and participant training).
- *Acceptance* – the willingness of participants to use a particular technique (associated with its face validity).

To these criteria the usual psychometric criteria of validity and reliability can be added:

- *Validity* – does the measurement technique actually measure pilot workload?
- *Reliability* – does the technique produce the same estimates for workload in the same circumstances?

Furthermore, if the technique is purported to measure mental workload the method should also be insensitive to other task demands imposed by physical activity (Casali and Wierwille, 1984).

The validity issue is perhaps the hardest of the above criteria to address as there is no accepted definition of what workload is. As a result, it is difficult to assess if a measurement technique is successful in achieving its aims. However, all the previously described criteria for the assessment of workload measurement techniques should be borne in mind when evaluating the four general categories of approach. No one approach satisfies all of them.

Analytical Approaches

Analytical approaches are usually based upon a formal task analysis of some sort, often incorporating elements of both a behavioural, Hierarchical Task Analysis (HTA) and a Cognitive Task Analysis (CTA) described on some kind of task time-line. Behavioural forms of task analysis describe the tasks and sub-tasks that need to be performed and their organisation. Generally these are observable, discrete operations (e.g. push switch; read display; adjust thrust lever). Cognitive forms describe the underlying cognitive skills and processes for performing a task. Crandall, Klien and Hoffman (2006) suggest that cognitive task analysis comprises three distinct stages: knowledge elicitation (extracting information to provide information on events, structures or mental models); analysis of the data (to provide structure and explanation); and knowledge representation (to depict the nature of the relationships and to help inform design and operational processes). Many texts are available describing task analysis in its major forms in far more detail (e.g. Daiper and Stanton, 2004; Crandall, Klein and Hoffman, 2006; Hollnagel, 1993).

AWAS (Aircrew Workload Assessment System) is an analytical approach which utilises a description of piloting tasks arranged on a timeline. Each of these tasks are assessed for their individual demands on each of Wickens's MRT channels and conflicting demands (where two simultaneous tasks occupy the same information processing channel) are also identified (Davies, Tomoszeck, Hicks and White, 1995). The output from this analysis is a continuous output of predicted pilot workload. Wickens has also developed a predictive model of workload based upon the computational version of MRT (Wickens, 2002). This operates on a similar basis to AWAS. Any shared tasks identified in the cognitive task analysis are analysed with respect to the demands placed on them in four categories: Working Memory (spatial); Working Memory (verbal); response (manual) and response (verbal). Demands are estimated in each category simply on a scale of 0 (no demand) to 3 (heavy demand). The total workload demand is simply the sum of the individual demands but, more importantly, the resource conflict score is also established from a conflict matrix as a step in establishing the dual-task interference score, which is then allocated to one of the input tasks depending upon task prioritisation criteria.

These analytically-based approaches are useful in the design stages for a new system where there is a requirement to predict information processing conflicts and spot potential workload peaks early in the process. However, such predictive models take little account of a pilot's familiarity with the task, their skill and experience, and other complex interactions with the environment. Furthermore there is no universal metric of 'how much' workload is acceptable. In general, it is very difficult to distinguish these approaches from measures of task load. Most of the criteria for the adequacy of workload assessment techniques outlined previously are difficult to apply in this category: validity is difficult to establish and reliability is a product of analyst skill rather than the workload experienced. The same comment may be made about the technique's sensitivity and diagnosticity. In terms of implementation of the techniques the AWAS approach on the one hand requires a great deal of data and dedicated software programs to perform the predictive analyses; the MRT approach on the other hand requires psychological and domain expertise but only requires simple arithmetic (Wickens, 2005). However, participant acceptance is a non-issue!

Behavioural Approaches

Behavioural approaches can be further categorised into two sub-categories: primary task-based methods and those based upon secondary tasks.

Primary task-based approaches are predicated on the simple assumption that when you have to do more things per unit time you are working harder than when you have to do relatively few things in the same time period. Furthermore, when you can't do the task properly, then the workload is probably too high. High workload is inferred from poor performance.

Wierwille and Connor (1983) used this approach. They measured pilots' control inputs per unit time to the primary flight controls. The level of workload was manipulated experimentally by varying the wind vector, strength of gusts and the pitch stability of the aircraft. The authors claimed that this measure was sensitive to workload imposed in this manner. A similar approach was used by Zeitlin (1995) for estimating driver workload. It was observed that an index of driver workload derived from vehicle speed and the frequency of use of the brake pedal could discriminate between different categories of road (urban, rural, etc.).

The basic problems with this approach are self-evident. Firstly the observed behaviour from which the level of workload is inferred is probably a direct response to the requirements of the task (a windy day or a winding road) rather than the level of workload experienced (workload is the *experience* of task load). The measures are task specific and cannot be generalised outside either the domain or even the experiment itself. The only way to infer when workload is too high is when performance cannot be maintained (a cliff effect) although prior to this point performance can still be maintained but at an increasingly high cost in terms of the mental effort required (see Figure 3.1). However, this cost is invisible in terms of its effect on performance. Furthermore, as is the case with analytical approaches to workload assessment, these primary task measures are confounded by such factors as a pilot's familiarity with the task, their skill and past experience.

Using O'Donnell and Eggemeier's (1986) criteria, primary task measures are poor in terms of their sensitivity and diagnosticity for measuring workload, as they are completely confounded with measures of pilot performance. As such their validity as a workload measure is highly questionable. Sensitivity can only be applied to the level of task demand and it is almost impossible to address issues related to diagnosticity (in terms of the load imposed upon the pilot's individual processing resources). Issues relating to implementation are the same as those relating to the measurement of pilot performance. Reliability is also really only addressing issues in replicability of performance and not the level of mental workload experienced. The major strengths of these techniques are that they are non-intrusive and they have a high degree of participant acceptance.

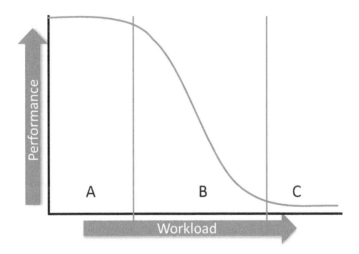

Figure 3.1   The hypothetical relationship between workload and performance. In region A pilot performance can be maintained as task demands (workload experienced) increases: task as demands do not exceed cognitive capacity. In region B cognitive capacity has now been reached. As task demands increase beyond cognitive capacity workload increases beyond acceptable levels and task performance suffers. Finally in region C task demands completely exceed the pilot's cognitive capacity. Workload is completely unacceptable and performance very poor

Consider the following which was actually observed during some flight control evaluation trials in an engineering flight deck. The 'objective' measures of performance were deviations from localiser and glideslope when performing an ILS approach. The evaluation pilot was flying two sets of flight control laws, one with US Department of Defense (DoD) level I handling qualities (desirable handling qualities) and one with DoD level II qualities (barely adequate). To everyone's surprise, pilot performance on the level II aircraft showed smaller deviations from localiser and glideslope than the level I aircraft: it was far superior! Looking at the 'objective' evidence it would seem that the level II aircraft had better handling qualities. However, on de-briefing it was found that if a concomitant measure of pilot workload had been taken it would have been noted that (perhaps) the level II aircraft occupied 90 per cent of the pilot's cognitive resources during the approach whereas the level I aircraft only occupied 10 per cent. The flight control laws with level II handling qualities were so poor that the pilot did not want the simulated aircraft to get away from the optimum approach path at any stage as it was so difficult to get back onto it. The pilot was constantly in the handling loop, making continual, small adjustments (hence the high workload). The level I aircraft was so easy to fly the pilot could let it get away from the localiser and glideslope and recover it back to the optimum approach path without even really trying! In other words, the pilot was just about within region A in Figure 3.1 with level II aircraft. You would wonder what performance would have been like if the pilot had to perform a task concurrent with flying the aeroplane. This leads neatly into a short discussion of secondary task measures.

Secondary task measures of workload are predicated upon the notions that workload is related to cognitive 'spare capacity' and that two compatible tasks may be performed simultaneously (i.e. they do not compete for cognitive resources – see Wickens's MRT) but only up to that point where the pilot's WM has sufficient capacity. Beyond this point performance on one task must suffer. The task of fundamental interest is designated the primary task. This is the one for which a workload measure is required and the participant is instructed that throughout the trials they *must* make all efforts to maintain their performance on this task. Any spare attentional capacity the participant is permitted to utilise is for performing a secondary, unrelated, task. As noted earlier, the nature of this secondary task will depend upon the nature of the primary task so that it doesn't compete directly for processing resources, but is usually something like a simple recall task, a choice reaction time task or a digit cancellation task (see Lysaght, Hill, Dick, Plamondon, Linto, Wierwille, Zaklad, Bittner and Wherry, 1989 for a comprehensive review). As task (input) demands in relation to the primary task are increased (for example in a manual flying task by varying the wind vector, strength of gusts and the pitch stability of the aircraft, as done by Wierwille and Connor, 1983, described earlier) performance on the secondary task suffers as the pilot is required to devote more cognitive resources to flying the aeroplane. As a result, workload is assessed by the decrements in performance of the secondary task. Ultimately, maximum workload capacity is delineated when performance on the primary task suffers with no attempt made to perform the secondary task (cf. primary task measures).

In general, secondary task approaches are quite rigorous and are sensitive to relatively small changes in workload but they are complex to undertake and are intrusive to perform (they can interfere with the primary task). They are also prone to the effects of experience (re-test reliability) and there is no standard performance metric (you can only infer 'spare capacity' from the measures applied to the primary and secondary tasks). As a result, they tend to be used in laboratory and simulator based studies and rarely in flight. The nature of the secondary task can also affect the participants buy-in to the study, especially if it is a non-flight related task that a pilot would never perform while actually flying (poor face validity). The actual validity of this approach can be quite high but is predicated upon the experimenter's view of the Human Information Processing system and their definition of operator mental workload. If implemented properly, though, using a battery of measures, sensitivity and diagnosticity can be high using secondary task approaches but may require a great deal of measurement equipment and subsequent analysis.

Physiological Approaches

There are a vast number of physiological measures that have been used to assess workload at some point. Probably those most often used relate to measures derived from an electrocardiogram (ECG), for example simple heart rate (or inter-beat interval), heart rate variability (sinus arrhythmia), height of the T-wave or more complex spectral analysis of the ECG output. Other related measures include: systolic and diastolic blood pressure; respiration rate; respiration rate variability; galvanic skin response; pupil diameter/blink response (the electro-oculogram – EOG); and other measures related to electroencephalograms (EEG).

The basic underlying premise is that when the pilot is under higher *mental* workload (note emphasis on the word mental) their heart beats more quickly (i.e. they have a shorter inter-beat interval, usually measured as the interval between R waves in the ECG – see Figure 3.2). This is because high mental workload (higher brain activity) requires a small increase in energy expenditure. The heart is under the control of the autonomic nervous system (i.e. not under conscious control) as is the control of the pupils of the eye. However, not only does mental workload increase heart rate (as suggested by Roscoe, 1993) it is also elevated by physical exertion in general, emotional factors (e.g. a high degree of responsibility or fear of failure – see Jorna, 1993), speech and high G-forces, any or all of which are encountered by a pilot in flight (Wilson, 1992). Other practical factors can also impede the use of any ECG-derived measure in flight, such as the pilot's straps rubbing against the backs of the electrodes resulting in spurious signals being generated. Moving the electrodes can avoid this but compromise the detail in the ECG wave form, hence limiting the analysis of its content.

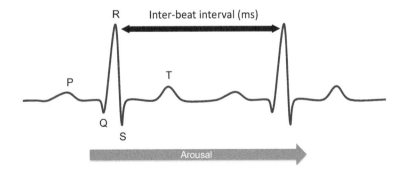

Figure 3.2   **Stylised components of the heartbeat wave form, with inter-beat interval indicated**

It has been suggested that a decrease in heart rate variability is more sensitive to increases in workload than heart rate alone (Zeier, 1979; Roscoe, 1992; Mulder, de Waard and Brookhuis, 2005). This is often measured with reference to the standard deviation of the inter-beat interval. However, all the above criticisms of heart rate can also apply, as the basic source of the data (and the factors that affect it) remains the same. Decreases in blood-pressure variability (another index of workload) are closely associated with decreases in heart rate variability.

More complex analysis of the ECG using frequency-based measures has also proved to be sensitive to fluctuations in mental workload, particularly in the mid-frequency band between 0.07–0.14 Hz, which is related to short-term regulation of blood-pressure, and in the high-frequency band between 0.15–0.50 Hz, which is associated with respiratory functions (Mulder, 1992). A decrease in power (the area under the wave form) in the mid-frequency and high-frequency components is associated with increasing mental effort and task demands (Mulder, 1992; Jorna, 1993; Veltman and Gaillard, 1993). Finally the height of the T-wave has also been found to decrease with increases in mental effort (Furedy, 1987). Jorna (1992), though, also points out that an individual's heart rate response is idiosyncratic and inferences about increases in workload can only be made with reference to an individual's resting ECG or their heart-rate response on a well-established reference task.

As noted before, high mental workload requires a small increase in energy expenditure hence the associated increase in heart rate. However, this also causes a small increase in respiration rate (Wilson and Eggemeier, 1991). Backs and Seljos (1994) observed that respiration increases as a result of increased temporal demands or WM load. But, as is the case with all ECG-derived measures, all measures associated with the rate of breathing are confounded by physical workload, stress and speech.

Moving away from ECG-derived measures of workload, electrodermal activity is measured by the Galvanic Skin Response (GSR) which reflects changes in skin conduction as a result of a small increase in the level of sweat production. GSR is sensitive to levels of arousal in general rather than mental workload *per se*. However, Kramer (1991) suggests that GSR is sensitive to information processing load. There are problems with measures of electrodermal activity, though. This approach tends to be sensitive to other factors, such as respiration, temperature, humidity, age, sex, time of day, season and emotional arousal. As such GSR is not a particularly selective measure of workload.

The EOG can actually be collected using the same equipment as the ECG – you just stick the electrodes around the pilot's eyes instead (although this can look somewhat unsightly and so is often unpopular)! Nevertheless EOG-derived measures can be sensitive to workload. There are three components to the EOG: blink rate, blink duration and the blink latency to a stimulus. Kramer (1991) observed that blink latency increases and blink duration decreases with an increase in task demands. However, blink frequency increases with fatigue and can also be affected by air quality, rather than workload. It has also been reported that eye pupil diameter increases concomitantly with information processing demands (Beatty, 1982) but this can also be affected by ambient illumination rather than workload, hence often restricting its use in an applied situation.

The final group of physiological measures of workload pick up signals directly from brain activity. EEGs pick up signals directly from the scalp, typically classified into four bands: delta waves (up to 4 Hz); theta waves (4–8 Hz); alpha waves (8–13 Hz) and beta waves (greater than 13 Hz). There is some evidence to suggest that alpha-wave activity decreases while theta-wave activity increases with an upsurge in task demands (Sirevaag,

Kramer, De Jong and Mecklinger, 1988). However, it has also been observed that increases in alpha-wave activity are more likely to be associated with decreases in arousal, rather than being directly related to workload (Wilson and Eggemeier, 1991). EEG is also difficult to collect and slightly intrusive for the experimental participant.

As a group of measures, physiological measures tend to be relatively easy to collect and do not intrude on task performance, however they are not particularly diagnostic and are often confounded by other factors not related to task demands, falling foul of Casali and Wierwille's (1984) criterion that any workload measure should be insensitive to other task demands imposed by physical activity. Interpretation of what these measures mean is often done in retrospect, limiting their utility and their predictive validity. Reliability is often poor but can be improved by baselining measures against resting measures (or reference tasks) and by the very careful collection of data. In general, physiological measures need to be supplemented with other measures of performance and/or workload to have any utility.

Subjective Approaches

Subjective approaches to the assessment of pilot workload come in two basic types: uni-dimensional measures of workload (e.g. the Bedford Scale by Ellis and Roscoe, 1982) and multi-dimensional measures (e.g. the Subjective Workload Assessment Technique – SWAT (Reid and Nygren, 1988); NASA Task Load Index – TLX (Hart and Staveland, 1988)).

Many uni-dimensional workload scales share a common ancestry. They are mostly descended from the Cooper–Harper aircraft handling qualities rating scale (Cooper and Harper, 1969). Examples include: the Honeywell Cooper–Harper rating scale (Wolf, 1978); the Modified Cooper–Harper rating scale (Wierwille and Casali, 1983); and the Bedford Scale (Ellis and Roscoe, 1982; Roscoe, 1984). An example of the latter is provided in Figure 3.3.

Uni-dimensional scales have the advantage of being very quick and easy to complete and can often be completed on-task, even in flight. Minimal training is required. In some cases, because they don't intrude unduly with the primary task, they can almost be used as a workload 'speedometer'. This approach is used by the ISA (Instantaneous Self-Assessment) technique, where the operator pushes one of five buttons, indicating low workload (1) to excessive workload (5) when prompted for a rating, which is usually about every two minutes (Kirwan, Evans, Donohoe, Kilner, Lamoureux, Atkinson and MacKendrick, 1997). In this way a workload profile is built up. This approach has been used to assess Air Traffic Controller workload and pilot workload when flying simulated novel approaches (Jarvis and Ebbatson, 2007). However, while the workload ratings may be meaningful, the best indication of high workload is not a rating of '5' but a complete failure to make a rating when prompted. This usually indicates that the pilot's brain is full!

Uni-dimensional workload scales based upon the Cooper–Harper scale have been used on many occasions in flight (real and simulated) and for many reasons, including for certification purposes demonstrating that the operating requirements of the flight deck do not impose excessive demands (Wainwright, 1987). Even so, it is still unclear exactly what uni-dimensional measures of workload actually measure. Vidulich (1988) suggests that subjective workload measures are sensitive to information processing demands in WM but are not sensitive to task execution demands. The Bedford Scale is supposed to measure

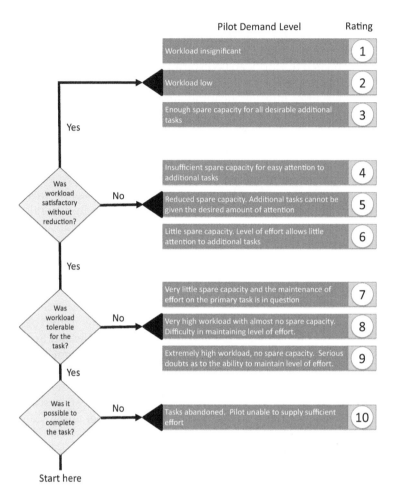

Pilot Demand Level   Rating

| Workload insignificant | 1 |

| Workload low | 2 |

| Enough spare capacity for all desirable additional tasks | 3 |

| Insufficient spare capacity for easy attention to additional tasks | 4 |

| Reduced spare capacity. Additional tasks cannot be given the desired amount of attention | 5 |

| Little spare capacity. Level of effort allows little attention to additional tasks | 6 |

| Very little spare capacity and the maintenance of effort on the primary task is in question | 7 |

| Very high workload with almost no spare capacity. Difficulty in maintaining level of effort. | 8 |

| Extremely high workload, no spare capacity. Serious doubts as to the ability to maintain level of effort. | 9 |

| Tasks abandoned. Pilot unable to supply sufficient effort | 10 |

Yes

Was workload satisfactory without reduction?   No

Yes

Was workload tolerable for the task?   No

Yes

Was it possible to complete the task?   No

Start here

Figure 3.3   Bedford uni-dimensional workload scale (adapted from Roscoe, 1984)

'spare capacity' in WM (see Figure 3.3) with the assumption that a great deal of spare capacity equates to low workload. Nevertheless it is unclear how someone completing the scale is supposed to introspect upon the information processing capacity they have available at any one point in time. This poses a significant problem for their validity. Uni-dimensional scales also only provide an overall assessment of the workload experienced, grossly limiting their diagnosticity. However, when used as a 'workload speedometer' they can be very useful in spotting workload peaks. Indeed, Wierwille (1988) suggested that momentary workload peaks (where task performance requirements exceed the pilot's capacity to perform) are of greater concern than averaged workload, particularly for design analyses. Being able to capture momentary work overload is a desirable property of any technique and uni-dimensional techniques may be the best option in this respect. One of the main failings of uni-dimensional techniques, however, is that their reliability tends to be quite poor. This may be related to a lack of sensitivity and diagnosticity.

Consider what is required of a pilot when making a workload rating using a uni-dimensional scale during a flight task. Using such a simple response format implicitly requires the rater to form mentally a composite score based on all the individual sources of workload in a particular segment of flight and then weight these components, relative to one another, with regard to the requirements of the task being flown, before finally deciding upon a workload score. Many implicit judgements and comparisons are required prior to making this final rating, hence there is a great deal of scope for variability in several aspects of the process that can contribute to this lack of reliability.

Multi-dimensional workload scales make this process explicit. Scales of this type require ratings to be made concerning the sources of the workload. For example, the NASA–TLX (Hart and Staveland, 1988) requires workload ratings to be made on three explicit dimensions: mental demand (how mentally demanding was the task?); physical demand (how physically demanding was the task?); and temporal demand (how hurried or rushed was the pace of the task?). In addition to these ratings concerning the sources of workload, three further ratings are also required: performance (how successful were you in accomplishing what you were asked to do?); effort (how hard did you have to work to accomplish your level of performance?); and frustration (how insecure, discouraged, irritated, stressed, and annoyed were you?). The latter group of ratings represent an interaction of the participant's performance with the task at hand (see Figure 3.4). The importance of each of these dimensions to the task being undertaken is then rated by making a series of 15 pairwise comparisons, the idea being that different sources of workload are more important in different situations. For example, when navigating using a map, stopwatch and compass in an old biplane, mental load may dominate as a result of few automated pilot aids being available. However, when re-configuring an aircraft for a last minute change of ILS approach from one parallel runway to another, while no one of the operations required is particularly difficult, there is a great deal of time pressure (temporal demand). The ratings for each scale are then multiplied by their associated importance weightings derived from the pairwise comparisons and summed to provide an overall workload score (software to perform the data capture process using the NASA–TLX is freely available from http://humansystems.arc.nasa.gov/groups/TLX/computer.php). However, the true power of any multidimensional scale resides in the interpretation of its individual sub-scales. The NASA–TLX can also provide a diagnostic workload profile for any particular task.

The SWAT methodology (Reid and Nygren, 1988) uses just three dimensions: time load (reflecting the demands placed by the time limits within which a task must be performed and the number of concurrent tasks that need doing); mental load (referring to the attentional demands imposed by the task); and stress load (which reflects the amount of stress imposed on the participant when performing the task, as evinced by their level of anxiety, confusion, risk or frustration). The SWAT approach is a two-step procedure. In the scale development phase, a number of activities, specified with respect to their task demands across the three workload dimensions, are rank-ordered by each participant. Using conjoint analysis these data are transformed into an interval workload scale ranging from 0 to 100. In the second phase the assessment of the workload associated with a particular activity is made. Each workload event is rated on a simple scale of one (low) to three (high) on each of the SWAT dimensions. Referring back to the results from the first phase, the scale value associated with this combination of scores is assigned as the workload score for that operation.

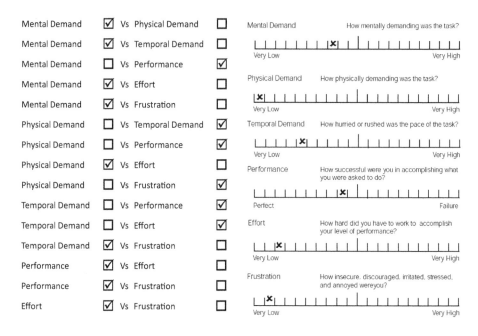

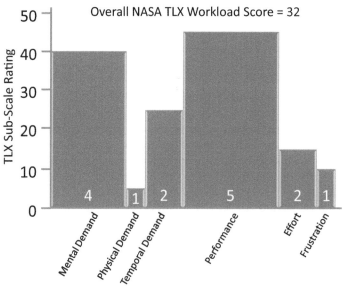

Figure 3.4   **NASA–TLX (Task Load Index) workload scale (adapted from Hart and Staveland, 1988). Source of workload (pairwise comparison of workload sources) is on the left: workload rating process using a 0–100 scale (in five-point graduations) is depicted on the right. Depiction of the resultant NASA-TLX output is shown below**

Both of these multidimensional techniques have been used in a vast range of studies of pilot workload (see Gawron, 2008) and both techniques have been shown to be reliable. In comparison with uni-dimensional techniques, the relative importance of specific, well defined aspects of workload for a particular instance are made explicit prior to performing the task. The pilot is then required to make workload evaluations using well-defined scales. Greater reliability is obtained as a result of making both more than a single rating and also through the manner by which the various aspects of workload are combined to produce an overall rating, which is now calculated through a mechanical and explicit process. Both scales also have content validity. In psychometric terms, for a scale to have content validity, its items must evaluate all major aspects of the domain to be assessed. Clearly, a uni-dimensional workload scale fails in this respect. The multi-dimensional nature of these scales also enhances their diagnosticity. However, both scales require detailed subject briefings and a great deal of preparation before they can be used. It is marginal if either approach can be used in real time during either simulated or real flight as a result of the number of ratings required. Both are likely to intrude on task performance hence their use tends to be restricted to the simulator or the laboratory. Furthermore, neither technique can be used as a 'workload speedometer' and both require skill to analyse and interpret.

A BRIEF ASIDE – MENTAL UNDERLOAD

This chapter has been concerned with the concept of mental workload and its measurement, however, in the aviation domain, the introduction of highly automated aircraft has generated a further concern: that of pilot underload. Automation on the flight deck has been criticised as vastly increasing mental workload demands on the pilots where the task load was already high (e.g. in the approach and landing phase) but decreasing further the demands on the pilot where the workload level was already low (e.g. in the cruise phase of flight). Many authors are now beginning to suggest that underload can be at least as detrimental to performance as overload (e.g. Parasuraman, Mouloua, Molloy and Hilburn, 1996).

In studies of driving it has been found that people susceptible to stress or fatigue may find their performance to be worse in conditions of underload (Desmond, Hancock and Monette, 1998; Matthews and Desmond, 1997). This was attributable to a failure to mobilise compensatory mental effort. Young and Stanton (2001: 2002) proposed a theory to account for why performance may be poor in very low workload (underload) conditions. They proposed that at very low levels of task demand attentional resources actually shrink (cf. other models of attention which assume a fixed attentional capacity). When task demands suddenly increase these now exceed the maximum attentional capacity normally available as these have shrunk due the very low levels of workload. This relationship is described in Figure 3.5.

Measuring underload in pilots though, is much more problematic. Braby, Harris and Muir (1993) developed an underload assessment process. It was found that the subjective experience of underload could be described in terms of an adjectival checklist reflecting aspects of mental effort and cognitive arousal. However, there was also a physiological component. An increase in work underload (if that makes any sense) was associated with a decrease in heart rate.

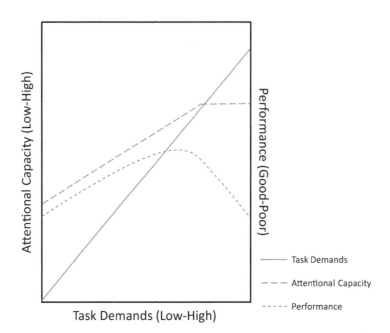

Figure 3.5   **The relationship between task demands and performance under a malleable attentional resources model. Underload conditions are at the bottom left of the figure (Young and Stanton, 2001)**

The major message is simply that work underload *is not* the same as very low workload. Different cognitive mechanisms are at play. Work underload seems to be associated with a shrinkage in attentional capacity, as opposed to low workload which is characterised by using only a little of the available attentional capacity. Furthermore, mental underload cannot be assessed using a workload scale: a response of '0.25' on the Bedford workload scale is meaningless. A bespoke work underload measurement approach is required.

APPLICATION GUIDELINES

All approaches to workload have their relative strengths and drawbacks. The question becomes which approach (or combination of approaches) should be used and in which circumstances?

In flight, the fundamental rule is that you must never interfere with the performance of the primary task in a safety critical situation: no obtrusive measures or ones that require divided attention on the part of the pilot should be used. This means that only simple, uni-dimensional scales can be used, and even then, only in carefully selected circumstances. You may have to wait for a workload rating until after a particular manoeuvre or flight segment has been completed. However, most physiological measures associated with the collection of ECG, respiration rate or blood pressure can be used at any time (although there can often be problems associated with locating and securing the equipment safely in the aircraft itself, especially in smaller types). In general, EEG, EOG and electrodermal responses are impractical to collect in flight. Consideration should always be made of employing a safety pilot.

In the simulator a far wider range of techniques can be used as the situation is not safety critical and you can freeze the simulation after every data collection point (although this can compromise the ecological validity of the trial). You can also afford to be a little more intrusive on the task. Any data collection activity you can perform in flight you can undertake in the simulator, plus in addition you also have the opportunity to collect good performance data and use more complex, multi-dimensional scales. To get pilot buy in, however, it is best not to 'push the envelope' with weird and wonderful secondary tasks (although you can embed them in an ecologically-valid manner at times, e.g. in the form of ATC instructions).

On the desk top in the laboratory, though, anything goes! You are not pretending it is an aeroplane in flight, so as long as the data collection technique fulfils the scientific requirements, and you have a willing guinea pig, any or all workload data collection methods are possible.

To help improve reliability of measures, particularly intra-rater reliability, consideration should be given to performing a series of workload ratings on a number of reference tasks (of varying levels of workload) to help 'calibrate' the pilots, prior to making the workload assessments on the tasks of interest. This also helps to avoid 'errors of severity' or 'errors of leniency' when making assessments. It is not unknown for participants to run out of scale length during data collection when an even harder (or easier) task comes along when the pilot has already used ratings at the extremes of the scale. All ratings tend to be made relative to the assessments made previously.

FURTHER READING

Very few comprehensive text books have been published in recent years on the topic of the measurement of mental workload. It would seem that workload research was very much a phenomenon of the 1980s and early 1990s. However, many of these older texts, despite their age, still contain valuable information about the theory and measurement of workload.

Hancock, P.A. and Meshkati, N. (1988). *Human Mental Workload*. Amsterdam: Elsevier Science.

Moray, N. (1979). *Mental Workload: Its Theory and Measurement* (NATO Conference Series 3: Human Factors). New York: Plenum.

Comprehensive overviews of many workload measurement techniques are available in either:

Gawron, V.J. (2008). *Human Performance, Workload, and Situational Awareness Measures Handbook*. London: CRC Press.

Or:

Stanton, N.A., Salmon, P.M., Walker, G.H., Baber, C. and Jenkins, D.P. (2006). *Human Factors Methods: A Practical Guide for Engineering and Design*. Aldershot: Ashgate.

There is also an annotated bibliography of workload measurement techniques potentially suitable for certification purposes freely available in:

Federal Aviation Administration (2005). *Advisory Circular AC 23-1523 Minimum Flight Crew.* Washington, DC: US Department of Transportation. Available at: http://www.faa.gov/regulations_policies/advisory_circulars/index.cfm/go/document.information/documentID/22118

4
Situation Awareness

All pilots like to be 'Situationally Aware'. Ask them what they mean by this and many will obligingly proffer a response along the lines of 'we like to be aware of our situation when we are flying …'. A slightly more helpful observation was provided by one pilot who commented 'you should never take your aeroplane somewhere where your head hasn't already been'. On a good day flight crew will often comment upon being 'ahead of the aircraft'; on a bad day they will lament about 'being behind the aircraft'.

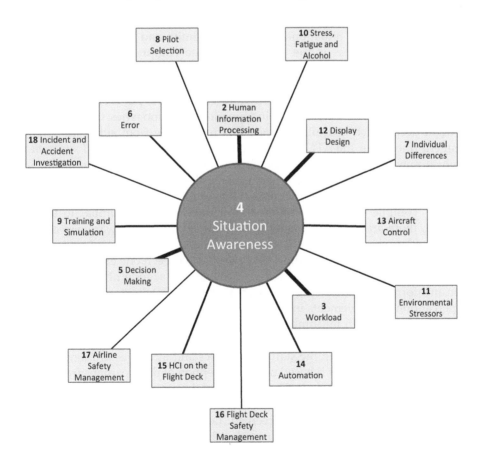

Situation Awareness lives in WM and is the precursor to all decision making. High levels of workload can impair Situation Awareness. Situation Awareness is also a product of the quality of the design of the aircraft's display systems. Error is a second order product of a lack of Situation Awareness.

Situation Awareness is required to perform the *Mission*. It may also be obvious to suggest that Situation Awareness applies to the hu*Man* aspects of any system. However it is also suggested that Situation Awareness can be resident in the *Machine* components if a distributed cognitive systems perspective is adopted. This chapter examines some of the definitions of Situation Awareness both from an individual and a crew perspective and also briefly introduces the reader to a dissenting voice, sceptical of the whole premise. The chapter concludes with an overview of the methods of assessing Situation Awareness and some application guidelines for these measures.

DEFINITION AND DESCRIPTION OF SITUATION AWARENESS

In the scientific literature there are various definitions of Situation Awareness (sometimes referred to as 'Situational Awareness'). Smith and Hancock (1995) proposed that 'Situational Awareness is the up-to-the minute comprehension of task relevant information that enables appropriate decision making under stress' (p. 59). Bell and Lyon (2000) suggest that, 'Situational Awareness could be defined as knowledge (in WM) about elements of the environment' (p. 142). Boy and Ferro (2004) put forward that Situation Awareness is a function of several quasi-independent situation types: the *available situation* on the flight deck which is available from data originating from either the aircraft, the environment (including the ATC) or other crew members; the *perceived situation* by the crew, which may be affected by various personal or environmental parameters such as workload, performance, noise and interruptions; the *expected situation* derived by the crew from their planning and decision-making processes; and/or the *inferred situation* by the crew, compiled from incomplete and/or uncertain data. Similar to the concept of workload, there are almost as many definitions of Situation Awareness as there are researchers, although as will be seen, some researchers reject the notion completely.

Mica Endsley was one of the earliest, and perhaps has been one of the most influential scientists working in the domain. Endsley (1995) developed a model of Situation Awareness based on an information-processing approach. Endsley (1995) defines Situation Awareness as 'the perception of the elements in the environment within a volume of time and space, the comprehension of their meaning, and the projection of their status in the future'. This model suggests that situation assessment begins with goal specification and includes at Level One the pilot perceiving the status, dynamics and attributes of relevant elements. At this lowest level awareness is based on the perception of data, perhaps from individual display elements on the flight deck. The fundamental premise is that if you have not got the basic building blocks (data) you cannot become situationally aware. At Level Two the pilot needs to comprehend what is perceived. Level Two Situation Awareness requires data to be combined in such a way as to provide a 'bigger picture' of the situation. The pilot needs to form a holistic picture of the situation and understand the significance of all the elements in it. At this stage data start to become information. Finally, Level Three Situation Awareness is achieved when the pilot fully understands his/her current situation and uses that information to project ahead to predict what is likely to happen in the near future. At this stage information starts to become knowledge. A model of Situation Awareness (and its importance in the decision-making process – see the following chapter) has been proposed by Endsley (1995) and is presented in Figure 4.1. Basically, Situation Awareness can be conceived of as 'living' in WM (see previous discussion of Human Information Processing) and its limitations are those inherent in the human cognitive system. It is also an active, not a passive process. Furthermore, Situation Awareness is not 'achieved'; it is a cyclical process that is ongoing throughout the flight.

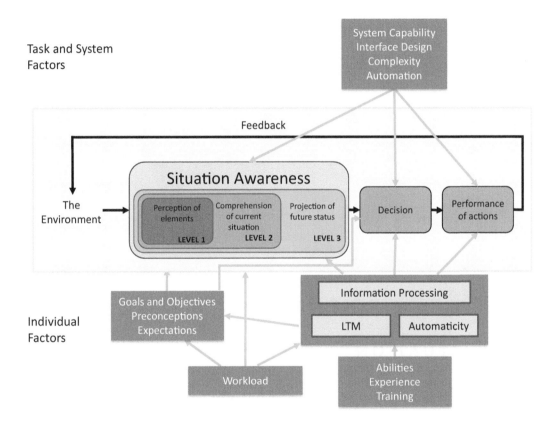

Figure 4.1 **Endsley's three-stage model of Situation Awareness (adapted from Endsley, 1995)**

Before going further it is worth making the distinction between Situation Awareness and Situation Assessment. Endsley describes Situation Awareness as a state of knowledge whereas Situation Assessment is the process (or processes) by which that state of knowledge is achieved, although not all researchers adopt this position. For example Smith and Hancock (1995) take a more process-oriented approach to Situation Awareness. Endsley also suggests that Situation Awareness is not solely a *product* of the Situation Assessment process; it also drives this same process in a manner akin to Neisser's (1976) perceptual cycle (see Figure 4.2). The perceptual cycle suggests that there is a reciprocal relationship between schemata (knowledge about the environment) which directs exploration of that environment (actions) which gathers information from the environment, which then in turn modifies the schemata. This is an ongoing process. This perceptual cycle describes how current knowledge, perception, action and the environment interact when engaging in goal-directed behaviour. With regard to developing Situation Awareness, a pilot's current awareness determines what she/he attends to which in itself dictates how the information is perceived, interpreted and what information is actively sought out subsequently (Endsley, 2000).

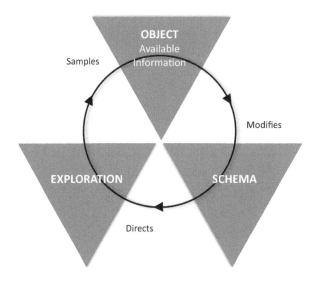

Figure 4.2   Neisser's (1976) Perceptual Cycle

CREW SITUATION AWARENESS

Up until this point, this brief discussion of Situation Awareness has concentrated on the individual. However, few pilots fly alone – most fly as a member of a crew. Even single-seat fighter pilots tend to fly as part of a formation of aircraft. This introduces the notions of shared and overlapping Situation Awareness (also referred to as team Situation Awareness).

The difference between shared and overlapping Situation Awareness is largely one of degree. In shared Situation Awareness all crew members have a common mental model and a complete shared understanding of the flight situation. An example of this is when undertaking an ILS approach with one pilot flying and the other pilot monitoring. To all intents and purposes, both pilots should have a common understanding of the situation which overlaps completely. However, during the normal conduct of the flight the crew will not have such a close (shared) appreciation of their situation. There will be some common elements to their Situation Awareness (e.g. where they are and what their immediate and longer term intent is – there many elements 'overlap') however it is likely that each crew member will be concentrating on their individual responsibilities associated with the role as Pilot Flying (flying the aircraft; dealing with navigation and general aircraft operation) or Pilot Not Flying (monitoring the flying pilot; monitoring aircraft performance; handling the radios; being responsible for monitoring the weather and for running the checklists). It can be seen that both pilots will be solely aware of many completely different things. It would actually be unproductive and inefficient for flight crew to attempt to achieve complete shared Situation Awareness throughout the flight.

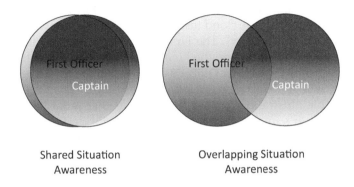

Shared Situation Awareness   Overlapping Situation Awareness

Figure 4.3   **Shared and Overlapping Situation Awareness**

Endsley (1995) suggests that team Situation Awareness can be defined as 'the degree to which every team member possesses the SA [Situation Awareness] required for his or her responsibilities' (p. 39). In this case each crew member requires Situation Awareness for those factors relevant for their duties as either Pilot Flying or Pilot Not Flying for that specific sector or phase of the flight. These components are role specific but are inter-dependent in meeting the overall goal of the flight. The major challenge to achieve good crew Situation Awareness (if it is assumed that this is overlapping Situation Awareness) is in the co-ordination of the crew resources.

Endsley and Jones (2001) developed a model comprised of four component parts to encourage shared Situation Awareness.

- *Requirements* – What information and goals need to be shared?
- *Devices* – What devices are available for sharing this information (e.g. verbal and non-verbal communication channels; shared visual or auditory displays)?
- *Mechanisms* – What mechanisms do crew members possess to aid in developing team Situation Awareness (e.g. shared mental models to facilitate communication and coordination)?
- *Processes* – What formal processes are used for sharing information, verifying understanding, prioritising tasks and establishing contingencies, etc.?

In the case of the individual pilot or even an aircraft's crew, Situation Awareness is deemed to reside in WM. However newer concepts of Distributed Situation Awareness are being developed in which the Situation Awareness is held both by the various human and/or machine components right across a socio-technical system (Stanton, Stewart, Baber, Harris, Houghton, McMaster, Salmon, Hoyle, Walker, Young, Linsell and Dymott, 2006). As Dekker (2004a) has observed, pilots nowadays have the role of active supervisors or managers who need to co-ordinate a suite of human and automated resources in order to get an aeroplane to fly. The automation is almost as much a part of the flight crew as the other pilot. Hollnagel (1993), in fact, suggested that the machine aspects of a system are all part of a joint cognitive system.

Extending Endsley's three-stage model of Situation Awareness (perception, comprehension, projection – which maps directly onto the tripartite input–process– output systems approach and Ackoff's data–information–knowledge categorisation) it is possible to apply this to a flight deck system, as illustrated in Table 4.1. In this case, the weather radar determines that the cloud formation ahead may pose a risk to flight (heavy precipitation and electrical activity). The First Officer, noticing the display commences in-flight re-planning activities. Before the First Officer communicates his concern to the Captain, the Captain notices the First Officer's activities and begins to become aware that a change of course will be required in the near future.

In this very simple illustration, output from the weather radar is input for the First Officer. The First Officer's unexpected activity on the flight deck is the Captain's input. However, two points need to be noted. Firstly, the 'knowledge' (not in the Ackoff sense) that underlies Distributed Situation Awareness is dispersed across the system. Second, there is *implicit* communication of information rather than detailed exchange of mental models. In this case all the agents were co-located on the flight deck, but in the aviation environment there is no reason why this should always be so. The agents (technological or human) may be separated by great distances (e.g. Air Traffic Control and the pilots on the flight deck). While an Air Traffic Controller may apparently be communicating routeing information to a pilot, flight crew often infer aspects of the Controller's situation from the manner of their communication and the tone of their voice (e.g. short, clipped communications are often taken as signs of a Controller under high workload). The pilots then comply accordingly, attempting to ease the Controller's task in any way they can.

Table 4.1   Distributed Situation Awareness as part of a response to bad weather

Agent	Perception	Comprehension	Projection
Weather Radar	Senses radar returns of storm clouds	Compiles picture of extent of cloud formation, distance and bearing	Displays information (along with projected tack) in appropriate colour to alert pilots
First Officer	Sees storm cloud formation on weather radar/ navigation display	Determines thunderstorm may present a risk to the aircraft	Needs to quickly determine new route to avoid the storm
Captain	Sees First Officer interrogate Navigation Display, Flight Management System and charts	Determines thunderheads present a risk to passengers and crew	Re-plans flight and initiates a diversion

Stanton, Baber, Walker, Salmon and Green (2004) proposed a set of basic tenets underpinning Distributed Situation Awareness:

- Situation Awareness can be held by both human and non-human elements in a socio-technical system.
- There are multiple views on the Situational Awareness of the same scene held by all the different agents, human or machine.
- Non-overlapping and overlapping Situational Awareness depends on the human or machine agent's individual goals: although they are part of the same system, the goals of the individual components comprising the system can be quite different (*q.v.* Endsley and Jones, 2001). In terms of Endsley's model it is also possible that that the different human and machine components in the system are representing different levels of Situational Awareness, rather than each being a microcosm of Situational Awareness in themselves.
- Communication between agents in the system may take many forms including the non-verbal behaviours of others in the system or even ingrained customs and practices.
- One component in the system (be it human or machine) can compensate for degradation in Situational Awareness in another agent.

The concept of Distributed Situational Awareness operates at a systems level, not at the individual level; it does not reside within WM. It must also be emphasised that Distributed Situation Awareness *is not* Shared Situational Awareness. Shared Situational Awareness implies the same collective requirements and purposes among the human and machine components in a system, all of whom share the same understanding of a commonly held 'bigger picture'. Distributed Situational Awareness implies different but compatible, requirements and purposes. Furthermore, the appropriate information/ knowledge relating to the task and the environment (held either by individuals or captured and processed by devices) changes as the situation develops (Stewart, Stanton, Harris, Baber, Salmon, Mock, Tatlock, Wells and Kay, 2007).

A DISSENTING VOICE

Some researchers, however, have criticised the entire concept of Situation Awareness (and indeed workload). Dekker and Woods (2002) and Dekker and Hollnagel (2004) have described the concepts of workload and Situation Awareness as 'folk models'. They argue that 'folk models' merely substitute one label for another rather than decomposing the construct into measurable specifics and that they are also immune to falsification. If, for example, Situation Awareness can be defined as a product of perception, decision-making, experience and knowledge structures in LTM, etc. then it would be better to de-construct it into its more definable and measurable component parts. If workload is WM spare capacity, then why not simply measure this? Dekker and Hollnagel (2004) illustrate their point as follows:

Take as an example an automation-related accident that occurred in 1973, when situation awareness or automation-induced complacency had not yet come into use. The aircraft in question was on approach in rapidly changing weather conditions. It was equipped with a slightly deficient 'flight director' … which the captain of the airplane distrusted. The airplane struck a seawall bounding Boston's Logan airport about one kilometer short of the runway and slightly to the side of it, killing all 89 people onboard. In its comment on the crash, the transport safety board [sic] explained how an accumulation of discrepancies, none of which were critical in themselves, had rapidly brought about a high-risk situation without positive flight management. The first officer, who was flying, was preoccupied with the information presented by his flight director systems, to the detriment of his attention to altitude, heading and airspeed control. (National Transportation Safety Board 1974)

Today, both automation-induced complacency of the first officer and a loss of situation awareness of the entire crew would most likely be cited under the causes of this crash. (Dekker and Hollnagel, 2004, p. 84)

Not surprisingly, such a posture has generated some robust counter rebuttals (e.g. Parasuraman, Sheridan and Wickens, 2008) that completely reject this argument. I merely point this out and move swiftly on.

MEASURING SITUATION AWARENESS

Measures of Situation Awareness have primarily been aimed at evaluating the success of new flight deck equipment in improving the Situation Awareness of the pilot. The majority of this work has been undertaken in the military domain, for example in the assessment of tactical displays. Measures of Situation Awareness are frequently used in conjunction with workload measures to enhance their sensitivity when comparing between design options, especially when in terms of performance, both design configurations produce similar results. Often the question becomes not 'which design produces the best performance' but 'what is the cost (in Information Processing terms) to achieve a certain level of performance'? Measures of Situation Awareness provide a measure of how successfully pilots can acquire and integrate information in a complex flight environment.

There are two basic approaches to measuring Situation Awareness: behavioural approaches involving experimentation and approaches using subjective rating scales. Distributed Situation Awareness cannot be 'measured' as such (as it is a property of the system); it can only be modelled. This requires a suite of inter-related analytical techniques to represent the knowledge elements, communication and information nodes within a system (see Stanton, Baber and Harris, 2008).

Behavioural Approaches

The most widely used behavioural approach is SAGAT (Situation Awareness Global Assessment Technique) developed by Endsley (1988). This technique uses a series of memory probes developed by subject matter experts that are applied during simulation scenarios. These questions are derived from a Situation Assessment requirements analysis for that mission. At random points during a simulation, the simulation freezes and the screens blank. The crew are then presented with a series of questions relating to Endsley's Level One, Two or Three Situation Awareness components. For example, a question relating to Level One may simply ask 'what is your current altitude'? Level Two questions would address issues such as identifying which of several radar returns is a potentially hostile aircraft. Finally, Level Three probes would encompass questions assessing enemy intent or the likelihood of diverting as a result of extreme weather. The pilot's answers are compared to the real situation (the ground truth) to provide an objective measure of Situation Awareness. This occurs several times during the simulation.

The main disadvantages with this technique are that it frequently interrupts the simulation scenario (although it has been suggested that even interruptions of longer than five minutes have little effect on performance – see Endsley, Selcon, Hardiman and Croft, 1998) and that it is merely a test of memory and not Situation Awareness (hence the criticisms levelled at the concept by Dekker and co-authors). Furthermore, the method for collecting the data is time intensive, requiring dedicated software support and the results produced are specific to the simulation scenario in which they are implemented, although when evaluating new interface designs, with careful planning using multiple representative scenarios the results should be more widely generalisable. With respect to O'Donnell and Eggemeier's (1986) other criteria (as outlined in Chapter 3 on Workload) the diagnosticity and sensitivity of the technique is very dependent upon the analyst's skill when performing the Situation Assessment requirements analysis.

SPAM (Situation Preset Assessment Method) developed by Durso, Hackworth, Truitt, Crutchfield, Nikolic and Manning (1998) used many of the same concepts as for SAGAT, however to avoid freezing the simulations participants were required to respond to real time probes for information. The measures taken were accuracy of the response to the probes and the time taken to respond, the rationale being that participants who were less Situationally Aware would take longer to find the relevant information and respond.

Rating Scale-based Approaches

Most approaches to the assessment of Situation Awareness using a rating scale have adopted a multi-dimensional approach. Jones (2000) alludes to several studies using uni-dimensional Situation Awareness rating techniques being unsuccessful as the scales were either insensitive or lacked diagnosticity, however few details are provided.

The Situation Awareness Rating Technique (SART – Taylor, 1990) is typical of a multi-dimensional approach to the measurement of Situation Awareness. It is comprised of three dimensions (Attentional Demand, Attentional Supply and Understanding) each of which contains several sub-scales (Attentional Demand: instability of the situation, variability of the situation and complexity of the situation; Attentional Supply: arousal, spare mental capacity, concentration of attention and division of attention; Understanding: information quantity, information quality and familiarity with the situation). Each scale is scored on a scale of 1 (low) to 7 (high). As a result of the number of ratings required, the SART tended only to be utilised post-task. A quicker to complete version of the scale (3D SART: Taylor and Selcon, 1991) uses only seven-point ratings on the three major dimensions. Using 3D SART it is possible to produce an overall Situation Awareness score from the simple formula:

Situation Awareness = Understanding – (Attentional Demand – Supply)

SART has been found to be sensitive to differences in display format in a simulated air-to-ground attack mission (Vidulich, Crabtree and McCoy, 1993) and it also showed some diagnosticity. However, it also demonstrated poor test–retest reliability.

The Situation Awareness Rating Scales (SARS) technique developed by Waag and Houck (1994) requires the rating of 31 behaviours regarded as essential for fighter pilots and their mission. These behaviours are grouped into eight major categories: communication, information interpretation, system operation, tactical game plan, tactical employment beyond visual range, tactical employment visual, tactical employment general, and general traits. However, unlike SART, Situation Awareness ratings are made by peers, rather than the pilots themselves. Evaluation trials demonstrated that Subject Matter Expert ratings of Situation Awareness were found to be highly correlated with independent measures of air combat skills performance in a simulated air-to-air combat mission (Bell and Waag, 2000). However, the SARS technique is specific to the air combat mission and is more akin to a rating of pilot Situation Awareness ability rather than a rating of Situation Awareness *per se*.

Perhaps the greatest drawback with any form of the self-rating of Situation Awareness, though, was elegantly summed up by one pilot who observed that: 'the problem is this: you only realise that you don't have Situation Awareness shortly after you have lost it …'.

APPLICATION GUIDELINES

As with workload, the approaches to the assessment of Situation Awareness have their relative strengths and drawbacks, although in this case the choice of technique is made somewhat easier as none of the self-rating methods are really suitable for application in flight (perhaps with the exception of 3D SART in limited, well defined circumstances). The main consideration that will guide the choice of technique concerns the objective of the assessment. All situations requiring an assessment of Situation Awareness are, almost by definition, complex, hence they will involve a reasonably complex simulation. All computer-based assessments involving freezing the simulation and administering memory probes involve a high degree of computing skill and Subject Matter Experts to formulate the questions. If the time or specific expertise required to do this is not available, this will dictate use of a more generic, rating scale-based approach (e.g. SART). Or, you could simply opt to measure the individual components of Situation Awareness!

It is important to realise, however, that Situation Awareness is not an end in itself. It can be an outcome of other factors, for example the design of the display systems on the flight deck (as will be seen later in this book) or it can be the pre-cursor for further action. This serves as a neat introduction to the subject of pilot decision making.

FURTHER READING

Two comprehensive text books are available that cover all aspects of the theory and measurement of Situation Awareness:

Banbury, S. and Trebblay, S. (2004). *A Cognitive Approach to Situation Awareness: Theory and Application*. Aldershot: Ashgate.

Endsley, M.R. and Garland, D.J. (2000). *Situation Awareness: Analysis and Measurement*. Mahwah, NJ: Lawrence Erlbaum Associates.

Overviews of many Situation Awareness measurement techniques are available in either:

Gawron, V.J. (2008). *Human Performance, Workload, and Situational Awareness Measures Handbook*. London: CRC Press.

Or:

Stanton, N.A., Salmon, P.M., Walker, G.H., Baber, C. and Jenkins, D.P. (2006). *Human Factors Methods: A Practical Guide for Engineering and Design*. Aldershot: Ashgate.

5
Decision Making

Aeronautical Decision Making (ADM) is defined by the FAA (1991) as 'a systematic approach to the mental process used by aircraft pilots to consistently determine the best course of action in response to a given set of circumstances' (Hunter, 2003). It needs to be stated right from the start that there is nothing special about ADM – it is only the context that is different from other aspects of decision making. The underlying cognitive processes are just the same as decision making in other complex, high risk, high tempo situations.

Decision making is a process within WM. Decision making is a direct product of Situation Awareness and Workload. Flight deck safety management involves a great deal of real-time decision making by the crew, but poor decision making results in error. The source of data/information for Situation Awareness (and hence decision making) is the aircraft display systems. The ultimate output of the decision making process is the control of the aircraft. In the end, this is what it is all about.

Decision making is all-pervasive throughout everyday life. We are making hundreds of decisions every hour. Admittedly, some are very minor – such as choosing tea or coffee – and in many cases there are no wrong decisions. However, you can make wrong decisions (which are defined by the context of the activity and the required outcome) and so there is an intimate link between the scientific study of decision making and that of human error. In fact, they are almost inseparable. This will become evident in this chapter and the following one. It was argued previously that there is also a close link between Situation Awareness and decision making, a theme that will be further expanded upon in the following pages. In recent years, the focus on human error investigations in aviation accidents has shifted away from pilot skill deficiencies toward decision making, attitudes, supervisory factors and organisational culture as being the primary factors (Diehl, 1991b; Jensen, 1997; and Klein, 2000a). As aircraft have become increasingly more reliable, human performance has played a proportionately increasing role in the causation of accidents.

Decision making in aviation is a joint function of the features of the flight tasks (the *Mission*) and the pilots' knowledge and experience relevant to those tasks (the hu*Man*). Decision making can also be either supported or inhibited by the flight deck systems, particularly the displays (aspects of the *Machine*). It is a complex cognitive process (or processes) affected strongly by situational and environmental conditions (Payne, Bettman and Johnson, 1988). In addition to making regular, everyday mission-related decisions pilots occasionally have to solve unexpected and ill-defined problems, for example in the case of a change of flight plan or in an emergency. In these cases these decisions are often made with incomplete information available and while under time pressure (Orasanu and Connolly, 1993). Diehl (1991a) estimated that decision errors contributed to 56 per cent of accidents in airlines and 53 per cent of accidents in military aviation. Similarly, Jensen and Benel (1977) reported that 51 per cent of fatal general aviation accidents between 1970 and 1974 were associated with decision errors. More recently Shappell and Wiegmann (2004) found that decision errors contributed to 45 per cent of accidents in the US Air Force and 55 per cent in the US Naval aviation; Li and Harris (2005a) observed 53 per cent of military accidents in the Republic of China (Taiwan) Air Force had an instance of decision making error within them and approximately 71 per cent of accidents occurring to Taiwanese commercial aircraft showed instances of decision error (Li, Harris and Yu, 2008). Put more succinctly, poor decisions are implicated in over half of all aviation accidents.

However, decision making occurs not just in the context of the flight deck but also in an organisational context. Orasanu and Connolly (1993) proposed that organisations influence decisions directly by stipulating standard operating procedures and indirectly, through the organisation's norms and culture. Therefore, as will be seen in later chapters looking at various aspects of *Management* on the flight deck and in the wider airline context, it is important to understand how the decisions made by the pilots at the sharp end on the flight deck are influenced by the actions of the people at the higher managerial levels in their organisations (i.e. the airlines). It can even be suggested that the *Medium*, in the guise of societal factors, can have a significant role to play.

This chapter looks at the two basic paradigms that have developed over the years that have characterised the study of decision making, followed by a brief review of two methods that have been proposed to improve ADM: training and the development of decision making mnemonics. The chapter closes with a brief examination of the problems of deciding what a decision making error looks like.

TWO PARADIGMS FOR INVESTIGATING DECISION MAKING

The early concepts of decision making were based on models derived from cognitive psychology. Collectively, these approaches to the study of decision making have been grouped under the generic label of Classical Decision Making (CDM). The CDM approach is based on normative models and it has usually employed an experimental paradigm as its chosen method of investigation. Studies are most often executed in the controlled conditions of the laboratory. More recently, though, the trend in aviation research has been toward conducting decision making research within the Naturalistic Decision Making (NDM) paradigm, which provides an alternative approach for understanding how decisions are made in complex, uncertain environments.

Classical Decision Making (CDM)

The CDM approach was at least partially derived from the normative decision-making models found in economics and statistics (Lehto and Nah, 2006). Implicit in these models was the concept of an optimal decision. The person making the decision was characterised as being completely rational, totally informed and infinitely sensitive (Edwards, 1954). Within the CDM approach the decision maker first formally recognises and describes what is known (and knowable) about the problem space, perhaps collecting more data as required. They then formally analyse options by applying principles of utility, uncertainty and risk to choose the optimal solution from the range of alternatives derived (Beach and Lipshitz, 1993). The alternatives under consideration include a set of potential actions, a set of events, a set of consequences for each combination of action and event, and a set of probabilities for each combination of action and event. Decisions outcomes are then represented and rationally evaluated in terms of these elements resulting in the choice of a logical course of action.

There are many different forms of CDM model (see Lehto, 1997 for a brief description of several of these theories). Subjective Expected Utility SEU theory is one of the most frequently cited (Savage, 1954). SEU is an attempt to model how people make decisions in the presence of risk. It has two subjective components to it: a subjective probability estimate relating to the likelihood of success (or failure) and a utility function. The utility function is related to the expected value of an outcome, which takes different values for different people and in different situations. The notion of utility can be explained very simply. If a £10 note is found by a beggar it has much higher utility (value) than if the same £10 note is found by Bill Gates. The expected utility of an outcome is derived from its utility weighted by its probability of occurrence. Thus a risk neutral decision maker will regard the expected utility of a bet coming to fruition to be equal to the utility of the stake's value; for a risky decision maker the utility of the expected value of the winning bet will be greater than the utility of the value of the stake (and vice versa for the risk averse decision maker). The SEU approach would suggest that a decision maker analyses all the various options and picks the one that has greatest expected utility.

While this approach makes sense in monetary terms (a great deal of CDM theory is based upon the rational, economic model of man) it becomes more difficult to employ directly in the aviation context. For example, it might be tempting to push on into bad weather and attempt to land at the final destination (the pilot may almost be at the destination and it is a long way to any alternatives). The utility of such a decision paying off could be high, in terms of time saved, flight pay bonuses or simply convenience. However, the metric upon which a failure of the gamble paying off could be quite different. How to you quantify crashing and any associated damage and injury? Multi-Attribute Utility Theory (MAUT) provides some answers to this problem (Keeney and Raiffa, 1976). In this approach SEU is extended to incorporate an evaluation of the potential expected utility of a decision on multiple dimensions (e.g. time saved, fuel saved, safety and/or inconvenience). Furthermore, some of these aspects may be deemed to be more important in terms of their outcome than others (e.g. safety).

The underlying rationale in CDM approaches such as SEU and MAUT are evident in Janis and Mann's (1977) seven-stage approach to decision making. In this method they describe what is expected of the skilled decision-maker.

The decision-maker, to the best of his ability and within his information processing capabilities:

1.   Thoroughly canvasses a wide range of alternative courses of action;
2.   Surveys the full range of objectives to be fulfilled and the values implicated by the choice;
3.   Carefully weighs whatever he knows about the costs and risks of negative consequences, as well as the positive consequences, that could flow from each alternative;
4.   Intensively searches for new information relevant to further evaluation of the alternatives;
5.   Correctly assimilates and takes account of any new information or expert judgement to which he is exposed, even when the information or judgement does not support the course of action he initially prefers;
6.   Re-examines the positive and negative consequences of all known alternatives, including those originally regarded as unacceptable, before making a final choice; and
7.   Makes detailed provisions for implementing or executing the chosen course of action, with special attention to contingency plans that might be required if various known risks were to materialise.'

Naturalistic Decision Making (NDM)

Some of the practical limitations of the CDM approach in the applied aviation context become apparent from even a brief consideration of the process described by Janis and Mann (1977). Firstly, it is implicit that within the CDM paradigm, the decision event is an end in itself. The approach assumes that the pilot has access to complete and reliable information concerning all aspects of the flight decision being taken and that they can process all of the data to define the problem so that it is ultimately clear and unambiguous. CDM also assumes they will be able to identify all possible solutions to the problem and be aware of their consequences. Finally, Janis and Mann's approach presupposes that they are capable of evaluating the outcomes by applying some kind of weighting to these options. There is also an implication that the decision maker is neither time constrained in undertaking this process nor under any degree of stress.

However, it has been observed that the key factors influencing pilot's decision making are time pressure and risk. Keinan (1987) established that under stress the range of alternatives and dimensions considered during the decision-making process are significantly restricted compared with normal conditions. Mjøs (2001) suggested that in emergency situations, uncertainty can outweigh time pressure as the key stressor having a bearing on decision making. Klein (1989, 1997) observed that decision makers in difficult situations and under time stress did not appear to use CDM-type strategies to make decisions even when they had been trained in that approach. Observations such as these contributed to the development of the alternative NDM paradigm for decision making research being developed.

As early as 1955, Simon argued that the information processing requirements of rational decision making exceeded limited human cognitive capacity. This necessitated the use of heuristics to reduce cognitive load and speed up decision making. As a result decision makers often sought *satisfactory* solutions rather than trying to make *optimal* decisions. Human beings have bounded rationality. When pursuing a satisfactory decision, a full review of alternatives is not made. The review of options continues only until a 'good enough' solution is identified (one that is satisfactory or sufficient). However, as a result, the downside to this approach is that the best option may never even get to the evaluation stage, let alone be selected.

There are a number of heuristics used when making decisions. Tversky and Kahneman (1974) in their classic paper identified three common principles to reduce the cognitively demanding task of assessing the probabilities underpinning decision events. These were the heuristics of representativeness, availability and anchoring. The representativeness heuristic is the tendency to judge someone or something according to how characteristic it appears to be of a particular category of event. The availability heuristic is the tendency to consider an instance or event as being more likely if it can be easily imagined. Finally, the anchoring heuristic is the tendency to give more weight to evidence that is consistent with the decision maker's initial hypothesis, rather than to contrary information which may subsequently be gathered. As will be seen in the following chapter, although these heuristics are cognitively efficient they often form the basis for certain types of human error as a result of the cognitive biases they induce (Reason, 1990).

The NDM research paradigm investigates how skilled operators use their experience to make decisions in the operational context (Zsambok and Klein, 1997). This approach originally had its roots in the military's need to understand the human dimensions of command and control but it rapidly found favour in both the civil and military aviation domain as a useful way of characterising and describing pilots' decision-making processes. Klein (1997) observed that unlike in the CDM-derived process suggested by Janis and Mann (1977) the expert decision maker does not generate alternatives and compare between them using pre-determined criteria to evaluate the likelihood of a successful outcome (cf. SEU or MAUT); their focus is on understanding the situation and judging its familiarity, not on the generation of a range of options to find a solution. In a later study (Klein, 2000a) it was found that a decision maker rarely considered multiple options but worked sequentially from the most plausible option generated. If the initial strategy was found as being unsuitable the next most plausible option would then be generated. Decision makers used their previous experience to frame the current situation. When a pilot found a good match between his/her situation and past experience, then that course of action would be pursued.

NDM research tends to put greatest emphasis on the investigation of the processes that underlie how an individual actually makes decisions and how they use their experience and training under demanding conditions, rather than stressing the importance of the outcome of the decision-making process (cf. CDM). NDM is characterised by 'dynamic and continually changing conditions, real-time reactions to these changes, ill-defined tasks, time-pressure, significant consequences for mistakes' (Klein and Klinger, 1991). All these characteristics are also often associated with decision making on the flight deck.

Orasanu and Connolly (1993) identified eight themes inherent in decision making in complex systems in a naturalistic setting. Every one of these is instantly recognisable when describing some of the more complex decisions required on the flight deck. They suggested NDM is characterised by:

- *Ill structured problems.* Unlike the assumptions inherent in the CDM approach, naturalistic decision problems rarely present themselves in a 'neat, complete form' (Orasanu and Connolly, 1993, p.7) which means that decision makers need to generate hypotheses about the nature of the problem before they can determine if a decision is actually required and before they can generate any response options.
- *Uncertain dynamic environments.* The naturalistic decision problem is often typified by incomplete, imperfect, ambiguous and/or unreliable information. Furthermore, in a dynamic, changing environment this information may rapidly become out of date.
- *Shifting, ill-defined or competing goals.* Outside of the laboratory environment, decisions driven by a single well understood goal are uncommon. More often, decision makers are confronted by multiple goals, some of which may not be clear and some of which may also directly oppose others. In such cases it may not always be clear what the trade-off should be. Furthermore, these conflicts and trade-offs may be complex by virtue of the dynamic environment in which they exist.
- *Action/feedback loops.* Unlike the CDM approach which focuses on a single decision event, it is more common in real life to be required to make a series of decisions to solve a problem. These may be iterative in nature, whereby the decision maker tackles a problem (or uncovers more information about it) with feedback loops serving to benefit the process.
- *Time stress.* Decisions are frequently made under significant time pressure. As a result, decision makers habitually experience high levels of personal stress, which often results in a poorer decision-making process (see Keinan, 1987 and Mjøs, 2001). Time stress may also lead to decision makers opting for less sophisticated reasoning strategies, evaluating only a limited number of options and/or making increased use of heuristics (with all their inherent problems).
- *High stakes.* Making a decision in a naturalistic environment may have particularly high stakes attached. While this will probably increase stress (which may be detrimental) it can also have the opposite effect in that the decision maker has a vested interest in the outcome thus making them more engaged in the task than perhaps they would in the laboratory environment.
- *Multiple players.* Many NDM problems involve more than just a single decision maker. A decision-making team may comprise a number of individuals who can act either co-operatively or competitively while attempting to achieve their goal. Hopefully, the small team on the flight deck (and if required off the aircraft) should behave co-operatively!
- *Organisational goals and norms.* Finally, as noted earlier, Orasanu and Connolly (1993) suggest that NDM often takes place in an organisational setting. Accordingly, the decision-making goals can also reflect the objectives of the organisation in addition to those of the individual decision makers.

Although these features considerably complicate the decision task (and the study of decision making) they are frequently absent in CDM research.

There are many NDM models of decision making. In fact one of the greatest criticisms of the paradigm is that there are as many NDM models as there are researchers and each model derived tends to be somewhat context-specific. This leads to a lack of generalisation. NDM models also do not predict behaviour, which as this is one of the ultimate goals of science, limits their utility (Duggan and Harris, 2001). NDM models tend to focus on process not product. When flying an aircraft the product of the decision is quite important!

The decision process model described in Figure 5.1 (from Orasanu and Fischer, 1997) is typical of many NDM models but is useful in this context as it was derived from studies of pilot's operational decision making. The model consists of two major components, situation assessment followed by choice of a course of action. Situation assessment (as opposed to Situation Awareness as described in the previous chapter) requires the definition of the problem, an assessment of the risk level and an assessment of the available time to make the decision. Jensen, Guilke and Tigner (1997) and O'Hare (1992) have also proposed that risk management is a key component of the decision-making process. They put forward that risk management also feeds into the decision-making process during the evaluation of potential courses of action. The time available is the principal factor for selecting subsequent strategies. If the situation that the decision maker is confronted by is not understood various diagnostic actions may subsequently proceed but only if there is enough time to do so. If both risk is high and time is limited then actions may be taken without a thorough understanding of the problem. Hence, a good decision-making process can be regarded as one in which the decision maker successfully accomplishes a series of sub-tasks (assuming the time is available). This involves the collection of information about a wide range of alternatives before making a careful assessment of the risks and benefits of each potential course of action and preparing contingency plans for dealing with the known risks.

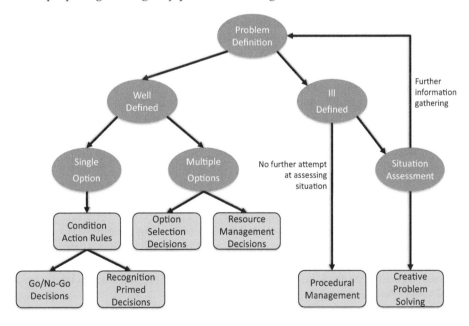

Figure 5.1   **Taxonomy of decision types (adapted from Orasanu, 1993)**

This NDM model developed by Orasanu and Fischer (1997) is also of interest as it begins to become apparent that there are several different basic types of decision (the bottom blocks in the decision tree presented in Figure 5.1) all of which require different types of information and skills (more of which anon).

- *Go, no-go decisions* – where an action is anticipated or in progress and a cue triggers a decision to terminate that action (e.g. abandoning a take-off as a result of a catastrophic engine failure).
- *Recognition-primed decisions* – where the decision makers interpret a cue pattern as being of a particular type and match it with an action according to standardised procedures (e.g. in a helicopter fuel flow increasing, EGT climbing and indicated torque increasing, suggesting a gearbox binding, necessitating an emergency landing).
- *Option/response selection decisions* – these occur when there are several legitimate options/actions which the decision maker needs to evaluate in terms of the requirements of the problem and from which one option needs to be selected (e.g. selection of a diversionary airport following an engine failure: response will depend upon factors such as: distance from destination; facilities available at alternative airports; weight of aircraft, etc.).
- *Resource management decisions* – in which the decision maker needs to co-ordinate several time-consuming tasks so their products are available when required (a classic instance of managing a failure on a flight deck through the prioritisation of actions and distribution of tasks between the crew members).
- *Procedural management decisions* – which are typically required when a decision maker faces ambiguous high risk situations requiring a diagnosis of the problem but can then subsequently act according to standardised procedures (e.g. the situation faced by the crew in the Boeing 737 at Kegworth where the crew were faced with smoke on the flight deck and heavy engine vibration, yet the engine was still producing thrust – there was little direct information about the root cause of the problem, which was the loss of a section of fan blade in the port engine: once the failure was diagnosed the course of remedial action was obvious, although unfortunately incorrectly carried out).
- *Creative problem solving* – requires the decision maker to first diagnose and define a problem. They then need to create an action to meet the needs of the situation since no specific guidance (i.e. standard operating procedures, manuals or checklists) are available to guide the decision maker: this is an even more complex version of the previous category.

Jenkins, Stanton, Salmon and Walker (2008) examined decision making from a related but slightly wider perspective. Their approach utilised all the characteristics and principles of the NDM method but they grounded their studies using Rasmussen's (1983, 1986) Skills, Rules and Knowledge (SRK) taxonomy and used the analytical principles in his decision ladder (Rasmussen, 1974). Similar to the approach proposed by Orasanu (and co-workers), Jenkins, Stanton, Salmon and Walker (2008) regard the decision making to consist of several stages:

- *Problem detection* – recognising that something has happened and that some action may be required.

- *Interpretation of information* – gathering relevant data and information about the situation and evaluating it in context.
- *Sensemaking* – fusing the available data and information to diagnose current and future system states.
- *Identifying options* – developing the viable options available to the decision maker.
- *Evaluation of competing goals* – determining the priority of competing goals.
- *Selection of an appropriate task* – to achieve the chosen goal, and
- *Development of a procedure* – to undertake the chosen task.

The relationships between these processes are described in Figure 5.2.

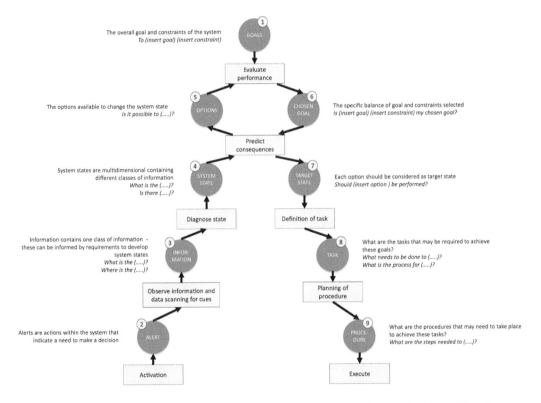

Figure 5.2   **Description of Rasmussen's (1974) decision ladder. The boxes indicate data processing activities, circles indicate resultant states of knowledge (Jenkins, Stanton, Salmon and Walker, 2008)**

Rasmussen's (1983, 1986) SRK taxonomy describes human cognition in terms of three basic types of behaviour: skill-based, rule-based or knowledge-based. Skill-based behaviour proceeds without conscious attention where responses are automatic and invoked by environmental cues (e.g. when flying, compensating for a gust of side wind or applying the brakes after landing). Rule-based behaviours operate on a range of IF … THEN rules stored in LTM. These are available to consciousness (e.g. IF traffic light is on red THEN stop the car). These rules can be acquired from direct experience or vicariously. They may also be quite complex requiring multiple IFs and containing exceptions. However when a problem does not map directly on to a stored rule knowledge-based behaviour is employed which depends upon the pilot's knowledge of systems, operating principles and procedures etc. In this case novel solutions need to be developed after the diagnosis of complex problems. These levels of processing relate to levels in the decision ladder, with rule-based behaviours associated with complex decision making at the top and skill-based behaviours at the bottom. Orasanu's decision types are also evident in this approach, with go/no-go decisions at the bottom (skill-based) level and creative problem-solving decisions at the top (knowledge-based) level. When a pilot is unable to formulate an intuitive response to a situation they are likely to search for a rule-based heuristic; if no such heuristic is available they will be required to resort to knowledge-based reasoning. However, the use of Standard Operating Procedures (SOPs) in the Quick Reference Handbook (QRH) tries to limit the pilot's requirement to indulge in complex, high-level diagnosis and problem solving. When operating at the knowledge-based level, the ratio of error to opportunity for error is at its highest. As a result, SOPs try to reduce most common and predictable non-normal situations to a set of IF … THEN rules, where the opportunity ratio for error is not so high.

IMPROVING DECISION MAKING

Aeronautical knowledge, skill and judgment have always been regarded as three basic faculties that pilots must possess (Diehl, 1991a). These are imparted as part of the academic and flight training programme and are subsequently evaluated as part of the pilot licensing process. In contrast, ADM has usually been considered to be a trait that good pilots possess innately or an ability acquired as a by-product of flying experience (Buch and Diehl, 1984). Means, Salas, Crandall and Jacobs (1993) pointed out that when pilots are under stress the likelihood of making serious errors increases and they are more likely to ignore relevant information, make more risky decisions and perform with less skill.

There have been many attempts to train ('improve') ADM. These have met with mixed degrees of success. Means, Salas, Crandall, and Jacobs (1993) advised that it was important to give pilots pattern recognition exercises during training. However, Zakay and Tsal (1993) found that practice without time pressure did not enhance decision making when under time constraints and suggested that if decision making is likely to be required under time pressure or other stressful conditions, practice should also include task performance under those conditions. A flight simulator can offer great advantages for training for pattern recognition by presenting simulated situations based around scenarios that are significant when making real-time decisions. Simulators can provide scenarios that occur both frequently or very infrequently – situations that pilots would never normally experience. Furthermore, this can be done safely. The simulator can provide many practice problems, designed to build up the recognition of patterns. Positive transfer of training can be expected the more the simulator duplicates the operation of the aircraft. Buch and Diehl (1984) found that judgment training produced significantly better decisions among civil aviation pilots in Canada. Connolly, Blackwell and Lester (1989) advised that pilots' decision-making skills could be significantly improved through the use of judgment training materials along with simulator practice.

There are a number of strategies (often embodied in mnemonics or acronyms) that have been developed by researchers and used by pilots to guide and structure decision making. The common aim of these techniques is to form a systematic approach to decision making that should be less affected by human nature and which should also reduce the cognitive workload on pilots (O'Hare, 1992). For example, the SHOR mnemonic (Wohl, 1981) consists of four steps: Stimuli (data: gather data; filter it; aggregate it; and store or recall it); Hypotheses (perception alternatives: create hypotheses about the situation; evaluate them; select one); Options (response alternatives: create response options; evaluate them; select an option); and Response (action: plan, organise and execute the response). SHOR was originally developed for use by US Air Force to aid decision making under high pressure and severe time constraints. The FOR-DEC mnemonic was developed in the civil aviation domain as a product from a Lufthansa CRM-course. It comprises of six steps: Facts, Options, Risks and Benefits, Decision, Execution, Check (Hörmann, 1995). It incorporates components addressing gathering data (situation assessment), analysis of risk and benefits (including the effects of time pressure, continually changing conditions and distraction) and having incomplete information. Each step of the FOR-DEC process uses a guiding question to focus the pilot's attention on a sequence of essential steps for effective decision making.

- *Facts* – What is actually going on here?
- *Options* – What are the choices we've got?
- *Risk and Benefits* – What is there to be said for and against the application of the different options?
- *Decision* – What shall we do?
- *Execution* – Who shall do what, when and how?
- *Check* – Is everything still all right?

Note that FOR-DEC implies a decision-making cycle rather than a discrete process with a beginning and an end. The DESIDE mnemonic method (Murray, 1997) was developed on a sample of South African pilots. It comprises of six steps: Detect, Estimate, Set safe objectives, Identify, Do, Evaluate (see Table 5.1).

Table 5.1   The DESIDE decision-making acronym (Murray, 1997)

	Description
D	Detect change *Are there serious risks if no action is taken?*
E	Estimate the significance *Are there serious risks if only the most available alternative or protective action is pursued?*
S	Set safe objectives (beware hazardous attitudes) *Is there a realistic possibility of finding a better solution?*
I	Identify options *Is there sufficient time to make a careful search for further gathering and evaluation of information and advice?*
D	Do the best option
E	Evaluate the outcome and continue to apply the DESIDE model if there are further changes or if the decision made is not producing the desired result.

Murray followed up a sample of pilots using the DESIDE method. Analysis of the questionnaire responses showed that it produced a positive attitude change in a lot of respondents but, like the application of many mnemonic methods, there was little empirical evidence to suggest that it was actually effective in improving ADM.

Orasanu (1993) has also pointed out that there was no evidence to support the development of *generic* training techniques to improve all-purpose decision-making skills. She attributed this to the fact that there were different component skills involved in making the six different basic types of decisions she identified. These skills are described in Table 5.2 (from Caird-Daley, Harris, Bessell and Lowe, 2007).

Li and Harris (2005b) undertook a study to identify the best ADM mnemonic-based methods for training military pilots' decision making. SHOR (Wohl, 1981) was rated as being the best ADM mnemonic in time-limited and critical, urgent situations. DESIDE (Murray, 1997) was regarded as superior for knowledge-based decisions which required more comprehensive considerations but when the pilot also had more time available to do so. These mnemonic methods were then developed into a short decision-making training programme which also encompassed Orasanu and Fischer's (1997) six basic decision types and their associated cognitive skills. The results of the training programme were formally evaluated using both pencil and paper tests (Li and Harris, 2006c) and in evaluations using a series of in-flight emergencies instigated in a full-flight simulator (Li and Harris, 2008). The results clearly showed significant improvements in the quality of pilots' Situation Assessment and risk management (the underpinning processes in ADM) as a result of this targeted decision-making training, although this was usually at the expense of speed of response.

Table 5.2　　Cognitive work requirements and training focus (from Caird-Daley, Harris, Bessell and Lowe, 2007 based on Orasanu, 1993)

Decision Type	Training Focus
Go/No-Go Decisions	• Develop perceptual patterns in memory that constitute the conditions for aborting an action. • Conduct under realistic time pressure and including borderline cases.
Recognition-primed Decisions	• Develop recognition of situational patterns constituting the condition side of a condition-action rule. • Learn the response/action side of the rule and the link between condition and action. • Develop evaluation skills (what will happen if I take/don't take this action, or is there a reason not to take this action?)
Option/Response Selection Decisions	• Train crew to use heuristics, e.g. Satisficing (Simon, 1955), elimination by aspects (Tversky, 1972) and dominance-structuring (Montgomery, 1983 and 1989).
Resource Management Decisions	• Acquire knowledge of the time required to complete various tasks, and the interdependencies among tasks. • Develop scheduling strategies.
Procedural Management	• Develop Situation Assessment and risk assessment skill.
Creative Problem-solving	• Develop Situation Assessment and risk assessment skill. • Develop skill in goal setting, planning, strategising and evaluation (e.g. case-based reasoning involving presenting many examples of others' experiences).

DECISION ERRORS

The NDM paradigm, however, is not the universal panacea for decision-making research and training. Klein (1997) points out that one of the greatest challenges for NDM research is accounting for errors. Unlike the researchers utilising the CDM paradigm who use controlled experiments to detect and study systematic biases in decision making, as NDM workers operate in an ecologically valid context 'there is often no clear standard of 'correctness' […and] there is a loose coupling of event outcome and decision process so that outcomes cannot be used as reliable indicators of the quality of the decision' (Orasanu and Martin, 1998, p.100). Baron and Hershey (1988) suggested that people tend to assess the correctness of their decision making based on the outcome of the decision. However, good decisions can lead to bad outcomes (and vice versa). Decision makers cannot infallibly be graded by their results (Brown, Kahr, and Peterson, 1974). All in-flight decisions are made under uncertainty. A pilot can apply the 'correct' decision-making process but still end up having an accident due to the vagaries of the weather or the behaviours of other people not under his/her control. Evaluating a decision as good or not must depend upon the stakes and the process, *not* on the outcome. Hence, studies of the evaluation of the effectiveness of decision making training employ assessments based upon using Situation Assessment and risk management measures and processes, rather than assessing the outcomes of the decisions made (Li and Harris, 2006c; Li and Harris, 2008). Accordingly, researchers using the NDM approach seek to move their understanding of error away from the cognitively-biased individual decision maker to the systematic causes underlying outcomes (Orasanu, Martin and Davison, 2001). 'Bad outcomes are thus caused not by the failure of individuals but by the failure of systems' (Lipshitz, 1997, p. 155). While this is laudable in many ways, the mechanisms underlying the final error in the chain of events still need to be fully understood to help derive suitable mitigating strategies.

Orasanu, Martin and Davison (2001) proposed two major ways in which a decision error may arise. Decision makers may make an error either in their Situational Assessment or in their action selection. Errors in Situation Assessment might arise through misinterpreting or ignoring cues. The root cause of errors in choosing a course of action can vary according to the type of decision strategy. Errors involving rule-based decisions might depend on failing to retrieve a response from memory (if indeed the correct action was actually known). Errors involving decisions where a choice is required among alternatives may be a product of failing to retrieve an appropriate response from memory or alternatively factors for determining the adequacy of the outcomes of the options derived may perhaps be unavailable. Creative decisions can be prone to error as a result of the absence of any support, requiring the decision-maker to develop a novel solution (cf. Rasmussen's SRK taxonomy (1983)). Klein (1993) suggested the root of these errors could ultimately be either a lack of experience on the part of the decision-maker, a lack of information or inadequate mental simulation.

This initial brief foray into the mire of human error in the decision-making context serves as a somewhat fortuitous introduction to a deeper consideration of this complex area which will be presented in the next chapter. This topic of decision error will be re-visited.

FURTHER READING

For a deeper consideration of human decision making the interested reader is referred to two classical texts on the issue:

Kahneman, D., Slovic, P. and Tversky, A. (1982). *Judgement under Uncertainty: Heuristics and Biases.* Cambridge: Cambridge University Press.
Zsambok, C. and Klein, G. (1997). *Naturalistic Decision Making.* Mahwah, NJ: Lawrence Erlbaum Associates.

For an applied perspective to decision making across a range of complex, high risk industries, try:

Cook, M., Noyes, J. and Masakowski, Y. (2007). *Decision Making in Complex Environments.* Aldershot: Ashgate.

6
Error

I have very carefully entitled this chapter simply 'Error'. Many people talk about and write about 'pilot error' as if it were something special. It is not. The errors pilots make are no different in their nature to those made by human beings in other walks of life. It is merely their context that is unique and unfortunately their consequences are potentially more severe.

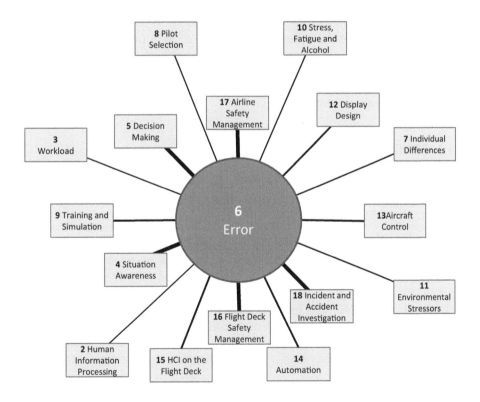

Safety management, be it on the flight deck or across the airline as a whole, is all about the error troika: stop error, trap error and/or mitigate its consequences. Ultimately errors can promulgate through the system to result in incidents and accidents. All errors are the product of a human decision at some point: if we don't make decisions we can never make errors, but all decisions are made on the basis of Situation Awareness. A failure in Situation Awareness can result in an error.

Human error is now the principal threat to flight safety. In a worldwide survey of the factors in commercial aviation accidents, in over 80 per cent of cases the crew was identified as a causal factor; in 66 per cent of instances the crew were implicated as the *primary* causal factor (CAA, 2008). The good news is that both of these figures are down slightly from the earlier analysis of accidents undertaken by the CAA a decade earlier (CAA, 1998). There are many other studies that suggest similar figures. In general, these only vary by a few per cent here and there. It is without doubt that error is the major contributory cause. But this in itself says very little. 'Human (or pilot) error' is not an explanation. It is merely the very beginning of an explanation. Dekker (2001) proposed that human errors are systematically connected to features of a pilot's tools and tasks and pilot error has its roots in the surrounding system. The question of human error or system failure alone is an oversimplified belief in the roots of these failures.

It could easily be argued that almost 50 per cent of this entire book is either directly or indirectly associated with making errors or avoiding errors, through either the good design of flight decks (*Machine*), appropriate selection and training of pilots (hu*Man*) or good design of the *Management* systems (either on the flight deck or in the organisation). There is certainly a massive overlap between the content of this chapter and those on safety management and incident/accident investigation. In something of an attempt to avoid a great deal of repetition, the majority of the content of this chapter will give an overview of the nature of human error and how to *identify* it, hence the reason why this chapter may be considerably shorter than one would initially expect for a topic of such central importance to Human Factors in aviation. How to *prevent* error is a theme that implicitly runs throughout many of the chapters.

DEFINING ERROR

Although 'error' is a term used practically every day it is not an easy term to define precisely once it is considered closely. Take a typical, if somewhat old, definition of human error, that proposed by Rigby (1970):

> Human error is any member of a set of human actions that exceeds some limit of acceptability. It is an out of tolerance action, where the limits of acceptable performance are defined by the system. (p. 457)

This is an adequate definition but it does ignore the fact that to fail to do something is also an error. It also only considers 'error' from one perspective, that of the 'system'. Hollnagel (1998) suggests three criteria must be fulfilled before any action can really be described as 'erroneous'.

- *Criterion* – There needs to be a specified performance criterion or standard against which performance can be compared.
- *Performance shortfall* – There must be an action that leads to a measurable performance shortfall.
- *Volition* – There must be the opportunity for the person to act in such a way that would not be considered erroneous.

The criterion problem has two facets to it. An error may only be termed an 'error' in one (or both) of two circumstances. If the action falls short of what is required from an external verifiable viewpoint (i.e. what is required by 'the system') then it is an 'error'. If the action executed is not what the person intended to do, then there is a failure to perform compared against some internal (but less verifiable) criterion. However, what may be an 'error' from an external perspective may not be an 'error' from an internal perspective. Formal safety requirements tend to view error only from an external (system oriented) perspective, which is ultimately what is required! However, simply viewing error in this manner can result in wrong conclusions about the aetiology of the 'error' being drawn and hence inappropriate remedial actions may be suggested as a result.

Errors are almost always associated with a negative outcome of some kind: a performance shortfall. Just occasionally though, people perform unintended actions (a failure against some internal criterion) that turn out to be for the better (in terms of the external, verifiable criterion for system performance). These serendipitous actions are rarely designated as 'errors', although from a psychological/internal perspective (as opposed to a system) they are clearly failures in Human Information Processing. For example, Alexander Fleming's discovery and isolation of penicillin in 1928 is often described as 'accidental' but is rarely (if ever) described as 'erroneous'! Taken from a human/system reliability perspective, however, outcomes are more important than processes hence such a fortuitous 'error' as this would not be regarded (from a system viewpoint) as actually being an error.

Finally, taking Hollnagel's third point, an 'error' can only be considered an 'error' if the person committing the 'error' has the possibility of avoiding it. If an action was unavoidable it cannot be considered to be an 'error' (although from a system perspective it may degrade the system – this is taking the external viewpoint). Flying into a hurricane, either deliberately or through negligence, could be considered an 'error'; flying into a microburst would not be considered a pilot error if either the warning came too late or if it was not detected at all.

However, volition is not an 'all or nothing' thing. This is where a socio-technical systems approach to the roots of human error enters the fray. In this approach the people at the 'sharp end' (and within the context of this book this is primarily the pilots) fall into traps left for them by the people at the 'blunt end' (those removed from day-to-day operations: the managers, designers, regulators, etc.). The idea implicit in this approach to the description of 'error' is that the performance requirements of the task (an objective criterion which is either met or not met by the operator at the 'sharp end') are either fulfilled or not fulfilled as a result of resources and constraints imposed by those at the 'blunt end'. That is, the operator does not have complete volition. This theme is very much evident in the current major models of organisational safety/accident causation that will be considered later in this tome (e.g. Reason, 1990; 1997).

As an example of defining what is an 'error', consider the American Airlines flight 587 accident to an Airbus A300 over New York in 2001, in which the principal cause of the accident was attributed to 'unnecessary and excessive rudder pedal inputs' on the part of the First Officer flying the aircraft (National Transportation Safety Board, 2004). As a result of these large rudder pedal inputs the aircraft's vertical stabiliser and rudder separated in flight. The engines then subsequently separated and the aircraft broke up killing all 260 people onboard. However, was this just pilot error or did the pilot (at the sharp end) simply fall into a hole that had been left for them by the aircraft's design and organisational flaws in the airline, both factors geographically and temporally remote from the accident? It was argued by the attorneys representing the estate of the First Officer that

American Airlines' crews operating the A300 had never been trained concerning extreme rudder inputs (a *Management* failure) and that the subsequent rudder failure was caused by a flaw in the design of the aircraft (a *Machine* shortcoming). Such rudder inputs should not lead to a catastrophic rudder failure in a civil aircraft. It could even be argued that this was a failure in the certification process (a failing of the regulatory *Medium*). Thus, pilot error is merely a starting point for an explanation concerning what 'caused' the accident: it is not an end in itself.

To round off this initial prologue illustrating the problems of defining what an 'error' actually is, the observation made by Woods, Johannesen, Cook and Sarter (1994) is worth considering. They suggest that 'error' is always a judgement made in hindsight:

> ... the diversity of approaches to the topic of human error is symptomatic that 'human error' is not a well-defined category of human performance. Attributing error to the actions of some person, team or organisation is fundamentally a social and psychological process and not an objective, technical one. (p. xvii)

While these debates are scientifically and philosophically interesting the net outcome can often be justifying why some out of tolerance action (or inaction) was not an error. From a practical perspective it is undeniable that people in the aviation industry occasionally do things that they shouldn't do or don't do things that they should do (whether they mean to or not). These actions and/or inactions degrade the system and can cause accidents. To be pragmatic, let's call these things 'errors' (taking very much a system perspective, as it is failures in the system that ultimately kill people) and consider a few ways that they may be mitigated. However, the terms 'error' and 'blame' should never be conflated. It has to be stated in the strongest possible manner that these terms *are not* synonymous.

ERROR CATEGORISATION

Reason's model of accident causation (Reason 1990, 1997) is perhaps the most widely used approach to the study and mitigation of human error in the aviation industry. It is basically a socio-technical systems approach to human error, however it is largely predicated on his earlier GEMS (Generic Error Modelling System) model (Reason, 1987). Reason's model of accident causation is described later in Chapter 17 on Airline Safety Management (do not confuse human error with accident causation: they are not one and the same). The GEMS model is best located within the psychological approach to human error. Errors are regarded essentially as failures of Human Information Processing (but which also happen to degrade the system). However, ultimately Reason's approach reaches out into the organisation, suggesting that these cognitive 'failures' have roots that are external to the operator, hence giving it a flavour of the socio-technical systems approach (see Lipshitz, 1997 and Orasanu, Martin and Davison, 2001).

However, before human error can be considered in any greater detail, a reiteration of the three basic modes of human operation described in the previous chapter is required upon which the GEMS model is predicated. These are the skill-based, rule-based and the knowledge-based mode (Rasmussen, 1983, 1986). In ascending order of cognitive complexity the skill-based mode is when the pilot is employing highly learnt, well developed behaviours which are essentially sub-conscious. Cognitively higher-order rule-based behaviours are stored in LTM and come from many sources (such as standard operating procedures that have been learnt or solutions encountered to previous problems) and are of the format of 'if X then Y'. The highest level of cognitive activity is that of knowledge-based behaviour. In this case there are no previously learnt rules to apply and solutions to problems must be worked out from first principles. These are essentially problems in higher order decision-making. Unfortunately, this is the level of cognitive activity that is most open to error and it is also the area in which the biggest mistakes can be made – due to combining pieces of information in wrong ways, knowing 'wrong facts', or not knowing all the pertinent facts, etc.

Reason (1987) derived an error classification scheme based upon these cognitive behaviour categories. GEMS contains just four major types of error:

- *Slips* – An extra step, a wrong step, an out of sequence step or a badly performed step is included in a series of well-learnt behaviours. These errors occur at the skill-based level of behaviour. Slips can be further sub-divided into sub-categories, for example:
- *Sequence error* – Doing all the right things but not in the correct order.
- *Selection error* – Attempting to do an incorrect thing as one of the steps in a sequence of steps.
- *Qualitative error* – Doing all the required steps but one or more of the steps is not completed to the required standard (defined by the system).
- *Time error* – Simply running out of time to complete all the required operations.

Slips are also collectively known as errors of commission.

- *Lapses* – Lapses occur at the same level of cognitive operation as slips but have no sub-categories. A lapse is simply omitting to perform one of the required steps in a sequence of actions. These are also known as errors of omission.
- *Mistakes* – Mistakes occur at both the rule-based and the knowledge-based levels of cognitive behaviour. Mistakes are errors brought about by a faulty plan or intention. They are associated with the correct execution of an inappropriate action (basically the unintentional misapplication of an IF … THEN rule). They arise from either applying the wrong principles or procedures, or from a lack of appreciation of the consequence of what (in retrospect) was an inadvisable course of action. Mistakes are made during conscious cognitive activities (although the person making the error will not necessarily be aware of making an error, otherwise they wouldn't have done it)!
- *Violations* – These 'errors' are quite different to the errors described in the previous three major bullet points. The previous categories were essentially defined as some type of failure in information processing. Violations are deliberate, conscious, inappropriate mis-applications of rules. They are an error in as much as they are an out of tolerance action when compared to what is required by the system. Reason defines two basic types of violation. *Routine* violations occur almost habitually and may be tacitly tolerated by the organisation or authority. The most typical example encountered in this category is that of 'corner cutting' (i.e. consciously omitting steps in a sequence of actions or performing a non-prescribed action in order to save time or effort). *Exceptional* violations are more unusual and much more extreme (they would not be tolerated at any level). The ultimate form in this category is

deliberate sabotage. However, successful attempts at sabotage, from an internal Human Information Processing viewpoint, certainly are not errors.

In Figure 6.1 it will be noted that the psychological 'roots' of the different types of error are suggested, be they failures in the attentional mechanisms, memory or in higher-order decision making. Skill-based errors are related to variability of sensorimotor co-ordination: rule-based errors are associated with misclassifying the situations leading to the application of a wrong rule from LTM (Reason, 1990). Figure 6.1 also describes the 'phenotypes' (the overt manifestations of the Generic Error Types) of the underlying error 'genotypes'.

Picking up from the previous chapter, decision making errors have different root causes to skill- and rule-based errors as a result of the cognitive processes involved (Rasmussen, 1993). At the knowledge-based level the decision maker is faced with a novel situation which may either require diagnosis (followed by the application of an already established procedure) or in addition, the formulation of a new procedure or development of new knowledge to address the issue. This involves situation analysis and formulation of appropriate goals (see the previous chapter). Hence, knowledge-based errors are associated with factors such as inappropriate information search or inadequate reasoning (Lipshitz, 1997). Reason (1990) suggests similar factors, such as incomplete or inaccurate information, but also adds bounded rationality (limitations resulting from cognitive capacity, particularly in WM and the finite amount of time available to make

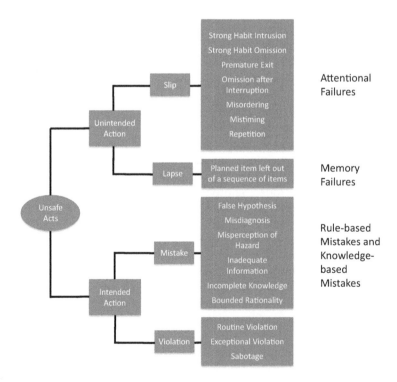

Figure 6.1   The underlying cognitive factors in the Generic Error Modelling
                System (GEMS) model (Reason, 1990)

a decision). Associated with bounded rationality are other cognitive biases including the availability heuristic (where undue weight is given to facts that come readily to mind – those not immediately thought of are discounted or ignored); confirmatory bias (where the decision maker decides upon one solution early in the decision-making process and then gathers further data/information to support this view, discounting contradictory interpretations and failing to regard negative information); and the overconfidence bias (where the search for data/information or other solutions is closed sooner than is really warranted).

Reason (1990) suggested that most errors were actually skill-based errors (slips and lapses. Skill-based errors accounted for 60.7 per cent of all errors, followed by rule-based errors (27.1 per cent) and knowledge-based errors (11.3 per cent). No mention is made of the missing 0.9 per cent. Perhaps this was a lapse … However, for these three types of error the subsequent detection rates were 86.1 per cent, 73.2 per cent and 70.5 per cent, respectively. Recovery from a slip or a lapse at the skill-based level has a greater probability than recovery from a mistake at the rule-based level or one at the knowledge-based level (Reason, 1990). Mistakes, especially those at the knowledge-based level, are likely to be aetiologically far more complex than slips or lapses as these originate at a higher level of cognitive activity. As a result, they will also be more difficult to detect subsequently.

The type of error is specific to the stage of cognitive processing (see Table 6.1). As there is often a period between the planning and execution of any action this short storage period is liable to a 'lapse', an error type associated with a failure of WM (see Reason, 1997).

Table 6.1   The relationship between cognitive stages and error type (Reason, 1990)

Cognitive Stage	Primary Error Type
Planning	Mistake
Storage	Lapse
Execution	Slips

This leaves us just with the tricky question of whether Reason's category of 'violations' is actually an error category. Again, it is a matter of perspective. Taking the perspective that errors are somehow failures in Human Information Processing, then the answer is 'no': in this case the pilot's actions proceeded as intended (even though she/he knew them to be wrong). From a system perspective, violations are certainly errors. If judged against an external criterion (e.g. the actions that are required to ensure safety) then there is certainly a shortcoming. As touched upon earlier, Reason (1990) defines two types of violation: *routine* (e.g. omitting seemingly unimportant items from a checklist when in a hurry) and *exceptional* (e.g. changing an aircraft's engine using a forklift truck instead of a hoist, as in the Chicago McDonnell-Douglas DC10 accident, or in the ultimate case, sabotage or terrorism). Violations certainly are a different category of error. The motivation to commit a violation is conscious and deliberate.

Reason (1990) further extended the GEMS model to incorporate a socio-technical systems dimension to the root causes of error. This is described in greater detail in Chapter 16 on Airline Safety Management in Part Four.

A SENSE OF PERSPECTIVE

Not all errors result in accidents. In fact many errors are not even detected either by the people making them or by the people watching them. For example, the effectiveness of the second pilot as an 'error checker' is questionable. Omission of action or inappropriate action was implicated in over 39 per cent of fatal civil aviation accidents and incorrect or inadequate procedures were implicated in a further 11 per cent. Becoming 'low and slow' (a failure to cross monitor the flying pilot) was a feature in 12 per cent of accidents (CAA, 2008). To be fair, what these data cannot show is in how many cases the second pilot trapped an error made by the other pilot and averted an accident. However, observational data obtained in routine flights supports the notion that the second pilot's role as an error checker is not all that it is made up to be. For example, Thomas (2003) reported that 47.2 per cent of errors committed by Captains during normal line operations involved intentional non-compliance with SOPs or regulations and a further 38.5 per cent were unintentional procedural non-compliance errors ('violations' in terms of Reason's taxonomy). Thomas also reported that in observations of line operations crews did not demonstrate effective error detection, with more than half of all errors remaining undetected by one or both of the flight crew.

Ruffell Smith (1979), in a series of full-crew simulator studies of representative high workload flight sectors, observed that the average error rate was between about 0.2 and 0.3 errors/minute (or an average of about 20 errors during a 90-minute flight). In low workload sectors the error rate observed was less than half of this figure. However, many errors made were quite trivial and were often unnoticed by the flight crew (e.g. missing call signs, not handing over control correctly or mis-setting radio frequencies). Several of the gross errors relating to navigation went unnoticed in the simulator but would have been picked up in a real flight by ATC. It is important to dispel the notion that all errors end in an incident or an accident – the vast majority do not. Decreasing the number of errors made on the flight deck, though, is desirable as it will decrease the overall risk. It is rare in modern high reliability flight operations for major errors to result in an accident. Accidents are often the result of unexpected combinations of seemingly minor errors coupled with poor working practices. A single minor error (or even a major one) should not result in an incident or accident in a well designed, high integrity operation. This was noted by Helmreich, Klinect, and Wilhelm (1999) in their model of flight crew error. Nevertheless, errors often beget errors (see Figure 6.2).

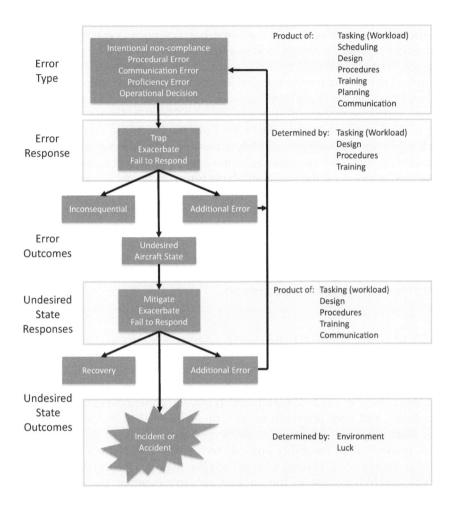

Figure 6.2   **Model of Flight Crew Error (adapted from Helmreich, Klinect and Wilhelm, 1999) overlaid with Reason's general failure types denoting the potential root of error (and mitigation)**

IDENTIFYING ERROR

There are alternative perspectives on the topic of human error in aviation that have a different root to them than the psychologically and organisationally-driven approaches described so far. Safety is all about the error 'troika': prevent error; trap error, or mitigate its consequences. Prevention of error may be achieved by either equipment design (as seen in many computer interfaces which will not accept an invalid input) or through procedural countermeasures (on the flight deck the processes of monitor and cross monitor, challenge and response) when setting up a flight deck system (e.g. the FMS/FMC). In the last case even if an incorrect entry has been made into the FMS/FMC, at each waypoint the crew should monitor the aircraft's actual behaviour against its expected behaviour from the flight plan, which should trap the error. Finally, continual monitoring of the aircraft's position should ensure that even if the error has not been prevented or trapped, the aircraft should not enter an unsafe condition (i.e. the error should be mitigated).

Helmreich, Merritt and Wilhelm (1999) suggest that the same set of CRM countermeasures may be applied to prevent, trap or mitigate errors, the difference simply being in the time of detection. However, there are other approaches that take an integrated human/system approach to error prevention. This approach is evident in formal error identification methods. Formal methods tend to deal with only the hardware and procedures on the flight deck rather than looking more widely across the organisation – a systemic approach. Reason advocates this latter approach to remove the underlying factors promoting error, suggesting that addressing the 'general failure types' in an organisation (which he lists as poor tasking, poor scheduling, poor design, poor procedures, poor training, poor planning and bad communication) is the most cost effective way of reducing error and hence preventing them promulgating throughout the system to result in accidents. This is described further in Chapters 16 and 17, respectively on Flight Deck Safety Management and Airline Safety Management in Part Four. Nonetheless it is worth having a very quick look at some basic error promoting factors associated with the design of both the flight deck hardware and its associated procedures.

Error Promoting Conditions

Performance Shaping Factors (PSFs) are conditions which substantially increase the likelihood of human error in a given situation. O'Hare (2006) divided PSFs into those external to the pilot (e.g. environmental conditions, equipment design, operating manuals and procedures, and poor supervision) and those internal to the pilot (including their emotional state, physical condition, stress, experience, and task knowledge). PSFs are inherent in many human reliability analysis techniques and human error identification methods. The Technique for Human Error Rate Prediction (THERP) developed by Swain and Guttmann (1983) for use in the nuclear industry contains the generic risk factors associated with the instance of error (see Table 6.2).

During the last decade 'design induced' error has become of particular concern to the major airworthiness authorities. In July 1999 the US Department of Transportation assigned a task to the Aviation Rulemaking Advisory Committee to provide advice and recommendations to the FAA (Federal Aviation Administration) administrator to:

> ... review the existing material in FAR/JAR 25 (Federal Aviation Regulation/Joint Airworthiness Requirements – part 25) and make recommendations about what regulatory standards and/or advisory material should be updated or developed to consistently address design-related flight crew performance vulnerabilities and prevention (detection, tolerance and recovery) of flight crew error. (US Department of Transportation, 1999)

These regulations have now been developed and were instigated in September 2007 in Europe by EASA (European Aviation Safety Agency) in Certification Specification (CS) 25.1302 – see Part Three in this book concerning the *Machine* aspects of the system.

Table 6.2   **Risk factors in the Technique for Human Error Rate Prediction –
THERP (Swain and Guttmann, 1983)**

Condition	Risk Factor
Unfamiliarity with the task	x17
Time shortage	x11
Poor signal: noise ratio	x10
Poor human-system interface	x8
Designer-user mismatch	x8
Irreversibility of errors	x8
Information overload	x6
Negative transfer between tasks	x5
Misperception of risk	x4
Poor feedback from system	x4
Inexperience (not lack of training)	x3
Educational mismatch of person with task	x2
Disturbed sleep patterns	x1.6
Hostile environment	x1.2
Monotony and Boredom	x1.1

Many of us have attempted to pull a door with a handle on it even though the label clearly says 'push'. The handle has an 'affordance' (Norman, 1988): its design encourages you to pull it despite instructions to the contrary. This is a typical instance of a design-induced error. On a flight deck the consequences of such errors can be dire. The crash of Air Inter flight 148, an Airbus A320, was attributed to such a design-induced error. The primary cause of the accident was that the crew left the autopilot in vertical speed mode when they should have selected flight path angle mode on the Flight Management and Guidance System console (see Figure 6.3). The crew entered '33' intending a 3.3° descent angle, however as the aircraft was in vertical speed mode this resulted in a rate of descent of 3,300 feet per minute (over three times the rate intended). The display of vertical speed or flight path angle was made on the same display element (at the right hand end of the panel) the only difference being the incorporation of a decimal place when in flight path angle mode. Mode selection was made using a switch in the centre of the panel (Bureau d'Enquêtes et d'Analyses pour la Sécurité de l'Aviation Civile, 1992).

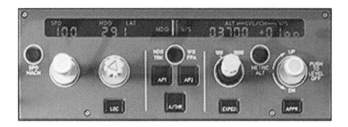

Figure 6.3   Airbus A320 Flight Management and Guidance System glareshield panel

Long lists of items in a procedure are especially prone to slips and lapses. With regard to checklists various axioms have been developed over the years. Reason (1987) observed that:

- The larger the number of steps, the greater the probability that one of them will be omitted or repeated.
- The greater the information loading in a particular step, the more likely that it will not be completed to the standard required.
- Steps that do not follow on from each other (i.e. not functionally related) are more likely to be omitted.
- A step is more likely to be omitted if instructions are given verbally.
- If steps are given in written form, the items towards the end of the sequence are more likely to be omitted.
- Interruptions during a task which contains many steps, (e.g. the implementation of a pre-flight checklist) are likely to cause errors in the form of either a slip or a lapse.

Formal Error Identification Methods

Most formal error identification methods work in a similar manner. They are often based on a task analysis followed by subsequent assessment of the user interfaces and task steps to assess their error potential. Hence, they are intimately related to the design of the flight deck and the standard operating procedures. SHERPA (Systematic Human Error Reduction and Prediction Approach – Embrey, 1986) has been assessed as being one of the more sensitive and reliable generic error prediction methodologies suitable for use on the flight deck (Harris, Stanton, Marshall, Young, Demagalski and Salmon, 2005) and is typical of this type of technique. The method operates by trained analysts making judgements about which error modes are credible for each task step based upon the analysis of the work activity involved. The technique proceeds through six basic steps:

- *Hierarchical Task Analysis* (HTA) – A hierarchical task analysis is conducted for the task under analysis. An example of such an analysis can be found in Figure 6.4.
- *Task Classification* – Each bottom level task in the HTA is taken in turn and classified into one of the five behaviours from the SHERPA taxonomy:
  - *Action* (e.g. pressing a button or pulling a switch): Errors in this category are classified into categories such as operation too long/short, operation mistimed

or wrong operation on right object, etc.

- *Retrieval* (e.g. getting information from a screen or manual): Errors in this category are classified into categories including: information not obtained, wrong information obtained or information retrieval incomplete.
- *Checking* (e.g. conducting a procedural check): Errors in this category are classified into categories including check omitted, right check on wrong object or check mistimed, etc.
- *Selection* (e.g. choosing one alternative over another): Errors in this category are classified into the categories of selection omitted or wrong selection made.
- *Information communication* (e.g. talking to another party): Errors in this category are classified into categories including: information not communicated or information communication incomplete, etc.

- *Human error identification* – After each task is classified into a behaviour, the analyst considers its associated error modes. A credible error is one the analyst judges to be possible. The nature of the error is noted along with its associated consequences, highlighting whether it is recoverable or not and suggesting potential remedial measures.
- *Consequence analysis* – The analyst considers the consequences of each identified error, which has implications for the criticality of the error.
- *Recovery analysis* – If there is a task step at which the error can be recovered this is entered next.
- *Tabulation* – The information gained from the SHERPA analysis is entered into a tabular output along with the ordinal probability of the error (simply low, medium or high) and its associated probability, assessed using a similar scale.

Several things will be noted from the manner in which the SHERPA technique proceeds. Error is regarded as a direct consequence of the design of the hardware and the operations employed. Although there is a strong behavioural element inherent in the methodology, the psychological root of the error is of little consequence. Secondly, error is defined strictly from a system perspective: unintended operations likely to degrade the system are categorised as errors. SHERPA also provides the facilities to incorporate the mitigation and management of errors into the analysis process (cf. the error 'troika'). Other error prediction methodologies employing a similar approach but developed specifically for the aviation domain (e.g. the award-winning(!) Human Error Template – HET – by Stanton, Harris, Salmon, Demagalski, Marshall, Young, Dekker and Waldmann, 2006) have demonstrated even greater degrees of sensitivity and reliability. An example of the output from HET (relating directly to the HTA example in Figure 6.4) can be found in Table 6.3.

It should be noted, though, that formal error prediction methodologies only really address Reason's skill-based (and perhaps some rule-based) errors within a fairly well defined and proceduralised context. Hence they can only help in protecting against errors of these types. They cannot cope with higher level decision-making errors, such as the knowledge-based errors that may occur when engaged in a non-diagnostic procedural management or a creative problem-solving task (Orasanu, 1993). They are quite limited in this respect. This has already been noted in the previous chapter along with the fact that these knowledge-based level errors are very difficult to detect, predict and recover from.

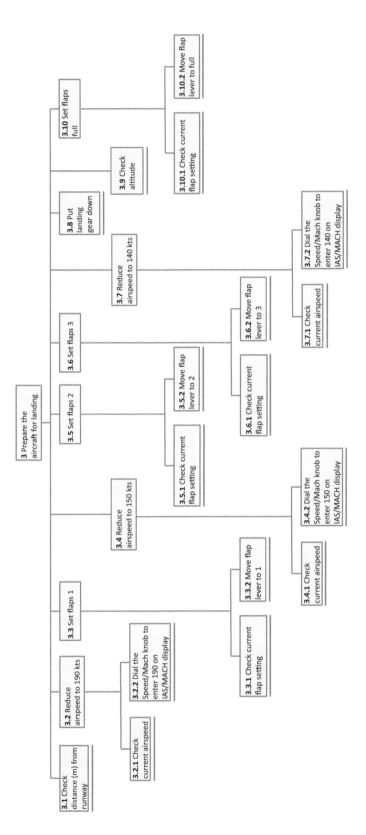

Figure 6.4 Section of an HTA for illustrative purposes – Land Aircraft X at New Orleans using Autoland system. A full copy of this analysis can be found in Marshall, Stanton, Young, Salmon, Harris, Demagalski, Waldmann and Dekker (2003)

Table 6.3 **Example of HET output (Stanton, Harris, Salmon, Demagalski, Marshall, Young, Dekker and Waldmann, 2006)**

Scenario: Land A320 at New Orleans using the Autoland system				Task step: 3.4.2 Dial the 'Speed/MACH; knob to slow down to 150kts on IAS/MACH display							
Error Mode		Description	Outcome	Likelihood			Criticality			PASS	FAIL
				H	M	L	H	M	L		
Fail to execute											
Task execution incomplete											
Task executed in wrong direction	✓	Pilot turns the Speed/MACH knob the wrong way	Aircraft speeds up instead of slowing down		✓			✓		✓	
Wrong task executed											
Task repeated											
Task executed on wrong interface element	✓	Pilot dials using the HDG knob instead	Aircraft changes course and not speed	✓				✓			✓
Task executed too early											
Task executed too late											
Task executed too much	✓	Pilot turns the Speed/MACH knob too much	Aircraft slows down too much	✓			✓				✓
Task executed too little	✓	Pilot turns the Speed/MACH knob too little	Aircraft does not slow down enough/too fast for approach	✓				✓		✓	
Misread information											
Other											

Design Options to Prevent Error on the Flight Deck

Once you have identified any potential error-producing situations it will be necessary to do something about it. This is all about the first element of the error troika: prevent the error occurring. There are some very simple design solutions that can be beneficial in helping to avoid simple errors if implemented in a carefully thought out manner.

- *Orient the controls so that accidental contact with them is not likely to activate them.* For example, if the natural, or most common movements of the pilot's arm in a certain area in the flight deck is in the up/down plane, ensure that all toggle switches are orientated in the left/right plane.
- *Provide sufficient control resistance to prevent accidental operation.* Put another way, make the control 'stiff' so that it requires a deliberate, sustained effort to operate it. Examples of this include overriding the activation of the stick-pusher.

- *Require a complex motion to activate the control.* Require a dual axis motion by the pilot when making a control input. Examples of this include having to pull and simultaneously twist on a rotary knob to change a setting or having to lift a collar before changing the position on a toggle switch.
- *Use interlocks.* Interlocks 'lock out' functions if the system is in a certain pre-designated state (e.g. operating the gear lever has no effect if the aircraft is on the ground with the system detecting weight on the aircraft's wheels).
- *Restrict access to the control.* This can be done by either placing the control in a recess, placing a physical guard around the control that must be removed first, or by placing the control outside of the normal reach envelope of the pilot. This option should only be used with care, though.

Computer-based error management includes using options such as lo-lighting menu items that are not available in a certain system states/modes or requiring confirmation before allowing the pilot to progress with that option (see later in Part Three which is concerned with flight deck design, including Chapter 15 dedicated to HCI on the Flight Deck).

Final Thoughts Concerning Design Aspects Underlying Error

Most errors have their root causes in a range of other aspects of the operation of the aircraft, not just flight deck design. This will become self-evident in the later chapters concerning Safety Management and Incident/Accident investigation. The design aspects of error (flight deck and procedural) are attractive from a regulatory point of view as they suggest the possibility of a remedial 'fix'. However, the systemic view of error is that it is a product of equipment design, procedures, training and the environment (Dekker, 2001). Error has its roots in the wider socio-technical system and can promulgate across organisational boundaries.

While the probability of design-induced error on the flight deck may be significantly reduced after implementation of the new rule alluded to earlier (CS 25.1302) described in more detail in Part Three, the level of overall risk in flight operations and the accident rate may only be marginally decreased. Simply adding a local regulation to fix one specific aspect of a wider system problem is unlikely to have a major effect. It will be argued in the final chapter that the management of error has to progress on a systemic basis, not a system-by-system, regulation-by-regulation basis.

NATIONAL CULTURE AND ERROR

Aviation accident rates differ across regions. Africa, Asia/Middle East and the Caribbean have higher accident rates than either Europe or America (Civil Aviation Authority, 2008). During the period 1997–2006 African operators had a fatal accident rate of 5.93 accidents per million departures; Asian/Middle Eastern operators a rate of 1.29 accidents per million departures; and the Caribbean airlines had a rate of 1.41 accidents per million departures. This is compared to just 0.17 per million movements for North American carriers and 0.34 for carriers from the European Union. Analysis of the underlying accident causal factors also shows differences between the regions. There has been a great deal of debate about the role of culture in aviation mishaps, however culture is only cited on relatively few occasions as having a role to play in the causation of accidents (Helmreich, 1994; Soeters and Boer, 2000).

Culture helps to fashion a complex framework of national, organisational and professional attitudes and values within which groups and individuals function. Hofstede (1984, 1991, 2001) proposed four dimensions of national culture: Power–Distance, Uncertainty Avoidance, Individualism and Masculinity. More individualist cultures (cultures which emphasise the rights of the individual over those of the group) such as the US show a lower probability of total loss accidents whereas collectivist cultures exhibit a greater chance of accidents. A high level of Uncertainty Avoidance (a preference for strict adherence to rules and norms compared to a tolerance for new ideas and flexibility) in a national culture has been found to be associated with a greater chance of accidents (Soeters and Boer, 2000). However, before progressing further it is essential not to conflate 'accidents' with 'error' (as emphasised at the beginning of this chapter). The two are quite different. Nevertheless …

It is essential to consider the cultural determinants of behaviour, including error. Li, Harris and Chen (2007) observed that the types of error implicated in aviation accidents differed between geographical regions. These differences were not related to the nature of the final errors committed by the pilots on the flight deck itself but were mostly concerned with contributory factors at higher organisational levels (see Chapter 17: Airline Safety Management). The analysis showed that Taiwan and India, both countries which have a high Power–Distance culture (cultures that value hierarchical relationships and which show respect for authority) had a higher frequency of events at the highest organisational levels compared to the USA (a much lower Power–Distance culture).

The corporate environments in Taiwan and India both prefer tall organisational pyramids with centralised decision structures and have a large proportion of supervisory personnel. In these cultures subordinates expect to be told what to do. On the other hand, the working environment of USA has low Power–Distance and is high on Individualism. It was suggested that the culture of the USA promoted greater efficacy in addressing safety-related issues through encouraging open discussion by all personnel. US organisational structures also allowed greater autonomy of action than in the Taiwanese or Indian cultures, which were less reactive as a result of their preferred 'tall' organisational configurations which served to discourage autonomy of action, especially at the lower managerial levels.

On the flight deck itself the only category of error that exhibited any differences between cultures was that involving decision making. The Taiwanese preference for collective (rather than individual) decisions may explain the over-representation of poor decision making being implicated in accidents from this country. Making decisions, implementing them and taking responsibility for their consequences is a central part of life in a low Power–Distance culture (such as the USA) and hence is well practised.

Unfortunately, there is one more level of cultural complexity associated with the determination of the root causes of pilot error on the flight deck. Independent analysis of the mid-air collision over Ueberlingen that occurred on 1 July 2002 (Bundesstelle Für Flugundfalluntersuchung, 2004) by groups of British and Taiwanese accident investigators demonstrated that there were cultural differences in the causal attributions underlying the active errors on the flight deck and in the Air Traffic Control Centre that led to the collision (Li, Young, Wang and Harris, 2007). At the beginning of this chapter it was noted that Woods, Johannesen, Cook and Sarter (1994) opined that an 'error' is a judgement made in hindsight. However, it would seem to be a judgement made in hindsight and from a particular cultural perspective; in an international business like the aviation industry, this is an observation that cannot be easily ignored.

A DISSENTING VOICE – AGAIN ...

It is worth rounding off this chapter by re-visiting some of the issues raised at the beginning. Some authors (e.g. Dekker 2004b) almost completely reject the notion of error. Dekker extends the argument from Woods, Johannesen, Cook and Sarter (1994) that 'error' is always a judgement made in hindsight even further. He suggests that we should be attempting to follow the line of reasoning that the pilot was following when an incident or accident happened. It's not a question of '... and then they made an error'. It is more a question of getting in the pilot's head and understanding their line of reasoning to understand why they did what they did (why did they think what they were doing was correct)? The 'error' is a misleading artefact identified and labelled with 20:20 hindsight. Dekker would argue that looking for an error doesn't necessarily help an investigator understand a sequence of events. He argues against the concept of dualism – the separation of hu*Man* and *Machine* (Dekker, 2004b) and subsequently looking for the broken human component (the 'error'). Dekker suggests a more holistic approach needs to be taken to the study of failures in systems. However, it is difficult to rationalise this conception of error (or rather the rejection of the notion of error) with the perspective employed by the system designer employing a formal prediction methodology to help avoid actions that will degrade the system. When considering human error, first of all pick your perspective then choose your label. But people do things that they don't mean to do (or shouldn't do). Honestly ...

FURTHER READING

For a deeper consideration of human (pilot) error try the classic:

Reason, J. (1990). *Human Error*. Cambridge: Cambridge University Press.

And for a complete explanation of the SRK taxonomy from which Reason's approach to human error was developed, see:

Rasmussen, J. (1986). *Information Processing and Human/Machine Interaction: An Approach to Cognitive Engineering.* Amsterdam: Elsevier/North-Holland

An alternative view of human error can be found in either:

Dekker, S.W.A. (2005). *Ten Questions about Human Error: A New View of Human Factors and System Safety.* Mahwah, NJ: Lawrence Erlbaum.

Or:

Dekker, S.W.A. (2006). *The Field Guide to Understanding Human Error.* Aldershot: Ashgate.

7
Individual Differences

Unlike many other aspects of psychology that attempt to isolate underlying universal principles of human behaviour which are invariant across people, the science of Individual Differences focuses on describing the manners in which people differ and the implications that these differences have for behaviour. Some individual differences are readily apparent such as age, sex and ethnicity; other aspects of individual differences require assessment using a quantifiable measure (e.g. social economic status) however many aspects of individual differences require measurement and inference about latent traits. Examples of the latter include personality, intelligence and other cognitive abilities. It is a fact that people differ in these traits, and the fact that some traits are desirable for the job of a pilot and some are not, underpins many aspects of the selection process (see Chapter 8 on Pilot Selection in Part Two of this book). This chapter is by no means a comprehensive view of the science of Individual Differences. It merely attempts to provide the reader with a flavour of a few key aspects of the underpinning science.

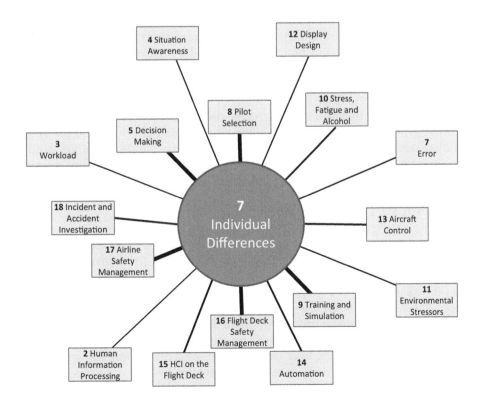

The study of Individual Differences provides the science base for the psychometrically-driven selection of flight crew. These differences in ability also underpin the training requirement both for flight skills and social skills (CRM).

The psychology of Individual Differences (see Kazdin, 2000) attempts to:

- Develop a descriptive taxonomy of how people differ;
- Examine differences in characteristics to predict behaviour; and
- Test the theoretical structure and dynamics of Individual Differences.

In this context the main emphasis will be placed upon the study of Individual Differences associated with personality, cognitive and psychomotor abilities. Personality and cognitive abilities cannot be measured directly, only inferred. Psychomotor skilled performance can be measured directly, though. Measurement of personality and cognitive aspects first requires the identification of the number and nature of their underlying dimensions. With respect to personality, these dimensions (or traits or types depending upon your perspective) are inferred from observation and self-report of behaviours, description of personal preferences, and reports of self-perception. In this category of psychometric test there are no 'right or wrong' answers. In contrast, tests of cognitive ability, of which intelligence is only one aspect (and there are many debates concerning what actually constitutes 'intelligence') are maximum performance measures, where the measure of a person's ability reflects the best a person can do. Tests of this type come in two broad sub-categories: performance scored in a limited time (speed test) or with no limited time (a power test) in which the object is to achieve the highest score possible on a series of questions of graduated difficulty.

There are fierce debates concerning the source of Individual Differences but basically this boils down to what is broadly termed the 'nature versus nurture' debate (e.g. Eysenck and Kamin, 1981). The 'nature' camp posits that differences in personality or intelligence stem from genetics, the effect of which is moderated by the effects of the environment. The 'nurture' side of the debate suggests that the new born is basically a blank canvas and that Individual Differences result from the manner and the environment in which they are reared. And of course, there is every shade of opinion between these polar opposites. The pragmatic view is that you cannot change the genes you are born with, so the only intervention available is that afforded by the environment in which you are raised. As will be discussed later, from a pilot selection perspective the source of job candidates' differences in personality or ability matters little. It is what they can do that counts.

PERSONALITY THEORY

There are many different philosophical or scientific camps from which the study of personality has emanated. The earliest school of thought (not surprisingly) was the psychoanalytic approach (from Sigmund Freud) who suggested that personality was basically the result of a three-way dynamic interaction between the Id, the Ego and the Superego. The Id is based on the pleasure principle, pursuing whatever it wants at a particular point in time with no consideration for the context of the situation or the feelings of others. The Ego is based upon the reality principle. The Ego's function is to service the needs of the Id but while taking into account the context of the situation and the feelings of others. The Superego is the conscience and operates by imposing moral and ethical restraints upon behaviour. Personality is a product of the balance between these three components. A dominant Id results in an impulsive, self-centred individual. A dominant Superego results in a judgmental and inflexible personality. The Ego is the conscious mind (containing the part of personality including defensive, perceptual, intellectual and executive functions). The Ego mediates between the Id, the Superego and the external world. In a 'healthy' personality (whatever that is) the Ego is dominant.

Type Theories of Personality

While this discussion of Freud's basis of personality may initially seem to be only of historical interest, offshoots of Freud's approach have found their way into widespread, mainstream use in the assessment of personality in many occupational settings.

Personality type theories classify people into distinct, discrete categories. Jungian (analytical) psychology, a divergence from Freud's psychoanalytic theory, provides the basis for the Myers-Briggs Type Indicator (MBTI). The MBTI is a multi-item, forced-choice personality test. There are various versions of the test, ranging from between (approximately) 100 to 150 items. These can be administered by computer or using the more traditional 'pencil and paper' method. The MBTI is composed of four basic dimensions which underlie the items on the questionnaire:

- *Extroversion–Introversion:* An extravert's flow of energy is directed outward toward people and objects whereas an introvert's is directed inward toward concepts and ideas. In a highly social situation the Extrovert draws energy from the situation where in contrast the Introvert has to expend energy to be sociable.
- *Sensing–iNtuition:* This describes the manner in which new information is understood and interpreted. Sensing personality types prefer information that is tangible and concrete and favour details and facts. People who prefer iNtuition err toward more abstract concepts and trust insight.
- *Thinking–Feeling* are decision-making functions in the MBTI typology. Those who show a preference for Thinking operate on the basis of a logical, rational tool set, whereas people who favour Feeling are more empathic and put greater emphasis (when making decisions) on considering the needs of others.
- *Judging–Perceiving* describes how a person interacts with the external world. A Judging typology prefers a structured lifestyle. Conversely, a person who is regarded as being Perceiving favours a flexible and adaptable approach to life.

(The Myers and Briggs Foundation. *MBTI Basics*.)

On the basis of their test responses a person's personality is then categorised into one of 16 categories (e.g. ENTJ – Extrovert, iNuitive, Thinking, Judging). For details of the characteristics about each of the 16 personality types, see www.myersbriggs.org/my-mbti-personality-type/mbti-basics.

There are other personality type theories, one of the most common being the Type A/Type B personality, developed by cardiologists Meyer Friedman and Ray Rosenman in the late 1950s/early 1960s (see Friedman and Rosenman, 1974). Type A personality people are often described as 'workaholics' who thrive on stress but who tend towards impatience and irritation. They have an intense drive to achieve goals and are highly competitive.

Some researchers have suggested that this may be linked to low self-esteem. Conversely, Type B personalities are portrayed as having exactly the opposite characteristics, being much more patient and relaxed.

Friedman and Rosenman originally hypothesised that Type A personalities were more prone to coronary heart disease as a result of their predispositions, although it has now been established that this link is statistically very weak. The Jenkins Activity Survey (Jenkins, Zyzanski and Rosenman, 1979) is one of the most widely used methods of assessing Type A/B personality, but in general results from the test are very mixed in predicting health and well-being. However, developments of the Jenkins Activity Survey place people on a continuum (from A to B) rather than describing them simply as Type A or Type B. This approach has more in common with trait theory. In general, type theories of personality have been criticised as being overly simplistic and insensitive to differences in human personality.

Trait Theories of Personality

Trait theory suggests that individual personalities are composed of several broad dispositions. They describe people on a number of continua, although how many is open to debate. Eysenck (1967, 1991) initially developed a model of personality based upon just two dimensions, later extended to three:

- *Introversion/Extraversion:* Someone highly introverted will be quiet and reserved, whereas an individual high in extraversion will be sociable and outgoing.
- *Neuroticism/Stable:* Neuroticism implies a tendency to become upset or emotional whereas stability refers to a tendency to be emotionally constant.
- *Psychoticism:* Individuals high in psychoticism may be antisocial, hostile, non-empathetic and manipulative.

In contrast, the OPQ 32 (Occupational Personality Questionnaire 32) from Saville and Holdsworth Ltd contains 32 dimensions of personality developed specifically for selection purposes which are designed to reflect an individual's preferred behavioural style at work (Brown and Bartram, 2009).

The EPI and the OPQ 32 probably reflect the extremes in the number of dimensions on the personality trait spectrum. Lying in between live tests such as the 15FQ+ (15 Factor Questionnaire) from Psytech International, the classic 16PF (16 Personality Factors) derived empirically by Ray Cattell (Cattell, 1946; Cattell, Cattell and Cattell, 1993) and the CPI (California Psychological Inventory) which comprises 18 underlying traits (Gough and Bradley, 1996, 2002).

The 16PF is worth a few further words in that it is somewhat different than the descriptions of the dimensions of personality derived from an underlying theory (e.g. the MBTI). Cattell derived his dimensions from adjectives describing personality traits found in the English language using self-report questionnaires administered to thousands of participants. These were reduced statistically using factor analytical techniques to form the 16 dimensions of personality. Actually, Cattell derived 15 primary personality traits. The sixteenth factor included in his personality inventory was intelligence (factor B – Reasoning) which he considered to be a trait rather than an ability. These are described in Table 7.1.

Table 7.1   Dimensions of personality as described in the 16PF

	Factor	Descriptors	
A	Warmth	Reserved	Outgoing
B	Reasoning	Less Intelligent	More Intelligent
C	Emotional Stability	Affected by feelings	Emotionally stable
E	Dominance	Humble	Assertive
F	Liveliness	Sober	Happy-go-lucky
G	Rule Consciousness	Expedient	Conscientious
H	Social Boldness	Shy	Venturesome
I	Sensitivity	Tough-minded	Tender-minded
L	Vigilance	Trusting	Suspicious
M	Abstractedness	Practical	Imaginative
N	Privateness	Straightforward	Shrewd
O	Apprehension	Self-Assured	Apprehensive
Q1	Openness to Change	Conservative	Experimenting
Q2	Self-Reliance	Group-dependent	Self-sufficient
Q3	Perfectionism	Self-conflict	Self-control
Q4	Tension	Relaxed	Tense

However, within personality trait theory there has emerged a general consensus of what are termed the 'Big 5' personality factors (Digman, 1990). These seem to be common underlying dimensions that have emerged irrespective of which personality test is used. The 'Big 5' are:

- *Openness to experience* (inventive or curious, versus cautious or conservative). A high score on this dimension suggests that the respondent is artistic, sensitive and intellectually curious. Lower scores are indicative of more conservative beliefs, with conventional, traditional interests and a distaste for abstract ideas.
- *Conscientiousness* (efficient or organised, versus easy-going or careless). A high score on conscientiousness is indicative of self-discipline or a high need for achievement. A low score indicates a preference for spontaneous behaviour but these people may have difficulty in motivating themselves.
- *Extraversion* (outgoing or energetic, versus shy or withdrawn). A high score on this dimension is indicative of people who are perceived as being enthusiastic and full of energy, and who enjoy the company of others. A low score on extraversion is usually referred to as introversion. Introverted individuals may be regarded as quieter and socially less exuberant.
- *Agreeableness* (friendly or compassionate, versus competitive or outspoken). Individuals with a high score on agreeableness are regarded as compassionate and cooperative, and willing to compromise to achieve harmony. A high score on this dimension is also associated with trustworthiness. People with low scores on this trait may be regarded as being unconcerned with others' well-being and they can

be viewed as being suspicious, unfriendly or uncooperative.

- *Neuroticism* (sensitive or nervous, versus secure or confident). High scores on neuroticism (also referred to as emotional instability) suggest that the individual is prone to negative emotions (such as anger or anxiety) and might also be prone to stress. A low score on this dimension (emotional stability) is indicative of calmness and even-temperedness and an individual who is less prone to interpreting evens in a negative context.

Use of the 'Big 5' personality factors (either directly or via other personality inventories) has been utilised in many personnel selection contexts. Conscientiousness has been shown to be a useful predictor across many professions. Extraversion has been shown to be a good indicator of performance in customer-facing occupations and ones which require teamwork (Barrick and Mount, 1991). Hörmann and Maschke (1996) found that personality variables reflecting high levels of neuroticism were more likely to be found in substandard airline pilots. Fitzgibbons, Davis and Schutte (2004) reported that the 'basic' personality profile for a commercial pilot was one of an emotionally stable individual, low in anxiety, vulnerability, angry hostility, impulsiveness and the depression aspects of the Neuroticism dimension from the 'Big 5'. Commercial pilots were also likely to be very conscientious; being high in deliberation, achievement-striving, competence and dutifulness (the Conscientiousness dimension). Furthermore, they were trusting and straightforward (Agreeableness) and were active individuals with a high level of assertiveness (Extraversion).

A Short Cultural Aside

All these theories of personality adopt a very Western European/North American perspective (a culture that is dominant across the globe in commercial aviation). The Chinese construe personality (and relationships) in a very different way, using dimensions that would be unrecognisable to the Western eye. However, these dimensions are equally as valid as those derived from the work of Jung or Freud. The most important lesson, perhaps, is that when evaluating Chinese personality types the dimensions should not be interpreted from a Western cultural perspective. For example, the concept of 'authoritarianism' is construed as a negative thing in Europe and America but not necessarily so in China. Chinese authoritarianism is a special type of authoritarianism. Jing, Lu, Yong and Wang (2002) and Jing and Yang (2006) use two basic dimensions to describe Chinese authoritarianism, each with further components to it:

- Inscrutable Power
  - Obedience
  - Respect
  - Fear of punishment
- Auspicious Expectation
  - Kindness
  - Taking care
  - Infallible superior
  - Family clan

They argue that Chinese authoritarianism cannot be understood completely by the simple concept of Power–Distance (see Hofstede, 1984, 2001). Chinese authoritarianism is a special type of authoritarianism that can be appropriately symbolised by the Chinese dragon. Aspects of 'dragonality' are desirable traits in a pilot, especially a Captain. The image of the dragon permeates deeply into the Chinese psyche: every parent wishes their child to become a dragon when they grow up. Becoming a dragon means being successful. 'Dragonality' additionally represents the power to punish people but also the power to bring wealth to people. It stands for rarefied power from heaven; the power people cannot resist but only obey and the power that they cannot understand but only accept (Jing, Lu, Yong and Wang, 2002).

I certain I haven't done this topic justice in the previous paragraphs, however I hope that this begins to demonstrate that Chinese notions and dimensions of personality are quite different from those encountered in the Western scientific literature. This will be re-visited when cross-cultural aspects of CRM are discussed in Part Four.

COGNITIVE ABILITY TESTS (INCLUDING TESTS OF PSYCHOMOTOR ABILITY)

There are many dimensions to cognitive ability, for example:

- Verbal ability
- Numeric ability
- Mechanical reasoning
- Spatial ability.

These are just a few of the more common facets. They may, or may not be considered as subsets of 'intelligence'. Intelligence is thought to be a stable characteristic and to differ between people. It is also assumed that the dimensions of intelligence can be quantified in some way using psychometric tests. However, unlike personality tests, cognitive ability tests are tests of maximum ability. As a reminder, two further broad sub-categories of ability tests may be defined. *Speed* tests require the candidate to answer as many questions of similar difficulty as possible within a given period of time. *Power* tests contain questions that vary in difficulty. Candidates sitting a power-type test are usually given few significant time constraints when undertaking the test, however even in these circumstances it is unlikely that any candidate will be capable of getting all answers correct.

For the sake of argument, although not strictly a cognitive ability test, psychomotor tests may also be included in this category. These tests assess motor skills and hand-eye coordination and are very commonly used in pilot selection, especially by the military. This category of test also falls into the category of a maximum performance test.

Intelligence tests, in the form of the revised version of the Binet-Simon test (often known as the Stanford-Binet test, developed by Lewis Terman) were used for the selection of US Army officers towards the end of World War I, onwards. The mean IQ (Intelligence Quotient) of the population was set as 100 (with a standard deviation of 15). Only recruits with an IQ in excess of 115 were deemed suitable as potential officer candidates. However, as the years have gone by, for selection purposes, pure intelligence tests have been found to be poor predictors of performance. More specific tests of selected aptitudes are generally more suitable. Nevertheless, most tests of cognitive ability do inter-correlate quite highly with each other leading to the notion of 'g', a generalised measure of intelligence (see also later in Chapter 8 on Pilot Selection). Not surprisingly, pilots tend to score quite highly in this regard.

Verbal ability tests come in several sub-types. Simple verbal ability tests examine skills such as spelling, grammar and the use of synonyms and antonyms. Verbal reasoning tests assess verbal problem-solving abilities. They make take the form of either verbal comprehension exercises or logical deductions to be made from complex statements. However, tests of verbal ability are influenced very heavily by the cultural and educational background of the job applicant.

There are what are claimed to be culturally fair IQ tests. These attempt to measure abstract analytical and reasoning ability without reference to knowledge of historical facts or the use of language. Perhaps the most well known of these tests is the Raven's Progressive Matrices which measures fluid intelligence (intelligence deemed to assess a subject's responses to new situations). This test requires candidates to complete a set of items, selecting an appropriate missing segment to complete a larger, abstract pattern. Many culturally-fair tests are administered under untimed conditions (power tests). Cultures have different attitudes towards time and pressure resulting from working against the clock and may therefore become anxious and not perform to their highest level.

Psychometric tests to evaluate the capability to work with numbers come in a variety of forms. Numerical ability tests may take the form of simple computation tests of the capacity to perform quick and accurate calculations using basic mental arithmetic. Numerical estimation items test the ability to make quick estimations without performing exact (time consuming) calculations. As a result these are most usually in the form of speed tests. Numerical reasoning tests examine a candidate's ability to determine the appropriate method to devise an answer rather than performing the appropriate calculations. Data interpretation tests require the subject to interpret information provided in tables, graphs or charts and then make appropriate decisions based upon these data.

Mechanical ability tests assess knowledge and reasoning about mechanical and physical concepts. They are tests of concrete rather than abstract reasoning. However, these are not pure tests of an underlying cognitive ability as they draw heavily on prior knowledge of things such as levers, gears, pulleys and springs.

Spatial ability can be either static or dynamic. Abstract tests of static spatial ability may involve things such as shape matching or the rotation of two- or three-dimensional shapes. More practically-oriented tests involve giving directions or map reading skills. Dynamic tests of spatial ability can involve assessing such skills as determining time to collision (or collision avoidance) or the management of arrival times. D'Oliveira (2004) has confirmed that dynamic spatial ability was a distinct dimension within the spatial abilities domain. In many ways dynamic tests of spatial ability can be conceptualised as one half of some types of psychomotor tests for assessing motor skills and hand–eye coordination, however there is only the judgement element present and not the control aspect.

Hand–eye coordination and psychomotor control tests have been used for the selection of pilots for over 70 years. The basic objective of the control of velocity test, a 2-D tracking task with various control lags incorporated within it which was used as part of the RAF selection process, was to assess candidates' basic psychomotor skills and measure hand–eye co-ordination. Various more sophisticated tests and test batteries have now been developed over the years to assess cognitive and psychomotor skills simultaneously in combination. One of these, the TBAS (Test of Basic Aviation Skills) – a component of the US Air Force officer selection process – is described in greater detail in Chapter 8 on Pilot Selection (Carretta, 1992; Carretta and Ree, 1993, 2003).

FURTHER READING

As was suggested earlier, this chapter provides only a very superficial introduction to a limited number of aspects of the psychology of Individual Differences. For a far more comprehensive treatment of this aspect of the science base, try:

Cooper, C. (2010). *Individual Differences and Personality*. London: Hodder Education.

And for an examination of intelligence-reacted aspects of Individual Differences, try his earlier work:

Cooper, C. (1999). *Intelligence and Abilities*. London: Routledge.

Finally, of course, there is the classic treatise on the psychometric assessment of individual differences:

Anastasi, A. and Urbina, S. (1997). *Psychological Testing* (7th edition). Upper Saddle River, NJ: Prentice Hall.

PART TWO

The (Hu)*Man*

A simple conceptualisation of Ergonomics is often presented as the process of either 'fitting the hu*Man* to the job' or 'fitting the job to the hu*Man*'. For a safe and efficient system both approaches are needed. The latter approach is covered in the chapters in Part Three of this book concerned with the *Machine*. The hu*Man* component in this book addresses the selection and training of commercial pilots and aspects of the occupational stressors unique to the aviation industry. These come in the form of psychosocial stressors (and their effects) and environmental stressors. The former have their roots in the domain of organisational processes (the *Management*); the latter need to be protected against, as far as is possible, by design of the aircraft and its systems (the *Machine*). In contrast to the material covered in the previous part, which although it had an aviation flavour to it was essentially a generic overview of the relevant parts of Applied Psychology, this part of the book is focused much more tightly on the commercial aviation system.

The pilot is at the very heart of Human Factors in aviation. The pilot is the hu*Man* around which all other aspects of the system are designed to operate. In stark contrast to Moore's Law, which suggests that the number of transistors that can be placed on an integrated circuit doubles every two years (hence it is used as a surrogate index of our speed of technological development) as Human Factors practitioners we are stuck with the 'Mark One' human being, which from a basic information processing perspective, has probably not evolved much for the last 100,000 years. As a result, in some ways the pilot can be considered as the ultimate system constraint in the safe and efficient operation of aircraft. However, some people are more likely to make safe and efficient (good) pilots than others. Selecting the people with the basic abilities required to fly (or is that manage?) a modern commercial airliner will considerably ease the demands on the training system.

The only option for adapting the hu*Man* component to have the skills, knowledge and ability to fly the aircraft is by training and education. These are the processes that ensure the pilot is equipped with the skills required to respond to the technical demands required of job. These are changing, evolving and increasing year-by-year. The design of the aircraft is imposed by the requirements of the *Mission* and this factor also dictates a large component of pilot training. There is an intimate relationship between the training required to make a commercial pilot and the specification and design of flight simulation facilities. However, simulators *do not* provide training; they merely support the training process.

The whole of the training process (curriculum content, delivery and the standards required of the training media) are heavily regulated by the airworthiness authorities (part of the Societal *Medium*). However, pilot selection methods are 90 per cent under the control of the *Management* of the airline – as long as the people selected can satisfy the medical, and in some countries the psychological criteria, laid down by the authorities.

However, the demands imposed on commercial pilots are not just technical in nature. The effects of repetitive sectors, operating across many time zones, working unsociable hours and being separated from family and friends are only a few aspects of the *Mission* that impact negatively on the hu*Man* in the system. Commercial aviation is a safety critical industry and the demands of having to constantly achieve high levels of performance are themselves wearing. The majority of crew training is aimed at emergency and abnormal situations (although it is suggested that this may need to change). Some say that flying is 99 per cent boredom and 1 per cent sheer terror. Thus may have been the case 50 years ago but it has now changed considerably. The operation of a short haul airliner is now 99 per cent high workload and 1 per cent sheer terror. This must be remembered when appraising the role of flight crew. Nevertheless, many aspects of pilot training will (hopefully) never be needed! You pay pilots as much for when things are going wrong as when they are going right. The economic *Medium* has made a significant transformation in the operation of airlines and has impacted on the human resources within them. Nevertheless, the broader mission remains the same: to deliver the passengers at the greatest possible speed and comfort while maintaining the highest possible standards of safety and economy. The hu*Man* component is absolutely central to this goal.

8
Pilot Selection

By definition, the pilot is the very heart of Human Factors. The pilot is the hu*Man* around which all other aspects of the system are designed to operate. The only option for modifying the hu*Man* component is by training, however as a result of all individuals being different in their abilities and knowledge, some people are more likely to make good pilots (after training) than others. The selection process analyses the requirements of the job (derived primarily from the *Mission* and the *Machine*) and attempts to match job applicants possessing the basic underlying abilities to the needs of the task (see Figure 1.2). The selection process is controlled by the airlines *Management* function, usually the Human Resources department.

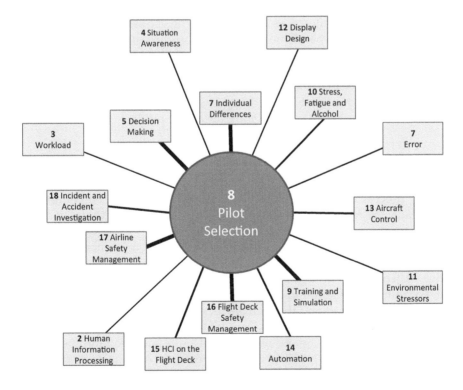

The selection of flight crew is based upon the science of psychometrics described in the chapter on individual differences. Pilot selection is also the feeder process for training and is highly related to issues in Crew Resource Management (Flight Deck Safety Management) and Decision Making.

Airlines can elect to select pilots from a pool of potential job candidates with a range of experience, skills and abilities. At one extreme they may choose to recruit highly experienced, qualified and type rated Captains for senior management pilot posts. At the other extreme they may wish to recruit *ab initio* cadets, possessing little or no flight experience, with the objective of training them to produce safe and competent flight crew, eventually capable of undertaking the First Officer's role on the flight deck of a modern commercial airliner. And, the selection process can apply to all situations in between these two extremes, for example the selection of First Officers for training and potential promotion to Captains or the selection of potential Training Captains. Selection and assessment principles can also be applied to the evaluation and *de*-selection of pilots *not* performing to the required standard during training. However, the selection methods employed differ considerably with respect to the skills and experience of the candidates to be recruited.

Training pilots is a costly business and poor selection decisions when recruiting *ab initio* pilots are expensive. Martinussen and Hunter (2010) quote figures from Goeters and Maschke (2002) who calculated in the year 2000 the cost of training a cadet pilot was €120,000. It was further calculated that cadets who failed training each cost the airline €50,000, against a figure of €3,900 for the cost of the selection process for each potential candidate.

Selection is largely based upon the psychological discipline of Individual Differences, a skimpy introduction to which was provided in Part One of this book. To recap slightly, the psychology of Individual Differences encompasses the assessment and measurement of constructs such as personality and intelligence. It is assumed that these traits are relatively stable but differ between individuals. It is also assumed that they influence behaviour. It is the fact that people differ in their personality and intelligence, coupled with the ability to measure these constructs, that underpins a great deal of the initial stages of pilot selection. Psychometric profiling has established that pilots with certain underlying abilities and personality characteristics tend to make superior airline pilots.

PRINCIPLES OF SELECTION

The selection process is driven by a job analysis. This process itself comprises of two separate threads: job requirements and person requirements (Martinussen and Hunter, 2010).

The first strand, *Job Requirements*, specifies the task competencies needed by job incumbents and the performance standards demanded of them. A competency is the knowledge, skill or ability needed by a successful post holder to perform to the standard required. For example, in addition to the technical skills required of a commercial pilot (the ability to fly the aircraft and the manage its systems) there is also a requirement for a successful crew member to be a good team player (e.g. they may need to exhibit good leadership/followership skills; good interpersonal skills and good intrapersonal skills – see the NOnTECHnicalSkills (NOTECHS) framework (van Avermaete, 1998) – outlined in the discussion of CRM in Chapter 16 on Flight Deck Safety Management). Such a collection of task competencies identified as being essential to undertake a particular occupation is referred to as a competency framework. The performance standards required in each of these individual areas forms the basis of a job description. Note, though, that all of these requirements are requirements of the job and *not* the person. The identification of these job requirements is usually carried out on the basis of some kind of formal analysis of, for example a task analysis undertaken using one or more of a variety of methods (see Annett and Stanton, 2000 for a description of a range of task analysis methods). The modern flight deck, however, as a result of the high level of automation incorporated into it, places considerable cognitive demands on the pilot, so it is more likely that a cognitive work analysis process (e.g. Vincente, 1999) and/or a cognitive task analysis (e.g. Crandall, Klein and Hoffman, 2006) will be a more appropriate basis for a job analysis. Non-technical skills may also be isolated on the basis of other analytical techniques, such as critical incidents technique (Flanagan, 1954) which was originally developed in the aviation domain. The technique was developed by the United States Army Air Force during World War II to produce objective definitions of effective or ineffective behaviours demonstrated by pilots for evaluation reports.

The second strand, *Person Requirements*, specifies the attributes required of the successful individual. To complement the basic psychomotor (stick and rudder) skills needed in a pilot certain personality characteristics may additionally be considered to be desirable (e.g. stability and outgoingness) as well as specific cognitive abilities (e.g. mathematical reasoning and spatial awareness). This is the realm of psychometrics and Individual Differences. In Chapter 7 on Individual Differences it was reported that commercial pilots were (among other things) emotionally stable, and low in anxiety, impulsiveness and depression. They also tended to be conscientious and strive to achieve, possessed a high level of assertiveness and were trusting and straightforward (Fitzgibbons, Davis and Schutte, 2004). Poor pilots had higher neuroticism scores than more successful pilots (Hörmann and Maschke, 1996). These attributes need to be operationalised into measureable quantities that can then be used in a selection context, for example, through the use of a suitable personality inventory. It should also be noted that the job competencies identified and the person requirement developed should 'fit'; the two aspects of the job analysis should complement each other, not conflict.

There are several methods of personnel selection commonly used, all of which have strengths (and weaknesses) depending upon the situation (see Arnold, 2005). These methods may be used alone or in combination with other techniques (largely depending upon the time and money available!). Frequently used methods include:

- Interviews;
- Personality tests;
- Biodata;
- Cognitive ability tests; and
- Work sample tests.

The relative strengths and weaknesses of each of these approaches is summarised in Table 8.1.

Table 8.1   **Advantages and disadvantages of various commonly used selection tests (adapted from Arnold, 2005)**

Advantages	Disadvantages
Interviews	
• Can test communication skills • Assess job knowledge • Interview can probe to gather important information • Allows the applicant to ask questions • Can introduce applicant to potential co-workers • Can be used with both experienced and inexperienced applicants	• Unreliable and low validity, especially unstructured interviews • Negative information given disproportionate weight in selection process • Decisions tend to be made within the first few minutes of the interview and based upon stereotypes
Personality tests	
• Can assess 'hidden' traits not readily apparent in interview • Can assess personality traits required for certain jobs • Can be administered in a group settings • Cheap and quick if the pool of applicants and number of posts available is large • Can be used with both experienced and inexperienced applicants	• Test needs to be selected carefully to assess required attributes • Training and experience may have greater impact on job performance than personality • Respondent may 'fake good' responding in the required manner to result in their selection • Only low predictive validity of some personality tests
Biodata	
• Cheap and quick if the pool of applicants and number of posts available is large • Allows structured collection of experiential data • Can be used with both experienced and inexperienced applicants	• Low predictive validity • Stronger predictors may be discriminatory (e.g. age)
Cognitive ability tests	
• Reliable with high validity for a wide range of jobs • May be used in combination with other tests to increase validity further • Can be administered in a group settings • Cheap and quick if the pool of applicants and number of posts available is large • Can be used with both experienced and inexperienced applicants	• Test needs to be selected carefully to assess required attributes • Can have negative impact on certain categories of applicant (e.g. racial minorities) – adverse impact
Work sample tests	
• High reliability and content validity if designed properly • Low adverse impact • Perceived as a fair process by applicants • Applicants cannot 'fake good'	• Costly and time consuming (often only one applicant at a time) • Can only be used with applicants with relevant job experience

The operation of personality tests has already been discussed in Chapter 7 however it is worth describing briefly the options for the remaining selection methods: interviews, biodata and work sample tests.

Interviews are a very commonly used selection technique even for the selection of pilots. However, interviews are one of the weaker selection methodologies in terms of their predictive validity (the degree to which an assessment measure accurately predicts subsequent job performance). There are, however, several things that can be done to enhance the success of the interview process. For instance, interviews should be conducted using several interviewers who all should be trained both in interpersonal skills and interview techniques. The interview board should be provided with a job description and person specification for the post and the questions asked of candidates should relate only to the job and should be standardised.

There are various basic types of interview. Instead of using an unstructured interview validity may be enhanced by using:

- *Situational Interviews* where the candidate is required to describe what they would do in certain job-related situations derived from the job analysis.
- *Behavioural Description Interviews* that ask applicants to describe the course of action they had previously taken in job situations similar to those they may encounter on the job. This may be further standardised by the interview board evaluating responses using structured behaviourally anchored rating scales.
- *Structured Interviews* that also ask the applicant how they would handle job-related situations (as before) but also aim to test specific aspects of job knowledge.

Huffcutt and Arthur (1994) demonstrated that predictive validity increased with an increase in the degree of standardisation of questions and their response scoring mechanisms. Campion, Campion and Hudson (1994), when comparing Situational Interviews (which are future-orientated) and Behavioural Description Interviews (past-orientated) found higher validities for the latter.

After completing any interviews, a structured process for comparing and evaluating all candidates can further aid the selection process. Assessors should avoid issues such as stereotyping applicants; giving too much weight in the selection decision to just a few characteristics (and giving disproportionate weight to negative information, i.e. reasons why an applicant *should not* be appointed); and basing selection decisions on first impressions.

Biodata are simply structured biographical data concerning a job candidate's qualifications, past activities and previous employment. These are usually elicited from application forms but may also be sought from previous employers in the form of a structured reference. The selection of biodata items is usually undertaken empirically, looking for strong relationships between applicants' past activities and job performance. However, this approach has been criticised as capitalising simply on chance relationships rather than having a theoretically-justified basis. For example, sports team membership (as a predictor of good job performance) may simply be a surrogate for personality factors such as conscientiousness or agreeableness, which are both associated with being a good 'team player'.

Work sample tests are based upon the assumption that the best predictor of job performance is the ability to do the job itself! The number and selection of tasks that job candidates are required to perform are based upon the job analysis. Work sample tests usually require the use of a simulator (although not necessarily a high fidelity simulator) and the candidate is presented with various job-relevant situations. Their responses are assessed by trained observers. The main drawback of work sample tests (apart from their time and cost) is that they can only be used on competent job applicants. Work samples approaches tend to be used when the number of posts and the number of applicants is small, but the post is of a high value (e.g. selection of Training Captains). A variation on the theme of a work sample test for unskilled applicants is the trainability test. The applicant is given a job-related period of instruction and is then evaluated on a work sample. The validity of this approach, though, is lower than that of the normal job sample.

Assessment centres consist of the evaluation of a job candidate's behaviour based upon multiple evaluations using multiple assessors. They may use a combination of several of the above approaches, plus other selection methods, such as leaderless group discussions and/or role play exercises (depending upon the results of the job analysis). Assessment centres usually take several days and as a result are very expensive so they tend to only be used for recruitment into high value posts or during the final stages of the selection process after the pool of potential job candidates has been reduced to only the most promising applicants. Trained observers are used to evaluate performance after each component of the assessment centre process. Judgments about a candidate's behaviour are recorded. At the end of the assessment centre results are pooled with selections based upon the results from all exercises. Every decision made should be backed up with behavioural evidence recorded during the process. In general, the validity of assessment centres can be quite high but this is a direct product of the selection and implementation of the exercises involved. It should be apparent by now that any selection process is only as good as the job analysis that precedes it.

The methods by which the personnel selection process progresses depends on several criteria, for example the number of jobs available, the experience required to fulfil the vacant post(s) and level of skills, knowledge and ability in the target applicant group. For example, is it expected that the people being recruited will be raw (untrained) recruits with the potential to develop into successful, professional pilots, or is the requirement to employ already competent and experienced personnel?

If the objective is to undertake a large-scale recruiting programme of cadets it is likely that the pool of potential applicants from which to choose will be large and will be inexperienced, thus negating the use of a job-sample based approach. Initial screening will need to be based upon methods easily administered to large numbers of people (perhaps remotely). Such techniques include biodata templates, remotely completed personality scales and job references. The initial selection of candidates may be made on this basis. Applicants passing this first stage might then be invited to an initial round of selection tests (e.g. tests of psychomotor coordination or cognitive abilities) to further reduce the size of the pool of applicants. Only once the group of candidates has been reduced to a more manageable size should more time consuming and expensive selection methods (e.g. assessment centres) be used. For example, in a 20-month period during the late 1990s, 5,834 people applied to join the British Airways pilot training scheme. Of these 2,606 candidates were deemed eligible for selection and were invited to a series of assessment centres. Three hundred and seven were subsequently offered places in the training scheme (Rawlins, 2000). Furthermore, in this example, as the applicant pool was inexperienced, a work signs-based approach was required (see Wernimont and Campbell, 1968 or Robertson and Smith, 2001). Signs are aptitudes in the applicant pool (perhaps based upon psychometric tests or biodata results) that have been empirically established to be indicators of successful job applicants.

Over the years various other sophisticated tests and test batteries administered as part of an assessment centre have been developed to assess a wide range of cognitive and psychomotor skills in *ab initio* pilot candidates. These are very commonly used as part of the selection process by the military who recruit many trainee pilots each year from a large number of applicants. For example, for many years the US Air Force used the Basic Attributes Test (BAT) which has relatively recently been updated and is now known as the Test of Basic Aviation Skills (TBAS). This comprises of various computerised sub-tests, of increasing complexity and difficulty that measure psychomotor skills and cognitive aptitudes. These are:

- *Directional Orientation Test* – to measure spatial orientation abilities by testing the candidate's ability to determine their position relative to a target.
- *Three-Digit and Five-Digit Listening Tests* – which require the candidate to respond when they hear any of the three or five digit pre-specified target numbers.
- *Horizontal Tracking Test* – where participants are required to use rudder pedals to control a target in the horizontal plane. The level of difficulty increases as the task progresses.
- *The Aeroplane Tracking Test* – requires that subjects use a joystick to keep a gun sight on a target aircraft as it moves at a constant rate.

Having experienced each of these component tests individually, the difficulty (power) is increased by requiring the US Air Force pilot candidates to undertake the tests in combination.

- *Aeroplane and Horizontal Tracking Test* – undertaken simultaneously.
- *Aeroplane Tracking, Horizontal Tracking and Three- and Five-Digit Listening Tests* – performed simultaneously.

Finally candidates are presented with the *Emergency Scenario Test* in which they must undertake both the horizontal and the aeroplane tracking tasks while responding to emergency scenarios which they must cancel by typing a code using the computer keyboard.

The TBAS (and its predecessor, the BAT) is just one component of the battery of tests used in the US Air Force selection process (the AFOQT – Air Force Officer Qualifying Test). The other aspects of this test battery include cognitive ability tests and personality tests. The process has been extensively researched and validated in a military context (see, e.g. Carretta, 1992; Carretta and Ree, 1993, or 2003).

Issues in Validity and the Efficacy of Selection Criteria

A selection process is said to have *content validity* when the procedure samples significant parts the job. Job sample-based tests require applicants to perform certain aspects of the job to demonstrate their competence. These components of the job on which the selection test(s) are based are determined from the job analysis. Job sample-based selection tests have *criterion validity* when it can be demonstrated that there is a strong relationship between these predictors and the candidate's ultimate ability to do the job (although a lot of the time such measures of predictive validity are established with respect to training success, rather than ultimate job performance). A selection process that uses psychometric constructs (perhaps based upon personality or cognitive ability tests, rather than job samples) has *construct validity* when it can be shown that a strong relationship exists between such a test and the job construct it seeks to measure (e.g. the relationship between interpersonal skills – a competency from the job analysis – and the 'Big 5' personality dimensions of Agreeableness and Extraversion).

Touching on the concept of validity (of any kind) brings up one of the big issues in selection research (and also the evaluation of training). How do you evaluate and measure the required skills in a commercial pilot? There is no simple, single criterion measure. Among other things, a 'good' pilot needs high-quality flying skills to control the aircraft manually; first-rate flight planning skills, decision-making skills, interpersonal/ management skills; first-class technical knowledge of aircraft systems; adaptability, flexibility, and even humour! The list is almost endless. What do you compare the selection criteria against to establish their usefulness? A selection process is only as good as the criterion measures for pilot performance used to evaluate its efficacy.

In many cases, especially for *ab initio* cadets, commonly used job performance criteria range from a simple dichotomous assessment of pass/fail after training, to slightly more sophisticated measures, such as time to complete training or the number of aspects of the training programme that required re-taking. These are still quite crude measures, though, that can tell the researcher little about the relative effectiveness of the individual selection tests employed. In common with assessing training outcomes, selection tests should only be evaluated against those aspects of pilot performance that they are supposed to select for. As a result, the evaluation of pilot selection criteria should be undertaken against specific training outcomes. Multiple comparisons of selected predictors against targeted desirable behaviours are required rather than assessing all predictors against a universal measure of success.

Success in training is, however, only a small part of the story. It is a proximal measure of the efficacy of the selection process but the real objective is to ultimately produce highly competent commercial pilots after they commence the job for real. However, the problem with *distil* measures of pilot selection methods are that the relationship between on-the-job performance and selection test scores is frequently weak as a result of the many other intervening factors, including the training itself! Furthermore, operational assessments of pilot behaviour are often crude and not well matched to the selection methods used.

Finally, all evaluations of selection decisions and selection criteria are biased toward the candidates selected, be they ultimately successful or not. Little is known about how well candidates de-selected during the selection process would have performed in their chosen occupation. How many potential Chuck Yeagers have not been selected for pilot training?

COMMERCIAL PILOT SELECTION

Airlines originally tended to rely on the military for producing trained pilots. As a result the selection techniques employed assumed that candidates were trained and competent (e.g. interviews, reference checks and maybe a flight check – see Suirez, Barborek, Nikore and Hunter, 1994). However, in the 1970s and 1980s, particularly in Europe where there was less emphasis on recruiting pilots from a military background but where there also was an increasing demand for commercial pilots, greater importance was placed upon the selection processes for *ab initio* trainees. In this case it was vital to assess the candidate's potential to become a successful airline pilot (and not fail the training) which is very costly (see Goeters and Maschke, 2002, earlier in this chapter). Psychometric and psychomotor tests became more commonplace similar to the methods military selection boards had been using for some years. However, as the management role of pilots began to develop with increasing importance placed upon CRM, qualities such as judgement, problem solving, communications, social relationships, personality and motivation became as important as the technical skills involved in flying a large commercial aircraft (Bartram and Baxter, 1996).

The vast majority of studies of aircrew selection have been undertaken by the military. Relatively few studies investigating the selection processes for civil commercial pilots are available in the open scientific literature (Carretta and Ree, 2003). Although airlines still rely on the military for producing trained pilots, this trend is changing due to the downsizing of the armed forces and increasing retention of their flight crew. In the USA, Hansen and Oster (1997) reported that up to 75 per cent of new hire airline pilots were recruited after commencing their flying career in military aviation. The situation in Europe is different where considerably more 'pilots' are selected with little or no prior formal flight training, and they are trained by the airlines themselves. As a result, there is a great deal of variance in selection methods between the two continents. US carriers tend to rely on techniques that assume candidates are already trained and competent (i.e. a work samples-based approach). Suirez, Barborek, Nikore and Hunter (1994) reported that reference checks, other background checks and interviews were the most common methods employed, often combined with a flight check. Aptitude and psychometric assessments were not generally employed, since it was assumed that the candidate's past record successfully demonstrated their ability as a pilot. In Europe, with more *ab initio* trainees, selection emphasises the assessment of the candidate's potential to become a successful pilot (greater stress is placed upon aptitude testing – a work signs-based approach). After pre-screening, a typical assessment centre for *ab initio* pilot candidates will involve personality assessments, tests for verbal and numeric reasoning, tests of psychomotor skill, group discussions and structured interviews.

There are several reasons why the vast majority of pilot selection studies have been undertaken by the military. All branches of the military, especially in the USA, recruit large numbers of pilot candidates every year from a vast number of applicants. They also keep extremely detailed training records which allows for validation of the selection measures employed at many stages during training and throughout the pilots' following careers. Furthermore, there is a scientific research infrastructure within the military that allows for validation of the selection procedures and there is no commercial disbenefit from the publication of studies investigating the effectiveness of the selection process.

In contrast, commercial airlines have some problems substantiating their selection criteria. There is often little formal validation of the selection processes in airlines, as selection procedures are deliberately changed and/or updated every year. The number of cadet pilots selected every year by even the largest airlines is also relatively small. As a result, few studies have been published on the job performance criteria for commercial pilots. Cathay Pacific report using selection criteria in six main areas: technical skill and aptitude; judgement and problem solving; communications; social relationships, personality and compatibility with Cathay Pacific; leadership/subordinate style; and motivation and ambition (Bartram and Baxter, 1996). The main point to note here is that flying skills *per se* form only a relatively small component. A similar observation may be made of the Qantas selection criteria (Stead, 1995). For direct entry (fully qualified) pilots, in addition to the required licences, candidates were required to demonstrate their numeric, verbal and critical reasoning abilities, pass various psychomotor skills tests, and they also underwent personality tests to evaluate their maturity and reaction to pressure/stress, adaptability/flexibility, task orientation, crew orientation, decision making and command/leadership potential.

Besco (1994) has argued that personality assessments have no utility in pilot selection and are unrelated to skill and performance. Meta-analyses of pilot selection studies suggest that personality is a weak predictor of performance (e.g. Hunter and Burke, 1994 and Martinussen, 1996) although this may be for a variety of reasons. These analyses are dominated by studies of military aircrew, which may result in a bias towards the assessment of pilots for single-seat aircraft and an emphasis on technical performance in the criterion measures. When criterion measures more closely related to team performance on the flight deck are used, the results are different. As discussed in Chapter 16 on Flight Deck Safety Management: Crew Resource Management and Line Operations Safety Audits, Chidester, Helmreich, Gregorich and Geis (1991) found that better performing pilots scored higher on traits such as mastery and expressivity and lower on hostility and aggression. Hörmann and Maschke (1996) found that below standard commercial pilots had significantly lower scores on personality scales assessing interpersonal skills and higher scores on emotional stability scales.

Situational judgement tests (as opposed to psychometric personality inventories), which present candidates with a series of job-relevant situations and ask them to respond with what they regard as the most appropriate course of action, have been developed by the US Air Force to identify pilots who will make effective crew members in multi-crew transport aircraft (Hedge, Bruskiewicz, Borman, Hanson, Logan and Siem, 2000). Results indicate that this methodology produces higher validity coefficients than personality tests when assessed against CRM performance. But as noted before, the utility and validity of personality scales for pilot selection depends very much on the choice of the criterion measure applied.

Non-personality related selection tests (e.g. verbal and numerical reasoning tests) have shown much larger correlations with performance (see any of the meta-analyses of pilot selection previously cited). This probably reflects the job of the airline pilot, where flight administration and planning is as important as the psychomotor skills required for flying. These measures may all be evaluating a common underlying construct, though. Analysis of US Air Force aircrew selection methods (Ree and Carretta, 1996; Carretta and Ree, 2003) suggests that most of the tests measure some aspect of general ability ('g' – see Chapter 7 on Individual Differences). In a battery of 16 psychometric tests used in the AFOQT measuring a range of aviation-related abilities (comprising of verbal, mathematical ability and spatial ability; aviation knowledge and perceptual speed) 67 per cent of the common variance was accountable by a single underlying construct (Skinner and Ree, 1987). Given the nature of the batteries of selection tests used by airlines, it is likely these also measure the same underlying general cognitive ability.

Military selection procedures still continue to place a great deal of emphasis on spatial and psychomotor skills in the evaluation of aircrew candidates (e.g. Hunter and Burke, 1994; Martinussen, 1996; Griffin and Koonce, 1996; Burke, Hobson and Linsky, 1997; Carretta and Ree, 2003). In general, these tests are reasonably good predictors of training success. However, military aviation is very different to commercial aviation. While many civil *ab initio* selection programmes still use tests of psychomotor skills, their main purpose is to maximise the chances of candidates passing the initial stages of flight training, rather than their utility in ultimately selecting good airline pilots. Nevertheless, although the job of a commercial pilot now requires very little hands on flying it still remains a necessary skill but greater emphasis is now placed upon the cognitive abilities needed for the execution of their day-to-day job.

While the factors discussed up to this point may be of utility in selecting pilots from the point of view of the airlines, the most controversial aspect of pilot selection (or de-selection), and one closely associated with safety, is that of psychological testing related to flight crew licensing. The JAA had the capacity to require flight crew to undergo *psychological* assessment (in addition to any *psychiatric* assessment) as part of the class 1 and class 2 medical requirements (JAA, 1999):

> An applicant for or holder of a Class 1 medical certificate shall have no established psychological deficiencies (see paragraph 1, Appendix 17 to Subpart B), which are likely to interfere with the safe exercise of the privileges of the applicable licence(s) (JAR-FCL 3.240).

Appendix 17 states:

> A psychological evaluation should be considered as part of, or complementary to, a specialist psychiatric or neurological examination when the Authority receives verifiable information from an identifiable source which evokes doubts concerning the mental fitness or personality of a particular individual. Sources for this information can be accidents or incidents, problems in training or proficiency checks, delinquency or knowledge relevant to the safe exercise of the privileges of the applicable licences… The psychological evaluation may include a collection of biographical data, the administration of aptitude as well as personality tests and psychological interview.

With the instigation of EASA in 2003, there has been an extended period of rule transitions and harmonisation from JAA to the new regulatory body. At the time of writing, this requirement is now being evaluated for inclusion within the corresponding EASA Flight Crew Licensing regulation.

This requirement for the psychological testing of pilots has provoked considerable debate both for it (e.g. Goeters, 1995) and against it (Johnston, 1996). Sweden and Germany have required psychological evaluation for pilot's licences for some time. However, Johnston puts forward ten problems and risks with testing, including issues such as: a lack of reliability and validity in the tests used; only a relatively weak probabilistic association (on an individual basis) with actual performance; no internationally agreed battery of tests; unintentional (or intentional) abuse of the results; and cultural differences in selection criteria and behavioural norms. Johnston does accept, though, that psychological evaluation has benefits in reducing attrition in training. The FAA has yet to adopt the requirement for the psychological assessment of commercial pilots so it is likely that this debate will continue for some time.

The selection process provides the raw material for flight crew training, no matter if the personnel selected are *ab initio* cadets or highly experienced Captains. It is only the type and degree of training required that differs. This is the topic of the following chapter in this part.

FURTHER READING

Although a little old now, the only book dealing specifically with the selection and assessment of pilots is:

Hunter, D.R. and Burke, E.F. (1995). *Handbook of Pilot Selection.* Aldershot: Avebury Aviation.

For a book more on the general principles of selection try:

Cook, M. (2009). *Personnel Selection: Adding Value Through People*. Chichester: Wiley-Blackwell.

9
Training and Simulation

At the beginning of Chapter 6 on Error I said that 'error' was a recurring theme throughout the book and that the chapter was concerned merely with presenting a framework for the reader. Almost the same can be said of the training aspect of this chapter. Training is pretty much ubiquitous throughout this volume. References to training can be found in the CRM aspect of Chapter 16 on Flight Deck Safety Management; and also Chapters 5, 13 and 14, respectively on Decision Making, Aircraft Control and Automation, just to name the bits that most readily spring to mind. As a result, the emphasis in this chapter will be on providing an overview of the training design and delivery process and the use of simulation to support it. Training is the method by which the selected hu*Man* is modified to operate the *Machine* to perform the *Mission*. It is overseen by the *Management* and specified by the regulator (part of the Societal *Medium*). Training is at the very core of Human Factors and was one of the starting points for the discipline in World War II.

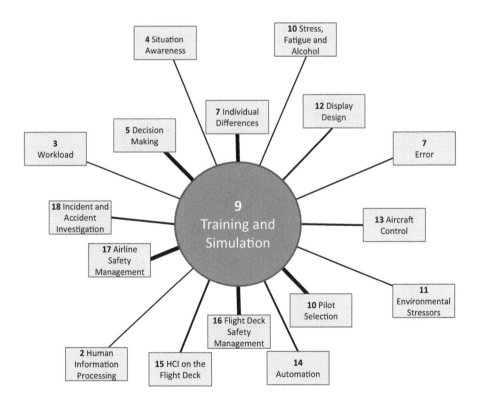

The crew selection process provides the basic material for subsequent flight crew training. Training is multifaceted and is associated with issues such as CRM (Flight Deck Safety Management), Decision Making and the psychomotor skills related to the Control of the Aircraft. Well designed flight deck interfaces minimise the requirements for training. The identification of a new (or modified) training requirement is also usually the first remedial action to be initiated as part of the Airline Safety Management processes following the investigation of any incident or accident.

Right from the outset it has to be emphasised that flight simulation *does not* provide training. Put an untrained person in a simulator and I guarantee that no matter how long they spend in there, they will not emerge as a competent pilot. Flight simulation supports the training process but simulation facilities can only be specified after the training objectives have been set, not *vice versa*. You do not set the training objectives with respect to the capabilities of the simulator!

Not surprisingly the training process, and the design and operational requirements for flight simulators are heavily regulated by the airworthiness authorities. It is not the role of this book to describe these requirements in any detail. The interested reader is referred to the regulations contained in: CFR title 14, Part 60 – *Flight Simulation Training Device Initial and Continuing Qualification and Use*; Part 61 – *Certification: Pilots, Flight Instructors, and Ground Instructors*; and JAR-FCL 1, *Flight Crew Licensing (Aeroplane)* – soon to fall under the remit of EASA. Interestingly, though, the airworthiness authorities are now beginning to relax their grip somewhat in this area to allow the airlines to be more responsive in designing their training syllabi to suit their operational requirements. Regulatory efforts are now placing as much prominence on the process of defining what should be trained and the manner in which it should be delivered, as on specifying what should be taught.

It almost goes without saying that pilot training is of paramount importance in maintaining aviation safety, but for a professional pilot, recurrent/assessment training is required as is the constant demonstration of the required levels of competence through regular checks. A failure to demonstrate continually acceptable skill levels may result in remedial training being required, loss of licence or ultimately, loss of employment. The provision of training courses and training facilities represents a huge investment for airlines. At the time of writing, depending upon their exact specification, a full flight simulator approved by the regulatory authorities for training and licensing purposes will cost between $6 and $12 million (and can cost up to $20 million). In a large airline, these facilities will be supplemented by cockpit procedures trainers, other part-task trainers, computer-based training facilities for aircraft systems and procedures, and 'regular' classroom facilities (see Farmer, van Rooij, Riemersma, Jorna and Moraal, 1999; Moroney and Moroney, 1999; or Kaiser and Schroeder, 2003). Furthermore, almost all training facilities and curricula in an airline will be aircraft type specific, as will be the assessments required.

Figure 9.1   CAE Airbus A320 FAA Level D Full Flight Simulator at the Shanghai Eastern Flight Training Company

As a final comment in this introductory section, although this book is dedicated to civil commercial aviation, it is worth noting the differences with military aviation at this point. With the exception of combat and regular transport operations, military flight crews spend most of their operational life training in one form or another. In civil operations, training costs time and money: it is a large financial overhead to the company and takes crew away from revenue producing operations. Crews in simulators are not generating money for the airline.

TRAINING

Training and Education are two different facets of learning. There is no well defined dividing line between the two. Training (in general) is normally undertaken to develop a specific skill or ability. The focus is on the trainee as a means to accomplish some well defined task. In contrast, education treats the development of the individual as an end in itself. This distinction becomes blurred when considering the demands on the crew imposed by the management of the automated systems on the flight deck. A basic knowledge of the manner in which the automated systems work is required if training is to be effective. The flexibility of automated systems means that a traditional approach to training along the lines of 'if this, then this' is inappropriate (see Chapter 14 on flight deck Automation in Part Three).

All training development proceeds through a series of generic stages. The Interservice Procedures for Instructional Systems Development (IPISD) has become a benchmark approach for training design, although it was originally intended for training design in much larger systems than those on the flight deck (Branson, Rayner, Cox, Furman,

King and Hannum, 1975). The IPISD has five major functional phases: Analyse; Design; Develop; Implement and Control.

- *Analyse* – This phase provides the foundations upon which the training design process is developed. It commences with a Training Needs Analysis (TNA). A TNA may use either qualitative or quantitative methods to identify the most important performance gaps which can be addressed via training. Gaps between the trainees' current levels of skills, knowledge and/or ability and that required to meet the requirements of their tasks need to be identified. The TNA can be developed from the job analysis undertaken for personnel selection purposes. All the tasks that need to be undertaken are identified and described along with their required performance standards (*q.v.* hierarchical task analysis and/or cognitive task analysis). It should also categorise them in terms of their difficulty, importance and frequency. However, a TNA can also identify deficiencies that should be addressed by interventions other than training (i.e. those that are not a result of knowledge or skills deficiencies). If the training objective is a development of an existing training syllabus, some of the required instruction may be already being delivered. Finally, all the tasks selected for instruction need to be analysed to determine their most suitable instructional setting. An options analysis should be undertaken to review the potential alternatives for training methods and the media required by these different options.
- *Design* – This phase begins to develop the output from the TNA into an instructional design. Each task selected for training needs to have a learning objective developed for it (i.e. what should the trainee be able to accomplish at the end of this instructional step?). These learning objectives are then subject to further decomposition and analysis to identify the steps necessary for their mastery. Tests are also designed to examine each of the learning objectives. These may be evaluated using a small sample of trainees (note how I am desperately trying to avoid saying a pilot study...).
- *Develop* – The instructional development phase classifies the learning objectives to identify and develop the guidelines required for optimum learning. The media selection process is also undertaken during this phase. In the aviation industry this often boils down to simply what sort of simulation facility should be used. However, simulation is not always the 'best' answer. The 'best' answer should also take into account factors such as: trainee characteristics (e.g. what stage of training are they at?); how many trainees are there?; what are the required/available simulation facilities?; are there other criteria relating to the setting of the training (e.g. distance)?; and is cost a major factor? On the basis of these considerations the instructional materials should be developed and tested. If the size of the pool of trainees allows, the content of the curriculum should be validated using empirical data obtained from groups of typical students before it is considered to be ready for wider delivery. Furthermore, in the case of a large organisation with many demands on its facilities and a large throughput of trainees, an instructional management system needs to be developed to allocate and manage resources. In the case of pilot training comprehensive record keeping is an essential facility of this system, particularly for the licensing and recurrent training requirements of the regulatory authorities.

- *Implement* – Before any syllabus can be implemented the personnel delivering the training also have to become familiar with both its content, the methods for evaluation of the trainees and the instructional management system. Opinions concerning the facilities to collect data should also be collected from all staff delivering the instruction with respect to aiding revisions to the syllabus and/or its method of delivery. Finally, the training should actually be delivered!
- *Control* – The training syllabus must be evaluated against targets and revised, if required. Internal evaluation involves the analysis of trainee performance (a proximal level of evaluation – see later). External evaluation requires an assessment of trainee performance on the job to determine if the course is producing graduates of the required standard (a distil measure). Data from the internal and external evaluations (and from trainer comments) are used as a quality control measure to assess and revise the training as required.

With greater freedom to determine the content of pilot training syllabi under programmes such as the AQP (see Advisory Circular AC 120-54; FAA, 1991) the design and evaluation process becomes even more important. Under the terms of the AQP the process of designing and delivering the training is as important as the competence of the pilots that it produces. Devolved responsibility for training design content and a regulatory 'light touch' requires robust processes to be in place.

The emphasis in the AQP is away from time-based training requirements to fleet-specific, proficiency-based requirements. Training needs are identified and developed directly from line operational requirements. The applicant (not the regulatory authority) develops a set of proficiency objectives based upon that airline's particular requirements for their specific type of operations. The AQP is based upon a rigorous task analysis of operations but with emphasis firmly placed upon the cognitive aspects of the flight task, such as crew decision making or the management of the aircraft's automation (see Seamster, Redding and Kaempf, 1998). Although the main emphasis in the AQP is biased towards CRM training (see also Chapter 16 on Flight Deck Safety Management) it is not exclusively targeted at this area.

The FAA describes the rationale underlying the AQP:

> Under the AQP the FAA is authorized to approve significant departures from traditional requirements, subject to justification of an equivalent or better level of safety. The program entails a systematic front-end analysis of training requirements from which explicit proficiency objectives for all facets of pilot training are derived. It seeks to integrate the training and evaluation of cognitive skills at each stage of a curriculum. For pass/fail purposes, pilots must demonstrate proficiency in scenarios that test both technical and crew resource management skills together. Air carriers participating in the AQP must design and implement data collection strategies which are diagnostic of cognitive and technical skills. In addition, they must implement procedures for refining curricula content based on quality control data. (http://www.faa.gov/training_testing/training/aqp/ – accessed 3 March 2010)

The first step in the AQP process requires an aircraft specific task analysis covering the full range of conditions internal and external to the aircraft (see also the description in Chapter 16 of the LOSA methodology). From this breakdown the required skills, knowledge and abilities are elicited, including the applicable performance standards. These may be generic skills or flight phase independent. The task analysis output can then be developed into a curriculum based upon a hierarchical series of enabling objectives to prepare crews for subsequent training in an operational flight deck environment. Individual training objectives are subsequently allocated to various curriculum components to construct the syllabus.

The AQP process is very similar to the elicitation, design and delivery of training described in the IPISD process. However, the revolutionary aspect of this process in the aviation industry lies in the releasing of the regulatory shackles when approving the content of the training programme. There have been some criticisms, though. For example, the complexity the AQP process means that considerable professional skills and resources have to be applied to the programme to gain the regulatory authority's approval, and there is a strong likelihood that only the major airlines with extensive resources will benefit (Maurino, 1999).

One of the interesting aspects of the AQP is that it encourages applicants to integrate the use of advanced flight training equipment, including full flight simulators, into the pilot training offered. It suggests that airlines should use *appropriate* technology for curriculum delivery based upon the analysis to the training requirements at a given stage of the process. By doing so airlines can significantly reduce the need for use of a full simulator in favour of lower-cost training facilities (see following section on Simulation).

The FAA emphasises the importance of quality control tools in the AQP process. It expects an AQP quality assurance programme to identify any necessary changes in the curriculum, courseware and/or equipment prior to any unwanted trends of reduced proficiency in graduates manifesting themselves. Airlines participating in the AQP need to evaluate continually all critical aspects of it, including the training programme, the instructors delivering it and the methods of evaluation. The quality assurance function must serve to critique performance, recommend changes and provide feedback to the training organisation at regular intervals (AC 120-54; FAA, 1991). The quality assurance programme must also take a data-driven approach. However, to take such a data-driven approach requires measurement of the 'output', i.e. the performance of the trainees produced by the system. As was touched upon in the previous chapter on Pilot Selection, this is by no means straightforward.

Kirkpatrick (1976, 1998) outlines four complementary approaches to the evaluation of trainees emerging from a training programme.

- *Reactions* (*proximal* measure) – The learner's affective reactions to the training ('happiness sheets'). Did the trainees like the training? This is not an actual measure of the competence of the participant but it does give some idea of how well the training is received. Training viewed more positively by candidates is more likely to be effective.
- *Learning* (proximal measure) – The measureable increase in skills or knowledge as a result of the training. This is usually in the form of some end of course evaluation (practical test or examination).
- *Behaviour* (distil measure) – The extent to which the skills or knowledge delivered during the training course and acquired by the trainee transfer successfully to the workplace. This requires a relevant 'on the job' measure of specific, targeted proficiencies.

- *Results* (distil measure) – The effect on the business as a result of the training received. This is the most difficult form of evaluation and usually requires some form of cost/benefit (utility) analysis. However, even if this step is performed causality is difficult to establish with any certainty (did the enhancement of crew competence resulting from the training actually result in a reduction of costs/increase in revenue?).

SIMULATION

The advantages of using simulators as a training device are manifold. Rolfe and Staples (1986) list several broad areas of advantage for using simulation as a training medium:

- *Increased efficiency* – Training is not affected by environmental or aircraft availability constraints; it is also possible to repeat manoeuvres (e.g. landing) without the time and cost of re-positioning the aircraft.
- *Increased safety* – Manoeuvres may be undertaken in the simulator that would be unwise to perform in flight unnecessarily (e.g. practising the response to engine failures at V1); non-normal and emergency procedures may also be trained.
- *Lower training costs* – Flight simulators are usually much cheaper to purchase and operate than large commercial aircraft (but this is not necessarily true for smaller aircraft; most aspects of flight simulator operation are independent of the size of the aircraft being simulated).
- *Reduction in environmental impact* – A very insightful comment in 1986!

To this list must also be added the benefits to the trainer and trainee of being able to log, replay and review performance, providing almost instantaneous feedback. Student progression may also be logged.

In the UK and Europe, Flight Simulation Training Devices (FSTDs) fall into one of four major regulatory categories:

- Full Flight Simulator (FFS)
- Flight Training Device (FTD)
- Flight Navigation Procedures Trainer (FNPT)
- Basic Instrument Training Device (BITD).

Both the FAA and EASA (previously the JAA) categorise FFSs into four sub-categories depending upon their capabilities (FAA, 1991; AC 120-40B – *Airplane Simulator Qualification*; JAA, 2008; *JAR-FSTD A: Aeroplane Flight Simulation Training Devices*). FFSs *must* simulate a specific type and variant of an aircraft.

Level A FFSs have the lowest level of complexity. They must have an enclosed, full-scale replica of the flight deck including simulation of all systems, instruments, navigational equipment, communications and caution and warning systems. The control forces and displacement must replicate the aircraft simulated and should respond in the same manner as the aircraft. Only basic motion, visual and sound systems are required, with the visual system providing a field of view of at least 45 degrees in the horizontal and 30 degrees vertical (per pilot). The lag following a control input must not be greater than 300 milliseconds more than that experienced on the corresponding aircraft. In addition to the all requirements for a Level A FFS, a Level B simulator must use validated test data for flight performance and systems characteristics and it is also required to include representative ground handling and aerodynamic effects.

A JAA/EASA approved Level C FFS must have a visual system providing each pilot with a collimated field of view of at least 180 degrees in the horizontal and 40 degrees vertical, capable of displaying daylight, twilight and a night visual scene. It also requires a six-axis motion platform and wind shear simulation must be provided. More sophisticated sound simulation is required that can represent environmental and significant aircraft noises, including that of a crash landing! The response to control inputs must not be greater than 150 milliseconds more than that in the representative aircraft. The requirements for a Level D simulator are almost identical to those of a Level C device, however it must have complete fidelity of all sounds and incorporate motion buffeting.

The capability of the FFS determines the training and licensing functions that may be undertaken on it. *JAR-FSTD 2A* (JAA, 2008) only permits procedures training, instrument flight training, transition/conversion training, testing and checking (except for take-off and landing manoeuvres), recurrent training and checking, and testing (type and instrument rating renewal/revalidation) on a Level A FFS. A Level B simulator also permits pilots to maintain recency of experience (three take-offs and landings in 90 days) and undertake type conversion training for take-off and landing manoeuvres. Aircraft type conversion testing and checking may also be undertaken in a Level B FFS, except for take-off and landing manoeuvres. An EASA/JAA-approved Level C simulator allows for the testing and checking of take-offs and landings for flight crew with a minimum level of experience defined by the airworthiness authority. In addition to all the above, a Level D FFS may be used for zero flight time conversions from one aircraft type to another for experienced pilots (see Figure 9.1).

The use of FFSs is now universally accepted as a method for providing training in civil aviation. There is little debate about its effectiveness. FFSs are based upon the notion proposed by Osgood (1949) that suggests that the greater the correspondence between the stimuli presented in the training environment and those in the live environment, the greater the transfer of training will be. If the simulator and the aircraft environments are identical then the required pilot responses should also be identical, hence the simulator should provide maximum training benefit. As a result, there has been a push for greater and greater fidelity in flight simulation, which some argue has led to the development of ground-based aeroplanes rather than training devices. This results in two interesting questions: what level of fidelity is actually required to produce effective training, and what actually is simulator 'fidelity'?

These are particularly interesting questions when it comes to considering 'lower' fidelity devices, such as FTDs, FNPTs and BITDs (using JAA parlance). FTDs are full-sized replicas of a specific type of flight deck. For EASA/JAA Level 1 approval (see JAA 2008: *JAR-FSTD 2A – Aeroplane Flight Training Devices*) the FTD may be either open or closed. It does not require either a force cueing system to the controls or a visual system but it must have the flight deck equipment and simulation capability to represent fully at least one system (this level of FTD is only approved as *part* of an approved training course for that/those given systems represented). A Level 2 FTD has all flight deck systems fully represented in a closed cockpit and should also include a navigation data base (sufficient to support the aeroplane systems). The aircraft flight dynamics and primary flight controls need only be representative of that class of aircraft (not type specific). This type of FTD can be used for system management training and CRM training.

Progressing further down the JAA continuum of simulator capability, a Type I FNPT has a generic representation of the flight deck environment of a certain class of aircraft. A Type II FNPT has more capability, in that it also incorporates a limited visual system providing an out of the window view. It simulates the flight deck systems of a generic multi-engine aircraft. A BITD represents only part of the pilot's position representative of a certain class of aeroplane. It may use facsimiles of aircraft instruments represented on a computer screen and spring-loaded flight controls. It is intended to provide a training platform for only the procedural aspects of instrument flight. The UK CAA suggest that five hours BITD credit may be allowed toward initial PPL (Private Pilot Licence) training and up to five hours credit against the Instrument Flying portion for a Commercial Licence. However, BITD time is not allowed to provide credit towards the requirements for an Instrument Rating (CAA, 2003: CAA Standards Document 18).

The US FAA uses a different system of designation and categorisation for these devices. The FAA classifies these lower level flight simulators as Flight Training Devices (FTDs). They fall into one of seven categories (although at the moment level 1 is not used). Level 2 and 3 FTDs are generic devices (they do not simulate a specific aircraft type), whereas levels 4–7 are type specific.

The reason why the issue of fidelity is of interest is essentially one of cost. These devices are much cheaper to purchase and operate than an FFS. It will also be recalled that the FAA AQP encourages applicants to use *appropriate* technology for curriculum delivery based upon the training requirements at a given stage of the process. This may not always be an FFS. So when should an applicant use a lower fidelity training device and what should they use it for?

Simulator Fidelity

Fidelity is not a uni-dimensional construct so it is difficult to describe any simulator as being of either low or high fidelity. Liu, Macchiarella and Vincenzi (2009) describe two major dimensions of fidelity: physical fidelity and psychological-cognitive fidelity. These dimensions themselves have sub-dimensions to them.

- *Physical fidelity* – The degree to which a simulator looks, sounds and feels like the environment that it replicates (Allen, 1986). This itself comprises:
    - Visual-audio fidelity
    - Equipment fidelity, and
    - Motion fidelity
- *Psychological-cognitive fidelity* – The degree that a simulator replicates factors such as workload, communication and Situation Awareness and demands the same degree of engagement with the task as when flying the real aircraft. This dimension comprises:
    - Task fidelity (fidelity of task requirements), and
    - Functional fidelity (how well the simulator reacts to inputs – Allen, 1986).

To these dimensions Brooks and Arthur (1997) added that of 'motivational fidelity' – the degree to which a training device engages the user and motivates them to engage in training using it.

Gross (1999) in a report from the Simulation Interoperability Standards Organization's (SISO) Fidelity Implementation Study Group, outlines the parameters by which the degree of fidelity of a simulation can be described and quantified. These may be applied judiciously to many of the parameters proposed by Liu, Macchiarella and Vincenzi (2009). These parameters are:

- *Accuracy* – The degree to which a parameter within a simulation conforms to reality.
- *Capacity* – The number of instances of an object that are represented by the simulator.
- *Error* – The difference between simulated values and their corresponding values in real life.
- *Fitness* – The capabilities required for a function or application.
- *Precision* – A measure of the rigour with which the computational processes are described and/or performed in the simulation.
- *Resolution* – The degree of detail used to represent aspects of the real world in the simulation.
- *Tolerance* – The maximum permissible error between the maximum and minimum allowable values in any component in the simulation.
- *Validity* – The quality that determines if an aspect of the simulation is rigorous enough for acceptance for a specific use.

In contrast, other authors have suggested that focusing on the physical aspects of a simulator is not an appropriate way of assessing fidelity. Steurs, Mulder and van Paassen (2004) argue that greater emphasis should be placed upon the psychological-cognitive aspects. They observed that as a result of technological constraints and a limited understanding of Human Information Processing, differences in visual and motion cues result in pilots behaving differently in a simulator compared to an aircraft. As a result they suggest that a better way to measure simulator fidelity is to quantify the extent to which a simulator induces pilot control behaviours comparable to those observed in the real aircraft.

Nevertheless, many aspects critical for pilots flying the line are not present in numerous flight simulation implementations. ATC communication has been identified as being highly desirable for pilot training (Longridge, Bürki-Cohen, Go and Kendra, 2001) which is an aspect of task fidelity, a psychological-cognitive component (see also Thomas, 2003 in Chapter 16 discussing CRM and Line Operational Safety Audits). Johnson, McDonald and Fuller (1997) demonstrated that CRM skills could be taught successfully using a desktop PC running a copy of Microsoft™ Flight Simulator. The physical fidelity of the simulation did not matter: what mattered was the complexity of the crew task, and the cognitive and communication demands placed upon the trainees. Nevertheless, the main foci in the fidelity flight simulation have been on the quality of the visual system and the effects of motion cueing, both aspects of physical fidelity.

Perhaps the most contentious issue in the flight simulation fidelity debate has been on the use of platform motion. At best, the results are somewhat mixed. Simulator motion seems to improve the acceptability of a flight simulator for its users (Reid and Nahon, 1988). It enhances its motivational fidelity, however it is debatable if motion actually improves trainee performance. Steurs, Mulder and Van Paassen (2004) observed that motion cueing enhanced pilot ratings of the realism of a flight simulator but did not contribute to enhanced performance. This supports earlier findings from research undertaken in a military training context (e.g. Waag, 1981). Longridge, Bürki-Cohen, Go and Kendra (2001) undertook a series of trials using a quasi-training transfer of training paradigm (see following sub-section) in a simulator using experienced crew members in both the PF and PNF role. Scenarios involving an engine failure on take-off (followed by with either a rejected take-off or continuing) were flown with the motion either switched on and with it off. The results showed that presence or absence of motion did not affect performance or the subjective perceptions of the PF or PNF, in terms of workload, comfort or the acceptability of the simulator. This again supports earlier work undertaken using commercial flight simulators, rather than simulations of more agile, military aircraft (e.g. Lee and Bussolari, 1989). These results are, however, in contrast to other findings that have observed the only major demonstrable benefit for the use of motion was in enhancing control behaviours for disturbance tasks (Hosman and van der Vaart, 1981; Hall, 1989).

With increases in the speed and power of computing (and the corresponding decrease in costs) issues in visual fidelity are not as important as they were 20 years ago. Two decades ago, with limited computing capability in simulator visual systems, care had to be taken when determining what to display and at what degree of resolution if the performance of the simulator was not to be compromised. These are still issues in many aspects of military simulation, where aircraft have bubble canopies (requiring a 360-degree visual) and which operate at high speed, frequently close to the ground (needing good resolution in a complex visual scene). The visual scene requirements for commercial FFSs though, are not quite so demanding.

One problem with visual scene fidelity is that the resolution available from computer projections is still relatively inferior to the acuity of the human eye. This is not a problem for images relatively near to the observer but images in the distance tend to be reduced in colour, contrast and resolution compared to the visual scene in the real world (Allerton, 2009). A second major issue is that of scene update rate. There is always a compromise between update rate and scene detail. Modern computers can achieve update rates in the region of 50 Hz, which is unnoticeable to pilots, however dropping below an update rate of 15 Hz is unacceptable.

Texture is an important cue for pilots, especially in the last stages of landing where it provides essential cues for height and speed. A failure to provide texture can result in inappropriate strategies being developed by trainee pilots. Mulder, Pleijsant, der Vaart and van Wieringen (2000) observed that in the absence of runway texture pilots judged height over the runway prior to flaring the aircraft by using the runway edge markings. With enhanced visual fidelity pilots used runway texture as a cue. Landing performance was improved when ground texture was added to the visual scene.

Collimation of the outside display in a simulator increases cost but it is desirable as it allows the pilot's eyes to accommodate to an outside scene focused at visual infinity, rather than on a scene, which although it purports to be an outside view, is actually on a screen a few metres away (at best). Collimated displays also give an enhanced experience of depth and distance perception. Non-collimated outside world views are often found on lower levels of simulator, for example FNPTs/FTDs. Perceived size and distance of images in the visual scene are significantly reduced when imagery is displayed in a non-collimated form compared to when it is displayed at or near optical infinity (Pierce, Geri, and Hitt, 1998). This may result in the distortion of features in the visual scene, like the apparent size of geographical features such as towns and rivers. This may be important for training visual navigation. Given the shortcomings resulting from the more cost effective display technology employed, the question becomes: 'by how much does this compromise training: is training still effective'?

Part Task Training

The basic function of many low-levels of simulation (including computer-based training systems) is of a device to support part task training. Part task training decomposes complex tasks (e.g. flying an aircraft) into smaller components each of which is subject to concentrated individual training, before being re-combined into the full task (perhaps in this case using an FFS). From a cognitive perspective, part task training reduces the magnitude of processing demands on the trainee when performing multiple complex tasks simultaneously, which inhibits the acquisition of each of these individual skills. Carlson, Sullivan, and Schneider (1989) suggested part task training is based upon the notion that complex cognitive skills are composed of a hierarchy of component skills; there are human cognitive processing capacity limits placed; and fluency on the component skills is critical to performance of the whole, complex task. With increasing complexity of the flight deck and the requirements of the flight task, part task training methods are becoming increasingly necessary. They may also be much more cost effective.

There are various methods by which a complex task can be de-composed for instructional purposes (Wightman and Sistrunk, 1987):

- *Segmentation* – partitions the whole task, based upon either its spatial or temporal dimensions. Trainees then practice one part, progressively adding more components until the whole task is assembled (e.g. flight path management using the facilities in the autopilot system, such as vertical navigation using vertical speed mode; using vertical speed mode plus adding in flight level constraints, then adding the further facilities and complication when incorporating speed constraints). If the tasks are segmented temporally, they may be instructed by either using *Forward Chaining* (which commences with the first required behaviour in the sequence, adding subsequent steps) or by *Backward Chaining* (which commences with the final required behaviour in the sequence and then adds the preceding steps).
- *Fractionation* – partitions two or more tasks that are normally executed simultaneously. Each are practised in isolation before being combined into the full task. As an example, when learning to fly the trainee normally just handles the aircraft. Navigation techniques are taught in the classroom (as may be radio procedures). These are later all combined into the full task of flying the aircraft.
- *Simplification* – decomposes a complex task by initially making it easier before slowly adding aspects of the increased the complexity of the whole task. For example, using a computerised training system the psychomotor skill of flying the aircraft may at first be confined to the horizontal plane; control in the vertical plane can then be added and once the trainee had mastered this yaw control using the rudders could be incorporated (perhaps followed by the disturbing effects of moderate turbulence).

Fractionation and simplification techniques are inherent in most training techniques (as previously explained, tasks are acquired one at a time when learning to fly; flying a light aircraft is simpler than flying a heavy aircraft) however segmentation can only be really exploited using computerised training or a flight simulator.

Part task training has been successfully used as a training technique in a number of well-reported experimental trials. For example, Wightman and Sistrunk (1987) reported that in part task approaches to train candidates for simulated carrier landings, task segmentation using backward chaining produced better trainee performance on the criterion task than did the same number of training trials on the whole criterion task. Task simplification through the enhancement of the simulated aircraft's response to throttle adjustments did not enhance transfer of training to the criterion task (see the following section for a description of transfer of training approaches). Goettl and Shute (1996) observed similar results, with backward chaining producing superior performance when training a flight task compared to whole task training. Lintern, Roscoe and Sivier (1990) found that transfer on a landing task was superior using part task simplification during training in the form teaching without a crosswind. However, simplification in the guise of reducing the control order in roll (see Chapter 13 – Aircraft Control) did not produce performance as good as that when using a normal control order in bank. Dennis and Harris (1998) demonstrated that use of supervised practice on a PC running a simple flight simulation program, even with non-representative flight controls, produced superior performance in *ab initio* pilots when transferring to flight.

Transfer of Training

Any training, irrespective of the medium and method by which it is delivered, should transfer to the real environment. Transfer of skills, knowledge and abilities acquired during training may be either positive or negative. Positive transfer occurs when performance on the aircraft flight deck improves as a result of training. Negative transfer occurs when the consequences of the training delivered (or perhaps the training medium) actually degrade performance in the live environment. It goes without saying that the analysis of the training system should serve to identify the features that ensure that positive transfer will take place and eradicate those that contribute to negative transfer.

Within a civil aviation context there are really no issues concerning the effectiveness of the higher levels of fidelity provided by FFSs. It is almost inconceivable that airlines would either recommence pilot training in the aircraft itself or even commission a controlled study into evaluating training transfer using an FFS. The main concern usually revolves around establishing if the required level of performance can be achieved more quickly or with less use of an expensive training resource, such as the aircraft itself or perhaps a high fidelity simulator. Can the training be done more quickly (or cheaper) using an FNPT/FTD? There is evidence that many cognitive tasks (as opposed to psychomotor tasks) can be trained very successfully using very low fidelity simulation (training) systems. For example, Pfeiffer, Horey and Burrimas (1991) found that instrument training transferred to performance in instrument flight. Schiewe (1995) and Johnston, McDonald and Fuller (1997) have both observed that low level simulation can be used to develop CRM skills. However, what methodological paradigm is most commonly used to assess the success of such training?

Transfer of training may be assessed by either (or both) evaluating the degree to which trainees' performance is improved via simulation training or examining the cost effectiveness of using a simulator. All transfer of training studies involve two groups of participants, one instructed using the device of interest (e.g. a simulator – the experimental group) and the other in either the aircraft itself or another, usually higher-fidelity simulator (the control group). In the latter case – which are often referred to as quasi-transfer studies – these are not true 'transfer of training' studies; true transfer of training studies can only be performed when the reference group is trained and evaluated using the aircraft itself. The idea behind such a quasi-transfer study is that the higher fidelity simulator is a further stepping stone in progression (e.g. a FFS following earlier training on an FNPT/ FTD). The applied training question tackled by such a quasi-transfer study is usually associated with evaluating if training can be done on a lower fidelity (cheaper) device than the one that it is currently being used.

Often it is not possible or very expensive to collect control group data. In this case, research using a design called the pre-existing control model, historical data provides the control group. The validity of this approach is based upon three assumptions: firstly, that the control group are identical in their composition to the experimental group; secondly, that there has been no change in the instructional syllabus over time, and; thirdly, the performance data were collected under the same conditions that would have applied to a control group being trained in parallel with the experimental group.

Table 9.1   Transfer of Training Models

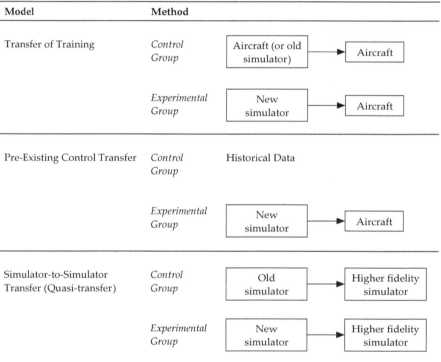

Model	Method		
Transfer of Training	*Control Group*	Aircraft (or old simulator) →	Aircraft
	Experimental Group	New simulator →	Aircraft
Pre-Existing Control Transfer	*Control Group*	Historical Data	
	Experimental Group	New simulator →	Aircraft
Simulator-to-Simulator Transfer (Quasi-transfer)	*Control Group*	Old simulator →	Higher fidelity simulator
	Experimental Group	New simulator →	Higher fidelity simulator

Liu, Blickensderfer, Macchiarella and Vincenzi (2009) list several general approaches for gathering data to assess trainee performance:

- *Trials to Criterion* – the number of trials required in the simulator (or aircraft) to reach the required level of performance.
- *Time to Criterion* – the amount of time required in the simulator (or aircraft) to reach the required level of performance.

- *Trainee Performance after Instruction* – on either the first, last or their mean performance over a number of trials after training is completed.

If suitable performance data can be collected then there are a number of methods for quantifying training transfer. Percentage transfer is one of the simplest and most common indices. It can be applied where performance measures have been taken from two groups trained in different manners (e.g. in the simulator and on the aircraft itself) using a performance measure elicited after instruction. It is defined as:

$$\% \text{ Transfer} = \frac{\text{Performance}_{(\text{Aircraft})} - \text{Performance}_{(\text{Simulator})}}{\text{Performance}_{(\text{Simulator})}}$$

Roscoe and Williges (1980) devised the training Transfer Effectiveness Ratio (TER). TER is the ratio of aircraft training time saved to the expenditure of training time in the simulator. A TER is calculated from simply:

$$\text{TER} = \frac{A_c - A_s}{S}$$

Where:

TER = Transfer Effectiveness Ratio.
A_c = Aircraft time required to reach criterion performance, without access to simulation.
A_s = Aircraft time required to reach criterion performance, with access to simulation.
S = Simulator time.

The use of a flight simulator can be cost effective even when the TER is small (much less than unity) as simulator time is far less costly than corresponding time in the aircraft. This is an issue associated with the training cost ratio (TCR) which is used to determine the cost effectiveness of a simulator. The TCR is simply:

$$\text{TCR} = \frac{\text{Cost of training in the simulator}}{\text{Cost of training in the aircraft}}$$

The cost effectiveness of a simulator is determined by multiplying the TER by the TCR. A simulator is cost effective if the product of this calculation exceeds unity (Wickens and Hollands, 2000). However, Roscoe (1971) demonstrated that transfer effectiveness diminishes as the number of trials (or hours) in ground-based trainers increases, and at some point they cease to be cost effective (see Roscoe and Williges, 1980 for a description of the calculation of incremental transfer ratios).

The products of formulae such as these are very appealing, in that they seem to provide tangible, quantifiable evidence of the effectiveness of a simulator or a training system. However, this is a little misleading. The production of the performance data for incorporation into these formulae can be difficult to quantify. ILS tracking accuracy, perhaps using root-mean square values for deviations from optimum flight path, may be a good measure of pilot psychomotor flight skills but how do you quantify less tangible performance, for example how do you measure empirically 'leadership' when assessing CRM skills? Furthermore, the product of each equation only reflects one parameter in a particular exercise. Any training package within a programme will contain many parameters on which the pilot will need to be evaluated simultaneously, each of which may result in a different TER or TCR. As will be seen in Chapter 16 when CRM is considered, a crew's success when undertaking a LOFT scenario in a simulator cannot easily be quantified from the sum of its component parts. Transfer of training formulae may only be applied to very straightforward, easily (and meaningfully) quantifiable training outcomes. These are few and far between in commercial flight crew training.

NEXT GENERATION PILOT TRAINING

Effective and efficient training can only be specified with respect to the characteristics of the aeroplane (*Machine*) and the operating *Medium*. It has already been stated that the pilot's task has changed considerably in the last few decades as a result of increasing levels of automation on the flight deck. Waiting for regulations for training requirements to keep pace with developments in technology is folly. However, regulations can potentially hold back developments in training, or impose requirements that are no longer valid. Professional pilots are required to undertake multi-crew co-operation courses but there has been no such corresponding advance in the training requirements for the understanding and management of advanced automation. Automation should not be looked upon as a separate 'add on' issue but should be regarded as central to the design and operation of modern airliners (Rignér and Dekker, 2000). Any inspection of the avionics architecture of a modern 'fourth generation' airliner (e.g. Airbus A330/340/380 or Boeing 777) would reveal this to be the case in the design of the aircraft, yet this is not reflected in the training of modern pilots.

Currently, the early stages of commercial pilot training are concerned almost solely with the control of the aircraft, followed by the development of communication and navigation skills. The new pilot then develops these skills further for use at night and in instrument conditions. This is followed by an introduction to airways flying. This initial training takes place in a low-powered, piston-engined aircraft with limited performance, simple systems and out-dated instrumentation. The basic syllabus has not changed radically since the 1930s. Until the 1950s this was acceptable. The technology in small aircraft cockpits was similar to that in large aircraft. Large aircraft flew differently simply because they were bigger and often they were slower than even the more advanced trainers. However, with the advent of the jet-engined commercial aircraft things changed. Flight deck technology developed to accommodate the enhanced levels of performance and the technology in the aircraft being flown by these new commercial pilots began to diverge from that in the training aircraft.

The transition to a modern highly automated 'glass cockpit' still occurs relatively late in the training of a new pilot, usually after they join an airline. It is also typically concurrent with being introduced to multi-crew and jet-transport flying. Several authors have recommended that to alleviate problems with this transition the introduction to 'glass cockpit' technology should be made earlier (e.g. Rignér and Dekker, 2000; Casner, 2003; Harris, 2009). Higher technology aircraft have been introduced into the early stages of flight training predicted on the basis that they resemble the future flight deck environment in terms of the type of instrumentation they contain. They also provide some of the automated functions found in advanced commercial airliners (Dahlström, Dekker and Nählinder, 2006) but this reasoning is over simplistic. The syllabus and training concept needs revision, not solely its means of delivery. A full TNA needs to be undertaken for the airline pilot operating a modern commercial transport to establish the best lead-in training. Simple evolution of technology and teaching is ineffective and inefficient. What is required is to map out the skill and knowledge requirements on a cognitive level and then implement them, as opposed to just updating the training technology hence 'making small aeroplanes fly like big ones' (Rignér and Dekker, 2000, p. 321).

Even a cursory analysis of current training shows many areas of limited utility, for example low-level visual navigation (most large transports don't even have VFR charts in them – and for what area should they carry them?). There is no need to learn the operation of a piston engine attached to a fixed (or even variable) pitch propeller. When transferring to jet transport aircraft with fly-by-wire (FBW) systems, as a result of the advanced flight control laws employed, the aircraft do not even respond in the same manner to stick inputs as a simple, light aircraft (see Chapter 13 on Aircraft Control in Part Three). Even the teaching of navigation using VOR/DME equipment may be questioned.

There have been some superficial evaluations to evaluate the training effectiveness of introducing higher levels of automation training earlier in the flight training syllabus but these have also addressed slightly the wrong question (Casner, 2003). For example, Wood and Huddlestone (2006) observed that the problem was not an issue in managing the automation interface but was rather an issue in understanding *what* the automation was doing and *how* it was trying to control the aircraft. This knowledge is required first before it is possible to 'manage' the automation. Teaching automation is not about teaching how to use its interface. It is what lies unseen behind the flight deck interface that is important.

Given the high fidelity of simulation currently available, the benefits of training delivered this way and the skills, knowledge and abilities required to fly a modern jet transport aircraft, the role of initial flight training in a light aircraft is questionable in terms of training content (although perhaps not in giving the trainee confidence in the air and assessing less tangible parameters). The objective here is not to provide answers but simply to provoke debate and begin to encourage exploration of the question 'could pilot training be done faster, better and cheaper?'

FURTHER READING

There are various excellent books available on training and simulation, for example:

Lee, A.T. (2007). *Flight Simulation: Virtual Environments in Aviation*. Aldershot: Ashgate, Farmer, E., van Rooij, J., Riemersma, J., Jorna, P. and Moraal, J. (1999). *Handbook of Simulator Based Training*. Aldershot: Ashgate.

Or:

Vincenzi, D.A., Wise, J.A., Mouloua, M. and Hancock, P.A. (2008). *Human Factors in Simulation and Training*. Boca Raton FL: CRC Press.

For a generic overview of training, there is also the classic:

Patrick, J. (1993). *Training: Research and Practice*. London: Academic Press.

10
Stress, Fatigue and Alcohol

This chapter is concerned with psychological aspects of stress; the environmental aspects of flight stress (high levels of 'g', hypoxia, etc.) are covered in the following chapter. Early models of stress were based simply upon the notion that it was a negative affect experienced when eternal pressures on the individual became excessive. Later characterisations of the condition were more sophisticated in that they did not regard the individual merely as a passive pawn but also recognised that it was the way in which people reacted to these external pressures that was an important factor in the experience. Some people react in a positive way, managing the sources of the stress and dealing with them; some people actually enjoy and thrive with a modicum of occupational stress (pressure). However, others choose a more maladaptive coping strategy, for example drinking excessive amounts of alcohol, hence the title of this chapter (although it should be stressed that the chapter is not solely about the consumption of alcohol in response to occupational stress). Personality has also been identified as a key factor in the experience of stress and the way that people cope with it.

Stress and fatigue are highly related to working practices which are highly related to organisational design, and hence Airline (and Flight Deck) Safety Management practices. Some people may be more susceptible to fatigue than others. A response to stress may be the increased consumption of alcohol.

One of the problems faced by all airline pilots is that they work irregular hours, sometimes on split shifts, and medium and long-haul pilots are subject to varying degrees of jet lag. In the case of long-haul pilots, periods of high workload at the beginning and the end of the flight may be interspersed with long periods of underload during the cruise, which can in itself be stressful and/or fatiguing. On arrival, crews will often then spend some time in the bar just to 'wind down' or to help foster a team spirit amongst the crew.

Stress and fatigue are excellent examples of how the demands of the *Mission* (and to some extent the *Medium* – as a result of the size of the planet and the distances between cities) can impact negatively on flight crew. This can be managed by sensible scheduling or its effects can be mediated by the provision of suitable support services in the airline. However, first we need to know what we are dealing with.

STRESS

Stress can be categorised into two basic forms: acute stress and chronic stress. Acute stress is of relatively short duration and is often experienced as high workload caused by high taskload (see Chapter 3 on Workload). This is not the focus in this discourse. This chapter focuses on the longer duration aspects of psychological stress in aircrew. Stress can result both from occupational sources and non-occupational sources (and an interaction between the two). As this book is essentially one about the occupation of flying, it is this aspect that will be concentrated on, but this is not to say that the organisation does not have a role to play in moderating the effects of other life stressors on its employees. If it is accepted that stress is deleterious to both performance and health, then it is within an organisation's best interests to assist its employees in coping with the problem. This is especially true in the case of a safety critical organisation, such as an airline. The UK National Work Stress Network (2006) puts it quite neatly: 'Work-stresses go home with the worker: home-stresses come to work with the worker' (www.workstress.net/whatis.htm).

The UK Health and Safety Executive (2008) define stress as 'the adverse reaction people have to excessive pressure or other types of demand placed on them' (HSE leaflet INDG424: *Working Together to Reduce Stress at Work: a Guide for Employees*). Everyone experiences and reacts to stress in a different manner. They may experience increased irritability, a deterioration of relationships with family and friends, sleep disturbances, depression or anxiety, headaches and indigestion. This can lead to an increase in stress-related work absences, increased smoking, drinking and/or drugs use and also reductions in performance (e.g. stemming from an inability to concentrate) and an increased unwillingness to work as part of a team or accept advice. Ultimately, stress can result in far more serious chronic health consequences, for example cardiovascular disease; impaired immune function; ulcers; an increased pre-disposition to cancer; musculoskeletal disorders; and psychological disorders such as longer term anxiety and depression. The International Labor Organization *Encyclopaedia of Occupational Safety and Health* (2007) goes somewhat further and suggests that in addition to these consequences there is evidence to suggest that stressful working conditions interfere with safe work practices and increase the likelihood of injuries at work. It also noted that some studies related stress to an increased risk of suicide.

The health problems related to chronic stress result from the mechanism whereby the human is programmed to deal with physical threats by invoking the 'fight or flight' response. However, this response is only intended to be short-term and is intended to prepare the person for conflict or escape. This is rare in the modern world but, nevertheless, the body still reacts to a stressful situation as if it was a life-threatening situation. As a result there is an increase in adrenaline which increases heart-rate, blood pressure and respiration rate and which puts the body into a heightened state of arousal. Continuous stress, irrespective if it originates in the home or workplace, keeps the body constantly in such a heightened state, which ultimately leads to the physical and psychological symptoms described earlier.

Certain personality types are more prone to stress than others and also are often poorer at coping. Not surprisingly, anxiety-prone people tend not to cope well in stressful situations (Warr and Wall, 1975). However, the selection processes and the very nature of the aviation industry tends to ensure that people with this type of personality are rare on the flight deck. No discussion of occupational stress would be complete without some mention of the Type A personality type (see Rosenman, Freidman and Straus, 1964). As noted in Part One in Chapter 7 on Individual Differences, the Type A personality tends to be competitive, impatient, aggressive, highly motivated and is often a high-achiever. They are also more prone to occupational stress and problems in sleeping.

People with an external locus of control have been found to be more prone to stress and clinical depression (Benassi, Sweeney and Dufour, 1988). The concept of locus of control was developed by Rotter (1954). People with an internal locus of control believe that they are in control of their life and have some control over events around them and involving them. Conversely, people with an external locus of control believe the opposite. They imagine that they are the puppet of other people, their environment or some higher power. However, it needs to be emphasised that these are not two discrete personality types: there is a continuum from internal to external. People with an internal locus of control tend to exhibit high achievement motivation (Rotter, 1966). They also believe that their own hard work will result in success. It may not come as a surprise to learn that pilots tend to have a much higher internal locus of control than the general population (Wichman and Ball, 1983).

To reiterate, emphasis herein will be on the workplace-related stressors encountered by professional pilots but when good performance on the flight deck is of paramount importance the stresses from a crewmember's personal life that they bring to work with them cannot be ignored. Good employers will recognise this and take appropriate actions to support the person and help ensure the safety of the operation. Furthermore, in the aviation business it can be difficult to separate workplace-related stresses from those encountered in the home. It is often very difficult to achieve an acceptable 'home–work' balance. As a result, it is worth taking a *very* brief look at the range of stresses potentially encountered in everyday life.

Stresses from Home Life

Over four decades ago Holmes and Rahe (1967) published the Social Readjustment Rating Scale (SRRS). This was the product of a research programme that linked common stressful life events with various physical illnesses. The SRSS examined factors such as death of a close friend or relative, divorce, moving house, marriage, serious illness, and even Christmas, on health. The contents of the SRSS life events scale is reproduced in Table 10.1 (however, it should be remembered that this was developed in 1967)!

Table 10.1 **Life events and associated life change event units from the Social Readjustment Rating Scale (SRRS). A score in excess of 300 is indicative of being at risk of illness; between 150–299 the risk of illness is moderate (reduced by 30 per cent) and a score of less than 150 indicates only a slight risk of stress-related illness (Holmes and Rahe, 1967)**

Life event	Units	Life event	Units
Death of a spouse	100	Trouble with in-laws	29
Divorce	73	Outstanding personal achievement	28
Marital separation	65	Spouse starts or stops work	26
Imprisonment	63	Begin or end school	26
Death of a close family member	63	Change in living conditions	25
Personal injury or illness	53	Revision of personal habits	24
Marriage	50	Trouble with boss	23
Dismissal from work	47	Change in working hours or conditions	20
Marital reconciliation	45	Change in residence	20
Retirement	45	Change in schools	20
Change in health of family member	44	Change in recreation	19
Pregnancy	40	Change in church activities	19
Sexual difficulties	39	Change in social activities	18
Gain a new family member	39	Minor mortgage or loan	17
Business readjustment	39	Change in sleeping habits	16
Change in financial state	38	Change in number of family reunions	15
Change in frequency of arguments	35	Change in eating habits	15
Major mortgage	32	Vacation	13
Foreclosure of mortgage or loan	30	Christmas	12
Change in responsibilities at work	29	Minor violation of law	11
Child leaving home	29		

Workplace Stress

NIOSH (National Institute of Occupational Safety and Health) in its 1999 publication 'STRESS... At Work' identified six basic factors that contributed to increased levels of occupational stress:

- *Design of tasks* – for example those that impose a heavy workload, have infrequent rest breaks, long working hours and shift work.
- *Management style* – including factors such as no participation by workers in decision making, poor organisational communication and a lack of family-friendly policies.
- *Interpersonal relationships* – including a poor social environment and no support from either management or colleagues.
- *Work roles* – conflicting or uncertain job expectations and too much responsibility.
- *Career concerns* – such as a lack of job security, few opportunities for advancement and rapid changes in working practices for which workers are unprepared.
- *Poor environmental conditions* – unpleasant or dangerous conditions such as noise, air pollution or other ergonomic problems.

To a greater or lesser extent, many of these factors apply to airline pilots.

However, just because these factors are present in a job or an organisation does not mean that the personnel will suffer from stress. NIOSH (1999) suggests that situational factors such as encouraging personnel to maintain a balance between work and family life, developing a supportive network of friends and co-workers and maintaining a relaxed and positive outlook all help the reduce the effects of stressful working conditions. Where the worker cannot cope themselves Employee Assistance Programmes (EAPs) can bring about considerable benefits through providing assessment, support services or referring personnel to other external, specialist services.

Transactional models (Cox and Mackay, 1976) hypothesise that the experience of stress results from a dynamic transaction between the individual and their environment. Central to this approach is the individual's perception of the job demands made on them and their perceived ability to deal with them. Stress is a product of the *perceived* demands exceeding the individual's *perceived* ability to respond (note: *not* the actual job demands or ability). The perceived ability to respond to the job demands are a consequence of, for example, experience, personality, the current situational demands and the existing level of stress. Models such as the Job Demand-Control model (Karasek, 1979) take this approach a little further. Stress and well-being at work is partly determined by the organisation via the interplay of job demand and control. Work situations where the employee has little control over their working environment but has high demands placed upon them have been shown to result in higher rates of sickness absence and cardiovascular disease, both symptoms of stress.

Stress is not just about the well-being of the individual. There are considerable benefits to the organisation if stress is managed properly. Stress pre-disposes people to making errors (Reason, 1997). This is undesirable in a safety critical industry where accidents are to be avoided at all costs. However, stress also causes organisations a great deal of money and this factor can often focus attention far more effectively and immediately than the desire not to have an accident. In 1990, it was estimated that in the USA the total cost of occupational stress (in terms of productivity losses and health costs) was somewhere between $50–150 billion dollars annually (Sauter, Murphy and Hurrell, 1990). Over 40 per cent of all employee turnover in the US is attributable to workplace-related stress (Bureau of National Affairs, 1987). The organisational symptoms of stress include high rates of absenteeism, high turnover of staff, poor industrial relations and poor quality control issues (Arnold, 2005).

Stress in the Airline Industry

There is no such thing as a typical airline flight. The problems and challenges faced by aircrew are complex, manifold and vary widely in nature. For example, the crew of a short-haul airliner may start early in the morning (or in the middle of the afternoon, finishing late at night) and fly up to eight sectors (take-offs and landings) in a single working day. Very little time will be spent in the cruise: most of the time will be devoted to the high workload phases of departure, approach and landing, and the turnaround on the gate. The problems associated with this type of flight are twofold: irregular hours of work and high workload from the large number of sectors (Gander, Gregory, Graeber, Connell, Miller and Rosekind, 1998). This can lead to disturbed sleep patterns. Unremitting periods of high workload with a great deal of responsibility, tight schedules and relatively little control over the situation (e.g. as a result of ATC delays, weather or missing passengers – see NIOSH, 1999) contributes significantly to stress levels.

However, while stress and fatigue may be high from flying many sectors and from working anti-social hours, jet lag is not an issue for short-haul crews (the topics of fatigue and jet lag/circadian rhythms will be considered more fully later). Long-haul pilots, on the other hand, may fly only one sector every three days, but will experience prolonged work underload during extended periods in the cruise. It has been observed that those susceptible to stress or fatigue may find their performance considerably impaired in conditions of underload (Desmond, Hancock and Monette, 1998; Matthews and Desmond, 1997). Excessively low mental demands can be as stressful as high levels of workload (Hancock and Parasuraman, 1992). It is often said that flying is 99 per cent boredom and 1 per cent sheer terror!

Not only do long-haul crews suffer from disturbed sleep and eating patterns as a result of time zone changes after arrival at their destination, their regular lengthy separations from family and friends can also be a source of stress (Graeber, 1986; Cabon, Mollard, Coblentz, Fouillot and Speyer, 1995). This leads to an interesting conundrum. While long stopovers may help to reduce the effect of fatigue and time-zone shifts on the crew's circadian rhythms (and hence have a beneficial effect on performance on the flight deck) these stopovers may further contribute to stress by increasing the length of separation from family and friends (see Sloane and Cooper, 1986). This is an excellent example of how it is sometimes difficult in the airline industry to separate a workplace-related stressor from those encountered in the home. There are particular challenges faced by both partners in an 'airline marriage' (Karlins, Koh and McCully, 1989). In an attempt to counter the effects such as those described, airlines are now actively engaging pilots' spouses in the management of stress, as they are a major source of social support and a significant factor in their ability to deal effectively with psychosocial stress. It should also be remembered that long stopovers by flight crew are economically 'inefficient'. Airlines require greater numbers of crew to be engaged to relieve those on stopover and furthermore, it also has to pay crew for doing nothing, while covering their accommodation costs!

On a related note, Cooper and Sloane (1987) observed that from a physical perspective pilots may be able to rest when away from home but they are often not able to relax psychologically. In studies of short-haul pilots (Gander, Gregory, Graeber, Connell, Miller and Rosekind, 1998) crews reported an average of 12.45 hours between duty periods but only obtained an average of 6.7 hours rest. Such factors lead to poor performance on the flight deck, increased likelihood of air traffic incidents, health complaints and lowered levels of psychological well-being (e.g. Winget, DeRoshia, Markley and Holley, 1984;

Haugli, Skogstad and Hellesoy, 1994; Loewenthal, Eysenck, Harris, Lubitsh, Gorton and Bicknell, 2000).

In the last 15 years or so, the birth of the 'low-cost' airline has emphasised the need for efficiency in the operation of commercial aircraft. As margins are low on each seat sold, load factors need to be high and turnarounds need to be swift. Furthermore, crews need to be utilised to the maximum which has resulted in higher levels of stress and fatigue (Bennett, 2003). As low cost carriers have taken market share from larger carriers, the major airlines have also needed to react and begin to operate in a similar manner. There has also been a raft of acquisitions, mergers and bankruptcies in the industry. This has promoted stress levels in both pilot and cabin crew populations, stemming from issues such as uncertainty in employment, new management practices and pre-merger 'group' memberships (Terry, Callan and Sartori, 1996). Little, Gaffney, Rosen and Bender (1990) found that worries about company stability and large numbers of last minute flying schedule changes (which are becoming more common as a result of flight crew being used more 'efficiently') were also factors related to stress and depression in pilots.

The effects of repetitive sectors, operating across many time zones, working unsociable hours and being separated from family and friends are only a few aspects of the *Mission* that impact on hu*Man* in the system. Commercial aviation is a safety critical industry and the demands of having to constantly achieve high levels of performance are themselves wearing. This must be remembered when appraising the role of flight crew. The economic *Medium* has made a significant transformation in the operation of airlines but this too has had a considerable impact on the human resources within them.

FATIGUE

The Federal Aviation Administration defines fatigue as:

> ... a condition characterized by increased discomfort with lessened capacity for work, reduced efficiency of accomplishment, loss of power or capacity to respond to stimulation, and is usually accompanied by a feeling of weariness and tiredness. (Salazar, 2007, p. 2)

Fatigue may be acute, occurring over a relatively short time period after significant physical or mental activity, or it may be chronic, developing as a result of insufficient sleep over a prolonged period of time (perhaps as a result of jet lag or shift work) or it may stem from continual physical or mental activity with insufficient rest periods. Chronic fatigue can also be an outcome of depression and stress or it can result from behavioural issues including excessive alcohol or caffeine use, working late into the night or simply being in unfamiliar or uncomfortable surroundings (Salazar, 2007).

The principal symptom of fatigue is tiredness (not necessarily sleepiness) but can also include headaches, general body aches and pains, and bleary eyes. It may well be a result of many factors including heavy mental or physical exertion; circadian rhythm disruption or boredom (underload), but it always results in a decreased ability to perform. More specifically, fatigue impairs concentration, complex cognitive tasks and manual dexterity. Graeber (1988) observed that fatigued aircrew exhibited a great deal more reliance on automated systems than did their non-fatigued counterparts. The Battelle Memorial Institute (1998) reported that fatigued workers were satisfied with lower performance and that errors went uncorrected; fatigued crews concentrating on one problem allowed other problems to develop. In the same report, the China Airlines Flight 006 mishap (where a Boeing 747 entered an uncontrolled descent from 31,000 to 9,500 feet as a result of the Captain focusing on a problem stemming from the loss of power in one engine) was attributed to this latter factor. Further contributory factors were a failure to monitor the flight instruments and an over-reliance on the autopilot (*q.v.* Graeber, 1988). The incident occurred during the low point in the Captain's circadian rhythms, approximately four hours after the normal time he had been beginning his period of sleep for the week prior to the flight (Lauber and Kayten, 1988).

Circadian rhythms operate on roughly a 24–25-hour basis. Not only do these rhythms dictate sleep/waking patterns they also control other processes, such as hormones, body temperature and appetite. People in 'normal' jobs wake up at approximately the same time every day and fall asleep at the same time. Typically, they will require between six to nine hours of sleep per day. This routine can be modified slightly, to little or no detriment if it is advanced by around one hour per day. However, periods in excess of this begin to cause some problems.

Various models to predict fatigue have been developed over the years. Gundel, Marsalek and ten Thoren (2005) produced a predictive model of fatigue based upon three components. The first two components of the model relate to alertness. These are a circadian process (C) and a homeostatic process (S) which put together result in a sleep–wake cycle (the third component in Figure 10.1). S increases during the hours of wakefulness until it reaches a circadian upper threshold when sleep is initiated. During sleep S decreases and when it reaches a circadian lower threshold sleep is terminated. To this a 'time on task' component is added (the final component in Figure 10.1 – NB: this is a hypothetical example!). The time on task component describes the fatigue that builds up during a task. The shape of this component will depend upon factors such as how demanding or monotonous the task is, its duration and the effect of taking a short break. The baseline for this component will also depend upon the initial level of fatigue that the person comes to the task with. What Figure 10.1 demonstrates is that it cannot be assumed that alertness (or performance) will be at a maximum early in the day (or on shift) and then subsequently decrease. It depends upon other biological factors some of which may exist in an antagonistic relationship.

During the 1980s the NASA Ames Research Center undertook an extensive programme of work to determine the extent of fatigue, sleep loss and disruption of circadian rhythms in flight crew and tried to establish what effect this had on their well-being and performance (Rosekind, Gander, Miller, Gregory, Smith, Weldon, Co, McNally and Lebacqz, 1994). Data were collected during regular flight operations, in full-flight simulators and in a series of laboratory experiments. Many findings accrued from the data, including:

- During nights away from home, pilots took longer to fall asleep, had about 1.2 hours less sleep and woke earlier compared to their pre-trip baseline measures. They also reported the quality of sleep to be poorer.
- Subjective fatigue and mood were worse during stopovers.
- Pilots consumed more caffeine on days when they were flying, primarily in the early morning to aid waking up and during the mid-afternoon to counteract the effects of fatigue.
- Pilots consumed more alcohol during a trip than either before or after the trip.

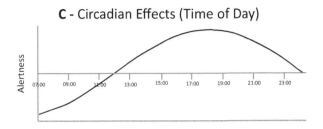

C - Circadian Effects (Time of Day)

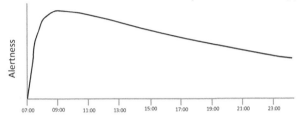

S – Homeostatic Process Effects (Time Since Sleep)

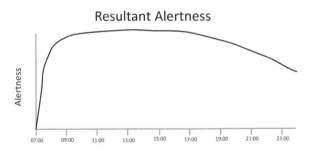

Resultant Alertness

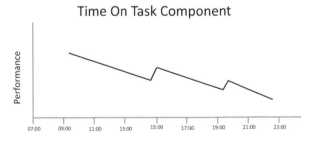

Time On Task Component

Figure 10.1   **Fatigue model. Cycle is based upon the period 07:00–23:00. Alertness and performance units are arbitrary (Gundel, Marsalek and ten Thoren, 2005)**

Chidester (1990) in a survey of airline pilots conducted over the course of typical, two-day short trips found that pilots generally experienced an increase in negative moods and tension. In common with other studies, on layovers, they reported sleeping less and having poorer sleep compared to being at home. It was also found that from Jenkins

Activity Survey measures of Type A personality, the impatience/irritability aspect was higher in pilots experiencing health-related problems on trips. Bourgeois-Bougrine, Cabon, Mollard, Coblentz and Speyer (2003) found that sleep before morning flights was significantly shorter than sleep taken before afternoon flights. This was further associated with the pilots taking morning flights experiencing higher levels of fatigue and workload compared to their colleagues flying in the afternoon.

Compensating for the Effects of Fatigue and Jet Lag

Sometimes efficiency and safety can be optimised and the well-being of flight crew can all be improved simultaneously. Rosekind, Gander, Miller, Gregory, Smith, Weldon, Co, McNally and Lebacqz (1994) observed that when pilots started duty periods earlier on successive trip days (requiring earlier wakeup times) it interfered with them obtaining adequate sleep, even when stopovers were long. This was as a result of the pilot's circadian rhythms preventing them from falling asleep earlier. They suggested that whenever possible, duty periods across successive days should either start at the same time or begin progressively later (moving with the natural tendency of the biological clock to extend the day). In a trial of an innovative flight rostering system implemented in a UK low cost carrier it was found that changes to shift patterns could bring about both safety *and* efficiency gains (Stewart, 2005). However, as a result of the regulatory *Medium* these benefits were not easy to achieve. Stewart noted that prior to the trial rostering practices were compliant with CAP 371 flight time limitations guidelines (CAA, 2004). In the normal flight roster, crews would work six days 'on' and three days 'off':

- Day 1: backward diurnal phase shift – starting 05:00.
- Day 4: forward diurnal phase shift – starting 13:00.
- Day 6: end work at 23:00–24:00 with an option to extend duty to 03:00.

This roster resulted in decrements in performance as the six days 'on' progressed. In the trial the company was granted a temporary waiver from flight time legislation by the UK CAA to evaluate a 'slow wave' shift pattern (five 'earlies'; two days off; five 'lates'; four days off). This revised shift pattern was found to:

- Reduce operational risk.
- Produce less fatiguing work patterns and reduce crew duty hours.
- Produce a reduction in insurance liability of the order of £4 million.
- Improve crew productivity by 7 per cent.
- Increase roster stability.
- Improve crew lifestyle and reduce sickness.
- Improve pilot retention and reduce training liability.

All of these factors increased both operational efficiency and safety. It needs to be noted, though, that the company had to be granted a temporary waiver from UK flight time legislations to realise these advantages. In this case the potential gains in efficiency (and safety) from the hu*Man* component in the system were bounded by regulatory structures (from the Societal Medium) which required changing before advances could be made.

The discussion of flight crew fatigue so far has pretty much entirely ignored the effects of jet lag experienced by all crew on long-haul flights. This only serves to complicate things even further. Jet-lag occurs when travelling across a number of time zones in a short period of time. As a result, the pilot's circadian rhythms, which are regulated by the day/night process, will be out of synchronisation with the local time at their destination. There are two principal biological drivers in sleep/wakefulness (and hence these are also implicated in jet lag). The hypothalamus in the brain regulates most of the functions that are disturbed in jet lag, such as sleep, appetite, body temperature, etc. It does this by releasing the hormone melatonin. When it becomes dark (normally after sunset) the hypothalamus releases melatonin, which promotes sleep. When it becomes light (normally after sun rise) melatonin production is inhibited. However, the hypothalamus cannot readjust its schedule instantly.

The degree of jet lag is linked to the number of meridians crossed not the length of the flight *per se*. A flight from Europe to America is (with a seven–nine-hour time difference) will be more disruptive than a longer flight to the tip of Africa, as there will only be a two-hour difference in local time. Furthermore, travelling Eastwards worsens the problem, as pilots have to effectively lose time (fall asleep earlier, which is difficult to accomplish) as opposed to travelling to the West, where the critical factor is to delay sleep. Adjustment to a new time zone is faster after a Westward flight, at about 1.5 hours per day, compared to an Eastward flight where adjustment is approximately one hour per day (Klein and Wegmann, 1980).

To compensate for the effects of jet lag, all airworthiness authorities have strict duty time limitations. A survey of these regulations by Bourgeois-Bougrine, Cabon, Mollard, Coblentz and Speyer (2003) identified 13 factors used in determining the rest periods between flights (e.g. number of legs flown; reporting time; adjustment to local time; duration of legs). However, no country used more than 11 of these criteria, and some used very few (only two or three). Worldwide, no two airworthiness authorities used the same set of criteria. These duty time regulations highlight the conflicting requirements of safety and performance. Only relatively few of the regulations surveyed took into account factors that have an influence on aircrew fatigue such as:

- *Hours of reporting for duty* – which impacts upon on the hours of work and consequently on pre-duty sleep time.
- *Crew adjustment to local time* – from a chronobiological point of view, the 'nocturnal' period (from 22:00 to 06:00) at a given geographical point may not cover the circadian low period if the crew has crossed several time zones.
- *The amount of jet lag* – for the purpose of calculation of rest duration.
- *The duration of rest periods*.

Although controversial, the use of short naps on the flight deck has been found to be effective as a short-term remedy to improve the performance of fatigued pilots. This approach, however, cannot aid in recovery from sleep debt and is not a substitute for proper sleep. Rosekind, Gander, Miller, Gregory, Smith, Weldon, Co, McNally and Lebacqz (1994) examined the effectiveness of a planned flight deck rest period as a fatigue countermeasure to maintain and/or improve crew performance on long-haul international flights. In a controlled quasi-experimental study in three-crew Boeing 747s, crew members were randomly assigned to either a rest or no-rest group. Pilots in the experimental trial group were allowed a 40-minute rest period during the cruise. Crew members rested one at a time on a pre-arranged rotation. The vast majority (93 per cent) of participants fell asleep and on average they slept for about 26 minutes. Flight crew allowed to rest were found to be able to maintain consistently good performance on a portable reaction time task and a psychomotor vigilance task at the end of flights compared to the decrements observed in the no-rest control group.

Petrie, Powell and Broadbent (2004) observed that over 50 per cent of Air New Zealand pilots used the cockpit napping procedure sanctioned by the New Zealand CAA. This was associated with lower levels of reported fatigue. This procedure allows a fatigued crew member to have a nap not exceeding 45 minutes while strapped into their seats. Before starting the period of sleep the other crew members (including cabin crew) are briefed and agree the time to wake the napping pilot. No course changes, altitude changes or fuel transfers are permitted during this period.

ALCOHOL

It has been reported that up to 12 per cent of professional pilots drink alcohol as a means of coping with stress and fatigue (Sloane and Cooper, 1984). Research by Canada Market Research (1990) and Ross and Ross (1995) concurs with this figure. These stressors may either be of a personal nature, a product of work-related pressures or an interaction between the two (such as the effects of long stopovers). Worries about the future of the airline, high workload and last minute changes in crew rosters are also issues that have been shown to be associated with drinking and flying behaviour (Maxwell and Harris, 1999).

In 1985, the US Federal Aviation Administration (FAA) established an upper blood alcohol concentration (BAC) of 40mg per 100ml of blood (also expressed as 40mg/dl – milligrams per decilitre). Above this BAC, it was expressly prohibited to act as a crew member of a civil aircraft (Code of Federal Regulations, Title 14; part 121.458 and 121.459). In Europe the permitted BAC is half his figure. JAR OPS (now EU OPS) Part 1 specifically prohibits a pilot to act as a crew member with a BAC of greater than 20mg/dl. An extensive meta-analysis of the research literature specifically concentrating on the effects of low BACs by Moskowitz and Fiorentino (2000) concluded that driving tasks, flying tasks and divided attention tasks were the types of task most sensitive to low concentrations of blood alcohol. These were followed by vigilance tasks, perception and visual functions, tracking tasks, general cognitive tasks, psychomotor skills and choice reaction time tasks. In several studies impairment was demonstrated at a BAC of less than 10mg/dl. In studies specific to aviation Smith and Harris (1994) observed that radio communication errors increased at BACs as low as 20mg/dl. Davenport and Harris (1992) found decrements in performance at BACs of 11mg/dl during a series of demanding simulated ILS approaches. Even with a BAC of zero, performance decrements as a result of a hangover may be detected. Yesavage and Leirer (1986) found that pilots who had ingested enough alcohol to achieve a BAC in excess of 100mg/dl were still showing significant performance decrements 14 hours later and at a time when their BAC had returned to zero. Bates (2002) examined hangover effects after an intervening night of sleep. The results showed that alcohol impaired performance 10 hours after reaching a BAC of 100mg/dl.

The reasons for drinking and flying are manifold. Harris (2002) suggested five basic reasons:

- *Promulgation of information* – Very few pilots were aware of the appropriate regulations. Rules that pilots are unaware of are ineffective.
- *Stressors* – Ross and Ross (1992) concluded that one of the fundamental reasons for drinking and flying was an inability of some individuals to control their use of alcohol. Maxwell and Harris (1999) also all reported that job-related stresses were frequently cited as a common cause of heavy drinking in professional pilots.
- *Socialisation* – Long-haul flight crew members commonly report drinking socially with other crew members after flights and also as a means of helping them to relax, especially after crossing several time zones (Sloane and Cooper, 1984). Other authors have suggested that pilots, and in particular military pilots, are from what may be characterised as a 'drinking culture', where great store is placed upon factors such as camaraderie and 'team spirit' (e.g. Anthony, 1988; Cuthbert, 1997). It is unlikely that a drink-flying event would immediately follow such a social gathering, however, it is more likely that after a heavy drinking session a pilot may fly the following day with an unacceptably high BAC (in addition to a hangover)!
- *Lack of knowledge* – Ross and Ross (1990) identified a fundamental lack of knowledge about the rate at which BAC declines as a function of time and the amount of alcohol consumed. The problem of accidentally transgressing the drinking and flying regulations in Europe was found by Widders and Harris (1997). In this study it was found that up to 24 per cent of UK pilot's licence holders could not determine when their BAC was likely to fall below 20mg/dl after drinking and may therefore be in danger of inadvertently infringing the regulation.
- *Attitudes, beliefs and opinions* – Attitudes, beliefs and opinions towards the use of alcohol all effect the likelihood of drinking and flying and in many different ways. Ross and Ross (1990) found that self-reported drinking behaviour was much more conservative when consuming whisky compared to wine or beer. Damkot and Osga (1978) earlier observed that 'hard' liqueur was perceived to be more dangerous to consume prior to flying than either wine or beer. Widders and Harris (1997) identified two potential groups of drinking and flying pilots: 'inadvertent drink-flyers', who did so as a result of a lack of knowledge and 'non-believers' in the regulation. The latter felt that they were safe to fly before their BAC had dropped below the 20mg/dl limit. Members of the 'non-believers' group of pilots tended to be older and more likely to possess an Air Transport Pilot's Licence.

Development of Effective Drinking and Flying Countermeasures

A simple, three-level model of deterrence of drinking and driving behaviour was offered by Vingilis and Salutin (1980). This approach can equally as well be applied in the control of drinking and flying. The deterrence model suggests remedial actions at three levels.

- *Primary Level* interventions are concerned with educating drivers about the effects of alcohol on performance.
- *Secondary Level* countermeasures are associated with enforcement of the regulations (i.e. increasing the likelihood, or perceived likelihood of apprehension of offenders).
- *Tertiary Level* interventions in the model are aimed to reduce the chances of recidivism through either punitive sanctions (aimed at suppressing offending

behaviour) or through counselling and rehabilitation (targeted at eliminating the root cause of offending behaviour).

In 1995 the FAA adopted a deterrence measure to reduce the incidence of drinking and flying. Every year, the FAA requires at least 10 per cent of all airline employees, including pilots, to be randomly selected for drug and alcohol testing and they may require up to 50 per cent of employees in safety-critical roles to be tested in any calendar year. In the case of a pilot producing a positive breath test with an indicated BAC of between 20mg/dl and 39mg/dl, that person is prohibited from flying until a repeat test indicates that their BAC is below 20mg/dl or at least eight hours have elapsed from taking the initial test. However, the US regulations also include provision for the suspension or revocation of an offender's licence. Lindseth, Vacek and Lindseth (2001) examined pilots' attitudes and opinions over a 10-year period regarding the effectiveness, adequacy and fairness of a random drug and alcohol testing programme. The results showed that respondents believed that alcohol use by pilots had decreased since testing was mandated and that alcohol and drug testing was more generally accepted by pilots than when it was first introduced.

Harris and Maxwell (2001) elicited opinions from 472 British private and professional pilots concerning the effectiveness of various countermeasures to reduce the likelihood of drinking and flying. Overall, punitive sanctions and tougher enforcement of the regulations were regarded as the most effective countermeasures (secondary level sanctions) although self-reported offenders and professional pilots thought these actions less effective than private pilots and non-offenders. The professional pilots in this study tended to regard counselling and rehabilitation more favourably than enforcement (tertiary level) which probably reflected the underlying reasons why some respondents in this section of the sample consumed alcohol. The results of an earlier opinion survey of US professional pilots undertaken by Ross and Ross (1995) also suggested that the provision of Employee Assistance Programmes would be the most effective approach to reducing the likelihood of drinking and flying. This was followed by (in decreasing order of effectiveness) education of pilots on the effects of alcohol and strengthening sanctions.

All airlines have Employee Assistance Programmes to help employees cope with their job-related demands, and by doing so promote safety. Figures are difficult to establish, but in 1983 one US carrier was reported to have over 120 recovering alcoholic pilots (Harper, 1983). The results of rehabilitation programmes for pilots with alcohol problems have shown considerable success with positive results in up to 85 per cent of cases (see Russel and Davis, 1995 and Flynn, Sturges, Swarsen and Kohn, 1993).

Being an airline pilot would appear to many people to be a very desirable job (and it is!). Nevertheless, it does carry a hidden cost in the stresses and strains it imposes upon the life of these professionals. At first, the obvious nature of these problems are related to the high workload (acute stressors) associated with certain types of operation and in certain phases of flight. However, it is the more subtle, chronic stressors that ultimately have the greater potential to do damage to pilot's health and well-being. With a little bit of thought, steps can be taken to alleviate most of these problems, potentially at relatively little cost, while also enhancing safety.

FURTHER READING

Relatively little has been written recently on the topic of stress in commercial pilots, however the interested reader might like to try:

Sloane, S.J. and Cooper, C.L. (1986). *Pilots Under Stress.* New York, NY: Routledge.

Or:

Stokes, A.F. and Kite, K. (1997). *Flight Stress*. Aldershot: Ashgate.

With regard to pilot fatigue, a practical perspective of this issue is available in:

Caldwell, J.A. and Caldwell, J.L. (2003). *Fatigue in Aviation*. Aldershot: Ashgate.

There are no volumes available specifically dedicated to the issue of the use and abuse of alcohol by pilots, however, there is an excellent review chapter available covering these issues in:

Harris, D. (2005). Drinking and Flying: Causes, Effects and the Development of Effective Countermeasures. In: D. Harris, and H.C. Muir (eds) *Contemporary Issues in Human Factors and Aviation Safety* (pp. 199–219). Aldershot: Ashgate.

11
Environmental Stressors

The previous chapter dealt with some of the more common sources and effects of the psychosocial stressors experienced by aircrew. This chapter is concerned with the physical stressors imposed on pilots. It has to be stressed right from the start, though, that the aeromedical aspects of these factors are only considered superficially (for a full treatment of these aspects see Rainford and Gradwell, 2006). This chapter is concerned with the human performance aspects of environmental stressors such as temperature and humidity, the flight deck atmosphere, noise, vibration and disorientation. Furthermore, the emphasis is firmly placed on the more regularly experienced values of these parameters rather than the effects of grossly abnormal values, as may be experienced in extreme circumstances or by military pilots in fast jets.

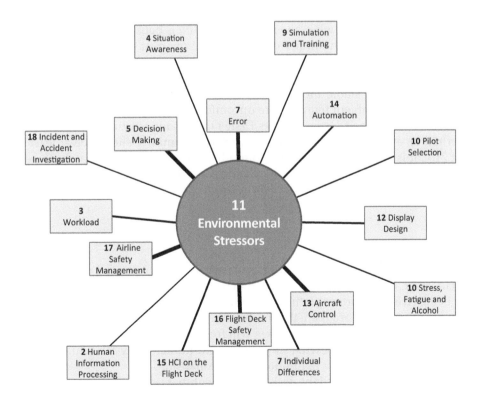

Environmental stressors severely affect the ability of pilots to perform all tasks, such as decision making and aircraft control. However, to some extent they may be moderated by the good design of the flight deck control and display systems and by the appropriate rostering of crews.

Not only is the flight deck the control centre of the *Machine*, it also has to protect the hu*Man* component from the hostile environment (*Medium*) outside the aircraft. It must provide a safe, comfortable, warm environment in which the crew can work for extended periods of time without becoming unnecessarily fatigued. Noise must also be kept within reasonable bounds to promote safe and efficient communication between the crew.

TEMPERATURE AND HUMIDITY

When it comes to a discussion of comfort and performance as a result of the environmental stressors of temperature and humidity it is very difficult to separate completely the effects of these two factors. Effective temperature (i.e. the thermal experience) is a product of several factors in addition to the ambient temperature, one of which is humidity. Konz and Johnson (2000) list seven factors that influence thermal comfort. Four of these are environmental aspects (dry bulb temperature, water vapour pressure, air velocity and radiant temperature) and two are individual factors (metabolic rate and type of clothing worn). The final factor is the length of exposure. As an example of how these factors interact, for a person wearing light clothing occupied in a sedentary task, such as piloting a modern commercial aircraft, at 25°C (adjusted dry bulb temperature) they will only (generally) feel comfortable when the relative humidity is between approximately 22 per cent and 60 per cent. If the humidity drops below 22 per cent the person is likely to feel cool; above 60 per cent they will generally begin to feel uncomfortably warm despite the actual air temperature remaining constant. It is essential that the humidity on the flight deck remains relatively stable and within reasonable bounds. High levels of humidity can make a comfortable ambient temperature uncomfortable.

The FAA (1996) in their *Human Factors Design Guide* suggests that to avoid throat and nasal irritation, and dry eyes, the minimum relative humidity should be at least 15 per cent. At 21°C it is suggested that relative humidity should be approximately 45 per cent. In general, relative humidity should not exceed about 70 per cent. High levels of humidity are often associated with stuffiness and drowsiness (Galer, 1987).

Comfort

The physiological effects of extreme cold are easy to quantify but need not concern us too much. However, for the record, if you ditch in water at a temperature of 0°C, the average human being (wearing indoor clothing) is unlikely to remain conscious for more than 15 minutes or survive for more than 45 minutes. At −20°C, with a wind speed of 5 ms^{-1} (18 km/h, 11.2 mph or 9.7 knots) exposed skin will freeze within 30 seconds (Allan, 1988). At the other end of the scale, somewhat surprisingly, the human body can actually withstand an air temperature of 68°C for 15 minutes before collapsing.

Commercial pilots need a flight deck environment that remains within a temperature/ humidity band that ensures their comfort for extended periods and does not induce drowsiness or fatigue. The ambient temperature must be neither so high nor so low that it adversely affects their cognitive functions. All discussions of optimal environmental working temperatures are dependent upon the nature of the work to be performed and the clothing being worn. Flying an airliner is essentially a sedentary occupation. For the purposes of this discussion it will be assumed that the flight deck crew are operating in light, indoor clothing. Expressed more formally, the insulation value provided by the clothing typically worn in a 'shirtsleeve' environment is approximately 0.6 clo[1] (Konz and Johnson, 2000).

While the airworthiness regulations (e.g. FAR 25.771) suggest that the temperature on the flight deck should remain within bounds so that the safe operation of the aircraft is not effected, this range of temperatures is not actually specified. For a sedentary person wearing light clothing, the vast majority of people will be comfortable in a temperature band between 20°C and 28°C with 50 per cent relative humidity (American Society of Heating, Refrigeration and Air Conditioning Engineers, 1997; Parsons, 1990). Ramsey, Burford, Beshir and Jensen (1983) demonstrated that safe working behaviour was most likely when the temperature in the working environment fell within the comfort bounds for the task.

Physiological Impairment

Some specific aspects of human performance impaired by non-optimal temperatures can be identified. For example, below 15°C (hand temperature) manual dexterity can be compromised, although the degree of impairment is also dependent upon exposure time. Tactile sensitivity drops when hand temperature drops below 8°C. On initial exposure to a cold environment dexterity is not really impaired but depending upon how low the ambient temperature is, manual dexterity can deteriorate very quickly (so next time that you ditch in a cold sea, make sure you do all the 'fiddly' things you need to do first and do them quickly!). Finger strength is also compromised by the cold. While these moderately low temperatures are not an issue during normal operations, in the very unlikely advent of a failure in the aircraft's environmental conditioning system, temperatures on the flight deck will drop compromising the pilots' ability to manipulate smaller rotary control knobs and toggle switches guarded by requiring a complex 'pull and click' action.

Moderately high ambient temperatures have no such effects on strength or dexterity (in fact they are improved when it is warm). The latest 'touch pad' cursor controls, as found on many portable computers and as an interface on some modern FMCs, can be compromised by sweaty fingers, as they work on the electrical conductivity of the fingertips. This reduces the possible control accuracy considerably.

Information Processing

The effects of cold and heat (especially cold) on the Human Information Processing system are difficult to demonstrate. Bensel and Santee (1997) provide an overview of research investigating the effects of low ambient temperatures on cognitive performance.

---

1   A measure of thermal insulation for clothing. One clo is the amount required for maintenance of body temperature in sedentary indoor conditions at 70°F/21°C with 50 per cent relative humidity and air movement of 10ft per minute/0.05 ms^{-1}. Body temperature is 98.2°F/37°C with metabolism at 50 kilocalories per square metre per hour.

In general, there would seem to be little indication of any impairment even at temperatures well below 0°C. Other authors (e.g. Oborne, 1987; Sanders and McCormick, 1987) concur with this view. This is not to say that keeping people in the cold doesn't adversely affect their mood, though!

Grether (1973) found evidence to suggest that human cognitive performance and vigilance begins to deteriorate above about 30°C. Hancock (1981), in a synthesis of studies of various different types of cognitive task, found that dual tasks suffered the most in high temperatures, followed by tracking tasks. Mental tasks were least impaired, however, these analyses only demonstrated impairments above approximately 33°C (and up to 46°C). These conditions would rarely be encountered in normal conditions on a flight deck, although it needs to be noted that due to the radiant heating of the pilots, they may be significantly warmer than the ambient air temperature.

ATMOSPHERE

It almost goes without saying that both passengers and crew must be provided with an atmosphere capable of supporting both their physiological requirements for survival while simultaneously maintaining their level of alertness. Not only does this require an adequate supply of oxygen, the air provided must have suitably low levels of carbon monoxide, carbon dioxide, ozone and other toxins.

Cabin Altitude

FAR 25 (section 25.841a) states that 'Pressurized cabins and compartments to be occupied must be equipped to provide a cabin pressure altitude of not more than 8,000 feet at the maximum operating altitude of the airplane under normal operating conditions.' Macmillan (1988) suggests that even though this is the maximum permissible cabin altitude in modern passenger carrying aircraft there is reason to suggest that prolonged exposure may induce mild hypoxia. He suggests that there is a growing acceptance amongst manufacturers that the maximum cabin altitude that is consistent with flight safety may be in the region of 5,000–7,000 feet. Once cabin altitudes rise above 10,000 feet then there is a considerable impairment of the pilot's cognitive abilities to perform flight critical tasks.

Ernsting, Sharp and Harding (1988) suggest that psychomotor tasks are the most resistant to the effects of hypoxia. Performance shows little decrement until the effective cabin altitude exceeds 12,000–14,000 feet. Cognitive tasks involving Short (Working) or LTM are less resistant to the effects of altitude. In some cases a deficit may be observed at 8,000–10,000 feet. At altitudes as low as 5,000 feet, there is some noticeable impairment of the sensitivity of the dark-adapted eye, although as the authors note, the impairment, is of little consequence for aviation safety. Although human performance in these various categories is impaired at the stated altitudes, it is not until somewhere between 16,000–24,000 feet (equivalent altitude) that unconsciousness will occur.

In the extremely rare advent of an explosive decompression the key factors are the speed of the decompression, the cabin altitude before the event and the aircraft's actual altitude at the time of the event. A rapid decompression from 8,000 feet to 40,000 feet in less than two seconds can result in the impairment of performance in 10–15 seconds and unconsciousness in 20 seconds (Macmillan, 1988). Less rapid losses of cabin pressure result in much longer times of usable consciousness. However gradual decompression is a potentially more insidious danger than a rapid decompression. The gradual loss of cabin altitude was implicated in the death of the pilots and crew of the Lear jet carrying golfer Payne Stewart in 1999 (NTSB, 2000). The aircraft was cruising at over 46,000 feet. It was concluded that the passengers and crew died from hypoxia as a result of failing to don emergency oxygen following a gradual loss of cabin pressure (the aircraft subsequently crashed when it ran out of fuel). However, the question concerning why they failed to don their emergency oxygen masks and initiate an emergency descent remains. With a relatively gradual rate of depressurisation, pilots do not recognise the problem developing. Hypoxia impairs judgement and reasoning (similar to alcohol intoxication) but may also result in euphoria and an enhanced feeling of well-being. This results in rapidly losing the cognitive functions essential to diagnose the problem and/or the abilities to don oxygen masks. This would be followed shortly after by a loss of consciousness. The crash of Helios Airways Flight 522 (Boeing 737) also resulted from a gradual loss of pressurisation, resulting in the crew becoming unconscious and the aircraft subsequently crashing when it ran out of fuel (Hellenic Accident Investigation and Aviation Safety Board, 2006).

Not only is it the degree of pressurisation that is important for maintaining the comfort and performance of the aircrew, it is also the pressurisation schedule (i.e. the speed at which the cabin pressure is reduced when climbing to the aircraft's cruising altitude and the rate that it is increased when descending). While this schedule has no direct effect on performance, it may indirectly impair the flight crew to some extent as a result of discomfort. Macmillan (1988) notes that the cabin altitude changes corresponding to high rates of ascent (in the order of 5,000 to 20,000 feet per minute) are tolerated with little difficulty or discomfort. However, when descending from a cruising cabin altitude (of around 7,000 feet) to sea level, increases in cabin pressure corresponding to a 5,000 feet per minute decrease in altitude will cause discomfort in the middle ear. The rate of descent of the cabin altitude in most passenger aircraft is usually much less than this, commonly of the order of 300 feet per minute.

Carbon Monoxide, Carbon Dioxide and Ozone

The airworthiness regulations specifically prescribe maximum levels for carbon monoxide, carbon dioxide and ozone. FAR/JAR/CS 25.831 specifies carbon monoxide concentrations in excess of one part in 20,000 parts of air are considered to be hazardous and that carbon dioxide concentrations during flight must not exceed 0.5 per cent by volume. FAR/JAR/CS 25.832 stipulates that cabin ozone concentrations must not exceed 0.25 parts per million by volume at any time above flight level 320 or 0.1 parts per million above flight level 270.

Carbon monoxide is a major product of combustion. Low concentrations of carboxyhaemoglobin (the product of breathing high levels of carbon monoxide) of around 20 per cent can cause mild headaches; higher concentrations (30 per cent) may impair vision or judgement. Ultimately, high levels of carboxyhaemoglobin (50 per cent) will result in severe headaches, confusion and ultimately, loss of consciousness. Such levels would only be achieved, though, after breathing carbon monoxide at concentrations of one part in 1,000 for over three hours. Carbon dioxide is carried in many aircraft as either a refrigerant and/or extinguishant. At concentrations of greater than 5 per cent (by volume) it acts as a narcotic and impairs vision and hearing. Concentrations in excess of 7 per cent will cause unconsciousness and ultimately death (Sharp and Anton, 1988).

Ozone occurs naturally in the atmosphere, especially above 40,000 feet. At very high altitudes, above the cruising altitude of current airliners (60,000 feet) ozone concentrations are approximately four parts per million. Ozone is an irritant to the eyes and respiratory tract at lower concentrations (one part per million) but can cause pulmonary oedema at higher concentrations (above 10 parts per million). Fortunately, ozone breaks down if heated to over 400°C, which is just the temperature of the bleed air from the first stages of the compressors on the engines from which the air conditioning system takes its supply. As a result, ozone concentrations in the flight deck and air cabin are rarely a problem.

FAR/JAR/CS 25.831 specifies the minimum ventilation requirements for the flight deck. Under normal operating conditions, the ventilation system must provide each occupant with an airflow containing at least 0.25 kg (0.55 lb) of fresh air per minute. This helps ensure toxic gases and vapours do not build up to significant levels on the flight deck. Maintaining a steady throughput of air also helps to reduce the effective temperature on the flight deck (see previous sub-section).

FLIGHT DECK NOISE

Effective and efficient communication between flight crew is essential for the safe conduct of any flight. Pilots also need to be able to hear and correctly identify any sounds generated by the aircraft warning systems. Excess background noise will interfere with both of these functions. Noise itself, especially if it is loud and/or prolonged, can also be stressful and debilitating to human performance. However, flight decks cannot be made into a silent workplace. It would be impossible to eradicate background noise altogether. Furthermore, the nature of the ambient noises changes with phase of flight and provides cues to the pilot. For example, at higher speeds high frequency noises generated by the window frames dominate: lower frequency 'rumbles' emanating from the high lift devices and the gear (especially on final approach) dominate at lower speeds.

Physical Properties of Noise

Background noise described in the form of a noise spectrum, with power (loudness) plotted on 'y' axis and frequency (in bands) on the 'x' axis. Not all frequencies in noise are of the same power. In the examples previously given higher frequencies dominate the ambient noise in the cruise whereas lower frequencies dominate on approach. A noise spectrum analysis also forms the basis for the design of auditory warning systems (see Chapter 12 in Part Three on Display Design). A simplified noise spectrum from a Boeing 727 during various flight phases is presented in Figure 11.1.

From a Human Factors perspective the most critical frequency band is between 600 and 4,000Hz where the human ear is most sensitive (the range of human hearing is about 20 to 20,000Hz). These frequencies should receive particular attention. From a physiological perspective sound deadening on the flight deck should also ensure that under no circumstances should the ambient noise on a flight deck exceed 120dB at any frequency. For general measurement purposes the dBA (decibels on the A scale) is used as a standard. The dBA mathematically weighs and combines noise measurements in several

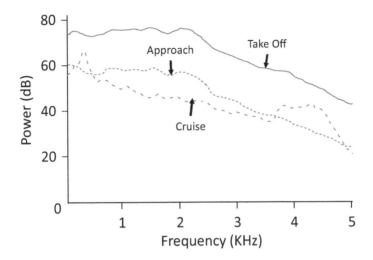

Figure 11.1 **Simplified flight deck noise spectra from a Boeing 727 during take-off, cruise and approach phases of flight**

frequency bands to produce a composite measure of noise. Greatest weight is given to sound in the 1,000–4,000 Hz bands, and least to sounds in the very low (up to 125 Hz) frequencies.

The effects of noise on the human are far more complex than can be described simply in terms of its physical properties. The apparent 'loudness' of a noise does not equate to its power (sound pressure level) measured in decibels (dB) as the ear is more sensitive at some frequencies than at others. The 'phon' is the unit of perceived loudness. A 60 dB tone at 1,000 Hz is defined as having a loudness of 60 phons. However, a 65 dB tone at 50 Hz produces a perceived loudness of only 40 phons. Unfortunately, the phon unit does not describe the relative loudness of sound; a 40 phon sound is not twice as loud as a 20 phon sound. This resulted in the development of the son. As a rough guide 40 phon = one son: every 10 extra phons doubles the number of sons. Thus two sons is perceived to be twice as loud as one son. However, it is sound pressure that impairs hearing but perceived loudness that causes annoyance and psychologically impairs performance.

There are two types of noise-induced hearing loss. The first type is of little concern in this context. In this case hearing loss results from acoustic trauma, for example an explosion, which results in permanent damage to the ear. Of more concern on the flight deck is noise-induced hearing losses which result from excess sound pressure levels at specific frequencies (e.g. noise resulting from the airflow going supersonic around the cockpit window frames in some aircraft). This is usually most profound at higher frequencies. The ear is full of small hairs (stereocilia) of differing length which are arranged on 'staircases' in the inner ear. Each length of hair is sensitive to a different frequency. Continual exposure to sound pressure at a specific frequency over about 85 dB can damage the hairs sensitive to that frequency. Short term exposure to loud noises can result in a temporary threshold shift which damages the structure of these hair cells but they recover after a couple of days. Continual exposure to the noise, however, results in the cells producing the stereocilia dying. The consequence of this is hearing loss and/or tinnitus. Lower and Bagshaw (1996) reported the general ambient noise levels on commercial flight decks was between 70 dBA and 79 dBA. It was concluded that such levels of noise would not be expected to result in hearing damage *per se*. Surprisingly, it was the noise levels from headsets that were the main concern. These were typically between 77 dBA and 89 dBA in order for pilots to hear clearly above the background noise levels. It was suggested that standard 'open' headsets should be replaced with the noise attenuating variety. However, these are not a universal panacea to the problems that flight deck noise imposes on hearing damage or communication. Several reports of in-flight incidents have been reported to NASA ASRS (Aviation Safety Reporting System) where pilots have failed to hear auditory warnings when wearing a noise attenuating headset. The headset attenuated the auditory warning played over the speakers on the flight deck, hence it never reached the pilot's ears!

Background Noise and Communication

Clear communication on the flight deck is absolutely imperative for flight safety. Speech (male or female) predominantly occurs in the 400–500 Hz band, although it also includes components right up to 5,000 Hz (Kryter, 1970). Kryter (1985) suggested that if the background noise exceeds 55 dBA use of items such as a telephone can become difficult. If noise levels exceed 63 dBA a raised voice is required to communicate with someone 1–2 m away (the typical distance between the pilot's seats in a commercial aircraft). At this distance, once the noise level exceeds 72 dBA, communication in anything less than a shout becomes impossible (Peterson and Gross, 1972). Unassisted communication on the flight deck becomes inadvisable once the background noise level exceeds about 55 dBA.

Annoyance Effects of Noise

The annoyance associated with noise is a product of its acoustic factors, its context and the personality of the listener. Sperry (1978) separated the annoyance effects of noise into two categories: acoustic and non-acoustic (environmental) factors. Acoustic factors included the sound level (and fluctuations in sound level), frequency (and fluctuations in frequency), duration, spectral complexity and the risetime of the noise. In the context of the flight deck, the non-acoustic factors include the pilot's activity, the predictability of the noise, the necessity of the noise and their personality. Noises that the flight crew perceive as being a product of poor design and hence avoidable (e.g. the loud, high frequency noise generated by the windscreen surrounds in some types of aircraft when in high-speed cruising flight) are regarded as more annoying than unavoidable noises. Kryter and Pearsons (1966) observed that higher frequency noise (over 2,000Hz) was also more likely to be annoying.

Noise and Human Information Processing

Most laboratory studies suggest that high levels of noise impair human performance. Poulton (1976, 1977) suggests that high levels of ambient noise mask the verbal rehearsal of items in the phonological loop component of WM (Baddeley, 1986) hence the interference often observed in these conditions in a letter and/or digit short-term recall task. Put another way, 'you can't hear yourself think'. Given the nature of such things as Air Traffic Control clearances, this is of some concern. Indeed any impairment of the pilot's WM is not desirable. Fortunately, impairment of this nature is only really evident above 70–80 dBA but this level of sound can be reached on a flight deck on occasions.

Broadbent (1976) also described the effects of high levels of noise on attention. He reported that there was a funnelling of attention onto only the most task-relevant information when performing in a noisy environment. On occasions, this may result in critical signals presented elsewhere being missed. Broadbent attributed this phenomenon to over arousal caused by this environmental stressor. Hockey (1970) also observed that in the presence of noise, certain aspects of a task that were perceived to be of higher importance were given more attention that those aspects regarded as being of lower importance.

VIBRATION

For the fixed-wing pilot, particularly those flying large turbofan-powered aircraft, vibration is usually not an issue of great importance, as vibration levels in the critical frequencies, are minimal. However, for smaller aircraft (particularly those with a high wing loading) and when operating aircraft in abnormal circumstances, there still remain certain times when vibration may be an issue. Turbulence can be considered to be a form of low-frequency, high-amplitude vibration, something with which all pilots will be familiar. Vibration was also implicated in the accident involving a Boeing 737-400 at Kegworth, UK. In this accident the port engine shed a small section of fan blade from the primary stage of the turbofan. This resulted in major vibrations being transmitted through the airframe making reading the displays on the flight deck difficult.

Stott (1988) describes the major sources of vibration that may affect the pilot. In order of increasing frequency, turbulence is usually of the order of one Hz, although it is usually of high amplitude (displacement). The pilot of a piston-engined aircraft may be exposed to vibrations resulting from the rotational frequency of the engine's crankshaft (which is typically somewhere between 2,400 and 3,600 rpm) corresponding to a vibration of between 40–60 Hz. The equivalent frequencies for a turbine engine are much higher, usually between 130–230 Hz. If the aircraft is a typical twin-engined turboprop, the blade pass frequency of the propellers (the propeller rpm multiplied by the number of blades in the propeller) will be approximately 100 Hz (assuming three blades per propeller) although vibrations of much lower frequencies may be produced as a result of harmonics resulting from the two propellers turning at slightly different speeds.

All vibrations have both a frequency and an amplitude component. Although it is the frequency of the vibration that may cause various parts of the human body to resonate (see later) it is the acceleration associated with these vibrations that is physiologically damaging to the body. If frequency remains constant, higher accelerations will occur at larger amplitudes. Fortunately, as frequency increases, amplitude tends to decrease. The ride in a large commercial aircraft is relatively smooth. Measurements taken in a variety of aircraft show that the typical range of accelerations (expressed as the root mean square value of 'g') varies between about 0.005 and 0.055 rms 'g' (compared to a car: 0.065–0.075 rms 'g' or a bus: 0.035–0.10 rms 'g' – Stephens, 1979).

Vibration is transmitted through contact of the human body with a vibrating surface, in this case the pilot's seat. The key frequencies are those at which the pilot's body (or parts of it) resonate. The human body as a whole has a natural resonant frequency of approximately 5 Hz. However, as humans are made of many different types of tissue of different densities, all of which resonate at slightly different frequencies, it is also important to consider the effects of various frequencies of vibration on specific parts of the body.

When considering the use of aircraft controls there are various key frequencies of concern. The first critical frequency band occurs between 2–6 Hz. In this region it is difficult to control an outstretched, unsupported arm. As a result, any manipulation of the small controls (e.g. those on the glareshield or overhead panel) is going to be considerably impaired. Between 4–8 Hz, control inputs using a joystick, even when the forearm is supported (as in the sidestick control inceptors in the Airbus 320/330/340/380 series aircraft) are impaired. However, this only really becomes a problem when the acceleration associated with the vibration exceeds approximately 0.20 'g', which is not particularly common in civil aircraft (Hornick, 1973). Control using a large control-column type of control inceptor is less likely to be affected at these frequencies as a result of the damping effect of the relatively large mass that the pilot is holding. The disadvantage of this latter type of control arrangement, though, is that it is likely that the pilot's arms will be more outstretched than when using a sidestick controller, making him/her more susceptible to the effects of vibration in the lower frequencies. In this same 4–8 Hz frequency band the muscles in the leg are also affected, which has obvious implications for inputs to the rudders (Stott, 1988). Two further bands of concern lie between approximately 4–5 Hz and 10–12 Hz. Both the head and the shoulder girdle resonate between these frequencies. In the lower band the amplification of the vibration is greater at the head than the shoulders and vice versa in the higher frequency band (Rowlands, 1977). At approximately 5 Hz, the amplification ratio of any vibration at the shoulder girdle may be of the order of three times the magnitude of the input. Again, this has implications for control inputs to the primary flight controls.

Careful design of the flight deck seating to reduce the transmission of critical vibration frequencies from the airframe to the pilot and the provision of armrests (e.g. when using a sidestick controller) is essential. However, somewhat counter-intuitively, matters may be made worse if the pilot straps in particularly securely when turbulence is encountered, especially if they tighten their shoulder straps. At certain frequencies, this may actually increase the transmission of vibrations to the shoulder girdle, making control more difficult in these circumstances.

Vibration and Viewing Displays

Visual performance may be degraded by vibrating either the display alone, the observer, or both. If the display alone is vibrated, at frequencies below 1 Hz the eye is able to track the target without problem as long as the amplitude of the displacement of the object does not exceed 40°/second at the retina. This tracking ability begins to decrease noticeably between 2–4 Hz. At higher frequencies than this the eye makes no attempt to track the object and instead the eye remains stationary and the target becomes blurred on the retina. If the observer is vibrated and the display remains static, reasonable visual acuity may be maintained up to about 8 Hz as a result of the vestibulo-occular reflex. At these frequencies the head tends to 'nod' (pitch) as much as resonate in the vertical plane. The vestibulo-occular reflex counteracts this pitching motion of the head and helps to keep the eye focused on the stationary object. At higher frequencies, vision begins to be compromised by the effects of resonances of the head, the eye-orbit complex and the eye itself. Concerning the resonant frequencies associated with the eye, there are two different peaks that are primarily responsible for inhibiting the use of visual displays. Below approximately 20 Hz the eye-orbit socket pair resonates. This begins to become damped above 20 Hz, however commencing at about 30 Hz the eyeball itself begins to resonate. This reaches a maximum somewhere in the region of 70 Hz (Stott, 1980).

Vibration and Human Information Processing

The effects of vibration on Information Processing are difficult to assess as vibration usually compromises the perception of the visual stimulus material and the speed and/or accuracy of any associated outputs. Thus it is difficult to assess if any performance deficit is a product of the mechanical properties of the human body or of the Human Information Processing system.

Several laboratory studies have shown that vibrating subjects has little effect on their reaction times while they are being vibrated but often their performance declines after the vibration has stopped. Hornick and Lefritz (1966), in a simulated military terrain-following flight task in which volunteer pilots flew a four-hour mission while experiencing vibrational frequencies between 1–12 Hz and associated accelerations up to 0.2 g, observed that after two and a half hours reaction times were almost four times longer compared to a control condition. Huddleston (1964) demonstrated that mental arithmetic tasks were unaffected at vibrations of up to 3.5 Hz, but performance decrements were evident between 4.8–16 Hz. However, in this study participants experienced accelerations in the region of 0.5 g, much higher than would be expected in a commercial aircraft. It has even been suggested that vibration at frequencies between 3.5–6 Hz may increase the arousal level (and hence vigilance) of participants engaged in a prolonged boring task, for example a trans-Atlantic or trans-Pacific flight! To summarise, it is difficult to establish if vibration at the levels or frequencies likely to be experienced by pilots of civil airliners will have any detrimental effect on their cognitive performance while flying.

DISORIENTATION AND AIRSICKNESS

The human was never designed to fly and as a result its sensory and perceptual system isn't really up to the job. Human senses are designed to operate in a 1 g environment in an upright position at speeds not exceeding about 40 km/h (and that assumes that you are Usain Bolt!). In the latest CAA review of accidents (CAA, 2008) worldwide, 'Flight Handling' was the second most common causal factor and of the 82 fatal accidents in this category, ten involved flight crew disorientation. Disorientation alone as the primary causal factor was implicated in four fatal accidents during this time. Pilot disorientation and airsickness both have their root causes in the inadequacies of the human senses for flight, most specifically the inner ear (or vestibular system) which is responsible for orientation and motion detection, the eye, and mis-matches between the sensory input from these two sources.

There are two organs in the inner ear responsible for the detection of angular motion and head position. The three semicircular canals are responsible for sensing angular accelerations (over about $2°/s^{-2}$). They are each arranged orthogonally to each other (see

Figure 11.2). The canal walls are lines with small hairs (cilia) and each canal is filled with fluid (endolymph). Thus, when the head is accelerated in any plane the fluid remains roughly stationary (relative to the Earth) but is in relative motion to the head, and hence also the cilia, which are connected to nerve fibres. Thus, when the head moves a signal is then sent to the brain from one (or more) of the canals depending upon the resultant forces. Note, however, that the semicircular canals are only sensitive to accelerations and not steady state motions. As a result, they do not detect any motion when the aircraft is established in a prolonged turn (or loop). In this case the cilia return to their datum position as there is no resultant acceleration.

The inner ear also contains two otoliths (again arranged at 90° to each other) comprised of particles of calcium carbonate on the end of small hairs. These are found in the saccule and utricle in the inner ear (see Figure 11.2). When the head moves their inertia causes them to stimulate the nerve fibres located at the base of the hair cells. Unlike the semicircular canals, the otoliths are sensitive to both motion and head position. Usually the brain deduces head position from combined inputs from the semicircular canals, otoliths and the visual system (particularly the peripheral visual system which is sensitive to strong horizontal and vertical features in the environment, such as the horizon). In this way it is possible to establish if the head is just tilted or if the whole body is in motion. However, the human sensory system is dominated by the visual system and it is when there is not a clear external horizon available that the likelihood of disorientation is increased, even when flying on instruments.

With regard to the physiology of the eye there are two independent visual systems at work. The foveal (central) visual system is employed for detailed, close up tasks. It operates when illumination levels are above 0.001 lux. The foveal region is a small region right in the centre of the retina consisting primarily of cone receptors that are colour sensitive and provide high visual acuity. It is the foveal visual system that pilots use to read the flight deck instrumentation, however it is not used to maintain orientation in space. This function is provided independently and simultaneously by the peripheral

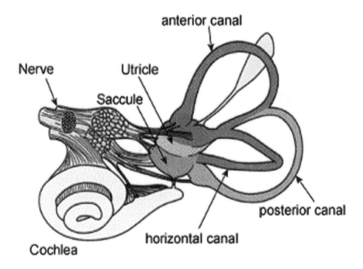

Figure 11.2   Inner Ear (Vestibular System)

visual system. The human being maintains their orientation in the environment by virtue of the mechanisms in the peripheral visual system which utilise strong horizontal and vertical features in the environment, such as the natural horizon. It also acts as a movement detector. The rod receptors in the peripheral visual system operate when ambient light levels are below 0.0001 lux. The rod receptors are more sensitive to light but have much poorer acuity and do not provide colour vision. Green and Farmer (1988) have suggested that one of the reasons pilots have difficulty in maintaining attitude when flying on instruments is that the artificial horizon found in most aircraft is quite small, hence pilots use their foveal visual system to read it. However this visual system is not designed to maintain orientation in space, which is a function of the peripheral system.

Disorientation

There are various general categories of pilot disorientation. Somatogyral illusions are false sensations of rotation which occur mainly in conditions of poor visibility. The most common of these illusions are the graveyard spiral and the 'leans'. As noted earlier, the semicircular canals are only sensitive to accelerations and not steady states. As a result, if a pilot enters a sustained turn in one direction they will initially perceive the aircraft to be turning as a result of the initial acceleration but this will eventually cease when the aircraft has been established at a steady rate of turn for a period of time (20–30 seconds); the cilia return to their datum position. As a result, the pilot may feel that their aircraft is no longer turning. If at this point they sharply level the aircraft it will produce the sensation that the aircraft is banked in the opposite direction to the initial turn, even though it is straight and level. If the pilot ignores their instruments they may re-establish the initial direction of turn in an attempt to counteract this sensation. The pilot now believes the aircraft is straight and level when it is actually turning. This can result in the aircraft losing altitude. Pulling back on the control column and applying power in these circumstances will actually only tighten the turn and the aircraft will continue to lose height.

The 'leans' occurs in a similar fashion to the graveyard spiral. In this case, as a result of there being no strong external horizontal reference, the pilot may unintentionally let the aircraft enter into a *very* gradual turn. The rate of angular acceleration in this case is below the sensitivity of the semicircular canals. If the pilot then notices the developing turn and returns the aircraft to straight and level flight at a rate of angular acceleration above the threshold of the semicircular canals, this can again cause the sensation of turning in the opposite direction. However, rather than entering a graveyard spiral, in this case the pilot may end up leaning in the cockpit in the direction of the original turn in an attempt to achieve the perception of the correct orientation.

The coriolis illusion occurs when the pilot moves their head sharply, in either a vertical or horizontal plane, while the aircraft is established in a turn. As a result, two or more of the semicircular canals are stimulated concurrently (along with the otoliths) resulting in the sensation that the aircraft is rolling, pitching and/or yawing simultaneously. This can rapidly result in profound disorientation and loss of control.

Somatogravic illusions are caused by linear accelerations which again take place mainly in conditions of poor visibility. In this case the otoliths are the sense organs mainly responsible. In general, the otoliths cannot discriminate between head position (e.g. tilted backwards) and prolonged linear acceleration in the same axis. For example, in the inversion illusion, the sensation that the aircraft is tumbling backwards may be provoked by an abrupt change from climbing to straight and level flight in instrument conditions. The head-up illusion is the sensation that the nose of the aircraft is pitching up when it is in straight and level flight. This is provoked by acceleration during level flight in restricted visibility. The pilot's response may be to push forward to pitch the nose of the aircraft down. This illusion can be stimulated during a night take-off from a well-lit airport into a dark sky. The head-down illusion is the opposite: this occurs with linear decelerations (e.g. when deploying the air brake) which can provoke the sensation of pitching down and hence the pilot's natural reaction may be to pull back on the control column. During a low-speed final approach this may result in the aircraft approaching the stall boundary.

Some of the visual illusions that may result in pilot disorientation have already been described in Chapter 2 on Human Information Processing. These illusions were specifically associated with issues in perception. However, there are other visual illusions that occur as a result of the physiology of the eye and the way that it works. This arrangement of the two visual systems in the eye can result in two particular illusions. The autokinetic illusion is caused by becoming fixated on a fixed single point of light at night against a featureless background. This can give the illusion that the light is moving and is on a collision course with the aircraft. The vection illusion may occur when taxiing. The motion detectors in the periphery of the eye detect the movement of other aircraft moving forwards (when the pilot's own aircraft is stationary) leading to the sensation that their own aircraft is actually moving backwards. This illusion is the same as that which arises when sitting in a stationary train when the one alongside it pulls out of the station.

Airsickness

Airsickness is simply the specific form of motion sickness experienced in aircraft. The most common symptoms of airsickness are nausea, dizziness, fatigue and pallor or sweating. Three basic types of motion sickness can be defined (Benson, 1988): motion sickness resulting from:

- Motion that is both felt and seen but the detected motions do not correspond (Type 1).
- Motion that is seen but not felt (Type 2i).
- Motion that is felt but not seen (Type 2ii).

It is hypothesised that motion sickness results from the brain being unable to resolve the sensory conflicts resulting from the visual, vestibular and perhaps also the proprioceptive systems when in motion (Reason and Brand, 1975). It is known that the inner ear is heavily implicated as individuals without a functional vestibular system are often immune to motion sickness. It is thought that the sensory conflict provoked by airsickness stimulates a response in the brain stem similar to that produced by some neurotoxins. This results in the release of a neurotransmitter that acts in the region of the Area Postrema in the brain stem, stimulating the nearby vomiting centre.

The aetiology of airsickness in the air cabin and on the flight deck is slightly different. The former case is an example of motion that is felt but not seen (Type 2ii) as a result of the small windows or the position that the passenger is seated in the aircraft. Without visual reference to the outside, the motion of the aircraft is perceived by the vestibular and proprioceptive systems, however the air cabin does not appear to move relative to the passenger.

Simulator sickness is an instance of motion that is seen but not felt as in the case of a non-moving base simulator (Type 2i). In the case of a full-flight simulator with a motion base (see Chapter 9 on Training and Simulation) the actual motion is fundamentally different to the motion in an aircraft undertaking the same manoeuvre. Aircraft simulators cannot reproduce strong or long-duration linear accelerations. These are simulated by tilting the simulator to create an illusion of being pressed into their seats. This is an instance of two sensory systems detecting motion that do not correspond. In a simulator there is an illusion of motion created by the virtual world but no motion (or inappropriate motion) detected by the vestibular system. Wertheim (1998) reports that simulator sickness is most profound in fixed-base simulators with large visual displays. In this case the large apparent movements of the visual environment create a strong sensation of ego motion that is not matched by input from the vestibular system. Simulator sickness is more profound in military aircrew who undertake more aggressive manoeuvres in the simulator. Kennedy and Fowlkes (1992) report that the majority of pilots have reported feelings of simulator sickness at some stage during their career. Corresponding figures for civil aircrew are much harder to come by. Anecdotal evidence, though, suggests that most pilots have experienced mild symptoms at some time. In the aircraft itself the coriolis illusion can result in airsickness as a result of a mis-match in sensory input between the semi-circular canals and the otoliths rather than a visual-vestibular mis-match.

FURTHER READING

There is one book that is mandatory reading covering the environmental stressors described in this chapter:

Rainford, D.J. and Gradwell, D.P. (2006). *Ernsting's Aviation Medicine* (4th edition). London: Hodder Arnold.

Although, to be honest I still prefer the second edition – the fourth edition is definitely better than the third edition! For a general book (non-aviation) covering environmental stressors try:

Konz, S. and Johnson, S. (2007). *Work Design: Occupational Ergonomics* (7th edition). Scottsdale AZ: Holcomb Hathaway.

PART THREE

The Machine

The *Machine* in question in this case is the aircraft and from a Human Factors perspective it is the flight deck that is of principal interest. The *Machine* is designed to undertake a *Mission* and the flight deck is designed to support the prosecution of that task. Fortunately for the flight deck designer, compared to a military aircraft the *Mission* of a commercial aircraft is relatively straightforward. However, there must be a good 'fit' between the hu*Man*, *Machine* and the *Mission* and this is what the flight deck design process needs to ensure. The pilots are the ultimate design forcing functions and the designer must operate within their abilities. A good flight deck should enhance pilots' Situation Awareness but it should not impose excessive workload. The opportunities to 'modify' the hu*Man* component are limited to what may reasonably be taught and practised during pilot training.

In the 1990s, as a result of a number of accidents involving highly automated aircraft the FAA commissioned an exhaustive study of the pilot-aircraft interface in 'third generation' ('glass cockpit') aircraft (FAA, 1996b). This highly influential report identified several major shortcomings and deficiencies in flight deck design. There were criticisms of the interfaces, such as pilots' autoflight mode awareness/indication; energy awareness; confusing and unclear display symbology and nomenclature, and a lack of consistency in FMS interfaces and conventions. The report also heavily criticised the flight deck design process itself, identifying in particular a lack of Human Factors expertise in design teams and placing too much emphasis on the physical ergonomics aspects and insufficient on the cognitive ergonomics. In total 51 recommendations came out of the report, including:

> The FAA should require the evaluation of flight deck designs for susceptibility to design-induced flight crew errors and the consequences of those errors as part of the type certification process. (FAA, 1996b, p. 97)

In July 1999 the US Department of Transportation tasked the Aviation Rulemaking Advisory Committee (ARAC) to provide advice and recommendations to the FAA administrator to:

> ... review the existing material in FAR/JAR 25 and make recommendations about what regulatory standards and/or advisory material should be updated or developed to consistently address design-related flight crew performance vulnerabilities and prevention (detection, tolerance and recovery) of flight crew error. (US DoT, 1999)

The European Joint Aviation Authorities (JAA – now replaced largely by the European Aviation Safety Agency – EASA) as a part of the airworthiness regulatory harmonisation efforts, also subsequently adopted this task. Since September 2007 the rules and advisory material developed from this process have been adopted by EASA as Certification Specification (CS) 25.1302 and AMC (Acceptable Means of Compliance) 25.1302. These rules basically require that the flight deck has an error-tolerant pilot interface and that the design process (and result product) employ good Human Factors. A good Human Centred Design is now effectively mandated on modern commercial flight decks. These rules apply to the Type Certification and Supplemental Type Certification processes for large transport aircraft certificated within the European Union. At the time of writing the same rule is soon to be adopted by the FAA in the USA.

However, if you are looking for the Human–Machine Interface on the flight deck as an entity in itself, you won't find it as it doesn't really exist! Figure iii.i (from Harris, 2010) shows why this is the case. On the Machine Output/Human Input side of the interface, the 'images' on the displays (hopefully) convey the required data/information to the pilot. These images are interpreted and their content should be transformed into knowledge and understanding that allows control of the aircraft and its systems. On the Human Output/ Machine Input side of the control loop, the pilot's control intent needs to be translated into the desired aircraft output. A good control system will translate the pilot's intent into the required change in the aircraft's flight path in the manner desired and with minimal workload (physical or mental). A 'high-quality' pilot–aircraft interface consists of a good 'fit' between the skills, knowledge and ability of the pilot and the controls and displays of the aeroplane.

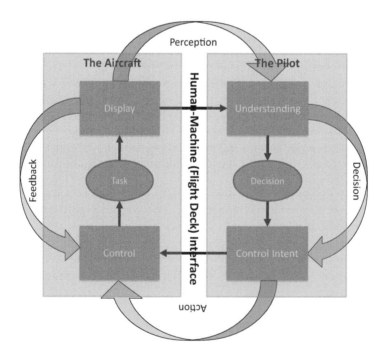

Figure iii.i   **The concept of the Human–Machine (Flight Deck) Interface superimposed over a representation of the classical 'Perception–Decision–Action–Feedback' control loop (adapted from Harris, 2010)**

In the following chapters in this Part the left-hand side of the diagram is considered in more detail (the flight deck controls and displays). The right-hand side has already been looked at in Part One. However, the system depicted in Figure iii.i is a classical manual control loop, and it is now rare that aircraft are flown manually for any protracted period of time. Most manual flight is confined to the take-off, and approach and landing phases; the remainder of the time the aircraft is flown automatically. With modern highly computerised aircraft flight decks the distinction between controls and displays is becoming increasingly blurred. Interactive displays, which are becoming a feature of many modern aircraft systems, are virtual controls and the flight deck is now one large, flying computer interface. Furthermore, interactions with a computer proceed in a different way to control using a traditional interface. This Part concludes with an examination of HCI (Human Computer Interaction) on the Flight Deck.

12
Display Design

The displays on the flight deck are the manner by which the *Machine* enhances the awareness of the hu*Man* controlling it. The displays convey data (or information) regarding the status of the *Machine* and the status/progress of the *Mission*. Displays are traditionally thought of as merely visual displays. In the case of a commercial aircraft these are the head-down displays (such as the Primary Flight Display – PFD; the Navigation Display – ND; and the systems displays) and increasingly the Head-Up Display (HUD). However, there are also auditory display systems, which are usually confined to conveying alerts and warnings to the crew and sometimes haptic displays. Haptic systems transmit information via the mechanoreception senses (kinaesthesia and proprioception). This is also touched on in Chapter 13 on Aircraft Control but haptic systems are also being experimented with as a means of conveying information to the pilot, liberating an additional information processing channel to those described by MRT (Wickens, 1984).

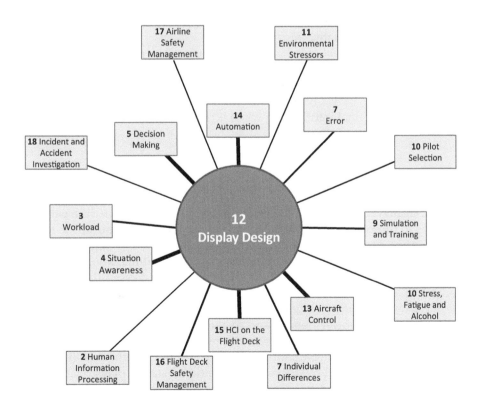

Displays are the primary means by which the pilots' achieve Situation Awareness, however modern displays are not the passive dials and indicators from the early years of aviation. Many displays are now multifunctional displays with an element of interactivity with elements of ubiquitous computing. Hence they are highly related to the aircraft's automation. Good flight deck design minimises training requirements and the number of opportunities for design-induced error while supporting crew decision making.

The ultimate objective of the display designer is to enhance the crews' Situation Awareness (SA), be it of their flight path, aircraft configuration or the functioning of their aircraft's systems. However, Dekker and Hollnagel (1999) have described highly automated aircraft cockpits as being rich in data but deficient in information. It should be the objective of display designers to convey information (and even knowledge) not simply to provide data (see the categorisation from Ackoff, 1989 outlined in Chapter 2 on Human Information Processing). For example, the airworthiness regulations require the display of fuel quantity and fuel flow rates. However, these *data* are of little use: what the pilot requires is *information* concerning what the remaining amount of fuel represents in terms of range or endurance.

Endsley (1988) lists 12 principles that may be used to enhance SA, five of which are directly applicable to display design:

- *Minimise divided attention requirements* – group information in terms of spatial proximity: avoid having many disparate display sources.
- *Encourage holistic, 'top down' processing* – present the pilot with the 'bigger picture'. Do not give them disparate components of data and expect them to assemble the 'big picture' (information) in their heads.
- *Filter information when required* – in times of high workload give the pilots only the most relevant, task-related information.
- *Display rate and/or trend information* – the Human Information Processing system is poor at predicting future states, especially in systems with either a high control order and/or a great deal of inertia.
- *Spatial information should relate to the situation* – for example, provide the pilot with track-up navigational information when it is important to know what is on the left or right.

VISUAL DISPLAYS

The flexible display technology now available on the flight deck has enhanced the opportunities to present information within its context and in a naturalistic form rather than as piecemeal, seemingly standalone, disparate nuggets of data as was the case in earlier, 'clockwork' cockpits using electro-mechanical instrumentation. Reising, Ligget and Munns (1998) argued that display technology has progressed through three eras: the 'mechanical' era; the 'electro-mechanical' era; and most recently, the 'electro-optical' era. The 'electro-optical' era (the 'glass cockpit' revolution) commenced with the use of relatively small, heavy, power-hungry and costly Cathode Ray Tubes (CRTs). However, the true revolution in glass cockpit aircraft was not in the type of displays used but was in the level of automation concomitantly introduced onto the flight deck (Billings, 1997). CRTs are now being replaced by active matrix liquid crystal displays (LCDs) which are thinner, generate less heat and consume less power. They can also be made much larger and in a variety of shapes and sizes. Their main benefit is that they offer the opportunity to use a wide variety of display formats appropriate to the nature of the data (or information) and the task at hand.

A Brief Diversion – The Anthropometrics of the Flight Deck

The best displays in the world are of no use if the pilots cannot see them. The spatial arrangement and the physical ergonomics of a flight deck, including the location and design of the display systems, begins at the Eye Reference Point (ERP). From this design eye position, internal and external vision, seat geometry and the arrangement of controls is all determined. From the ERP the next key design reference is the Seat Reference Point (SRP) where the seat back meets the seat cushion, when viewed from the side elevation. To keep the pilot's eyes in the required fixed position in three-dimensional space in the flight deck (i.e. at the ERP) the amount of seat height adjustment is largely determined by variations in the sitting eye height (see Lovesey, 2004 for an appropriate range of dimensions). The rudder reference point (RRP), the required point at which the pilot's feet should rest to operate these controls, is the final key anthropometric dimension. The range of adjustment required in the rudder pedals is determined by knee height and buttock–knee length (see Lovesey, 2004).

It is impractical to design a flight deck to accommodate all potential users (and remember aircraft are sold into a worldwide market). Designs are a compromise intended to accommodate around 90 per cent of the piloting population from the 5th percentile to the 95th percentile. This means that all but the smallest 5 per cent and the largest 5 per cent will be able to fit comfortably. Unfortunately, people are not the same percentile value in every critical measurement. For example, someone who is average (50th percentile) in stature probably may not be average in leg length or arm length (although these dimensions are highly correlated: Tiller, 2002; Pheasant and Haslegrave, 2005). Furthermore, stature is the critical dimension for flight deck design specified in CS/FAR 25.777 but is perhaps the least useful from a design perspective.

To emphasise, though, the ERP is critical: it determines visibility of displays (especially HUDs); indirectly determines the reach to controls; and directly determines the view out of the flight deck windows, perhaps the simplest but most important displays in the aircraft. FAR/CS 25.773 regulates external vision for the pilots. The primary considerations are the vision 'polar', which establishes a minimum standard for view of the outside world for safe manoeuvring and collision avoidance and the three-second rule, which dictates the amount of ground that can be seen ahead of the aircraft during the final stages of approach and landing (see Kelly, 2004). The primary flight deck displays should be situated between a horizontal datum taken from the ERP and 30° down and within 30° either side of it. Head-down displays should ideally be between 50 cm to 70 cm from the eye (although in some larger modern aircraft they are slightly further away than this).

The Basics

The character size required on a display is dependent upon the distance from the viewer. Opinions differ slightly but display characters should subtend an angle of between 17–25 minutes of arc at the eye (Pastoor, Schwarz and Beldie, 1983; FAA, 1996a).

To be easily readable any component on a flight deck display must have significant luminance contrast compared to its background. DEF-STAN 00-25 (part 7) suggests that the minimum contrast ratio should be 3:1 and should be user adjustable between up to an upper bound of about 10:1. *Advisory Circular 25-11 Transport Category Airplane Electronic Display Systems* (FAA, 1987) acknowledges the importance of these factors for display legibility yet merely suggests that these parameters should be 'adequate' or satisfactory' without actually specifying a value. LCD displays can achieve the same contrast ratio at lower luminance. Luminance contrast should always be used in preference to colour contrast to aid readability. This latter aspect is done by using a complementary colour pair (e.g. red and green). Colour should only be added to a display once the display format's effectiveness has been optimised in a monochromatic format. Colour alone is not a reliable method by which to convey information. It should be used sparingly and only as a redundant information code. Design standards strongly advise against using more than six or seven colours on an electronic display. Colours:

- should be consistent in their use;
- have only one meaning per colour;
- should be easily discriminated in all operating conditions;
- be consistent with conventions; and
- users should not be able to change the colour conventions.

Care should be taken to ensure that the use of red, yellow (amber) and green (and to a lesser extent, blue) is consistent with popular stereotypes and the conventions specified in the airworthiness regulations and associated advisory material for the use of warning and cautionary areas on instrumentation (e.g. CS/FAR 25 and AC 25-11).

Even with the advent of modern electro-optical displays, there are still only three basic formats for conveying quantitative data; digital counters; fixed-index, moving-pointer displays; and fixed-pointer, moving index displays (Figure 12.1). Each of them has strengths and weaknesses for transmitting certain types of data that depends upon the situation. These are summarised in Table 12.1. Basically, there are two general types of data/information that a display provides: rate information and state information. Rate information allows the pilot to estimate the rate of change in a certain value. State information conveys information concerning specific values at a given point in time.

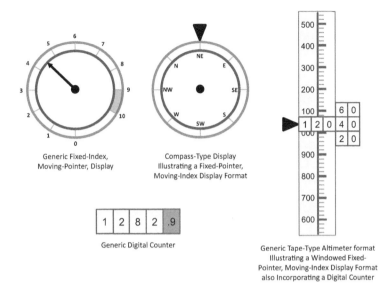

Figure 12.1   The basic display types for conveying quantitative information (Harris, 2004a)

Table 12.1   The relative merits of: digital counters; fixed-index, moving-pointer displays; and fixed-pointer, moving index displays (Harris, 2004a)

Situation	Preferred display format		
	Analogue		Digital
	Fixed-index, moving-pointer	Fixed-pointer, moving-index	Counter
Reading accuracy is very important	✗	✗	✓
Reading speed is important	✓	✗	✓
Values change quickly or frequently	✓	✗	✗
User requires rate of change information	✓	✗	✗
User requires information about deviations from a nominal value	✓	✗	✗
Minimal space is available	✗	✓	✓
User required to set a quantitative value	✓	✗	✓

Fixed-index, moving-pointer and fixed-pointer, moving-index instruments come in two common formats, circular dials and vertical or horizontal tapes. Fixed-pointer, moving-index scales can also come in either a full-scale or a windowed scale format. When designing a display system the first step is to define the data that you need to display and the relationships between the pieces of data (to make them into information).

Certain displays will be used in conjunction with others, for example when comparing the relative functioning of two (or more) engines. Other pieces of data will be used in isolation (e.g. cabin altitude). For each of these pieces of data it must be ascertained if the task requirements prioritise either rate or state information, if the values are likely to change slowly or rapidly, what degree of precision is required, and how quickly must the pilot read them. There is always a speed/accuracy trade-off. Displays that are designed to be assimilated quickly will be less precise; displays that are designed to convey accurate data/information will take longer to read. Finally, it needs to be established if it is going to be utilised to make a swift check reading of a value and if that value has a nominal value that can be incorporated. On this basis the appropriate display format can be selected.

With electro-optical displays, combined dials and counters can be presented, providing the best of both speed and accuracy. If the individual components on the display are arranged appropriately fast scanning 'check readings' are also possible (see Figure 12.2). With the older, smaller CRT-based displays (which could be as little as 160 mm x 160 mm) it was sometimes required to make compromises in the design of the display formats as a result of this lack of space. However this is becoming less of a problem with the use of larger, LCD-based displays. Counters take up little space, convey precise quantitative data but are slightly slower to read than a fixed-index, moving pointer display. Many of them on the same display area also lead to clutter and long search times to locate the appropriate piece of data. Fixed-index, moving-pointer dials are the best compromise between speed and accuracy, especially if they also incorporate a digital counter, however to be readable they take up a relatively large amount of display area. A circular dial format for this type of display is also good at conveying trend information. Tapes are poorer in this respect. But one advantage of tapes is that they take up less display space than a dial.

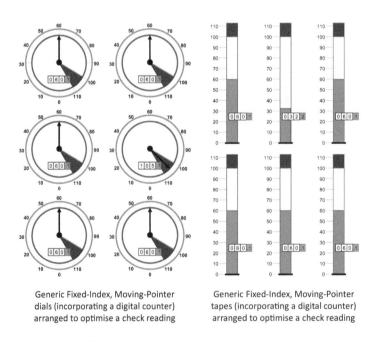

Generic Fixed-Index, Moving-Pointer
dials (incorporating a digital counter)
arranged to optimise a check reading

Generic Fixed-Index, Moving-Pointer
tapes (incorporating a digital counter)
arranged to optimise a check reading

Figure 12.2   Circular dial and tape formats of fixed-index, moving-pointer displays arranged to optimise a check reading (Harris, 2004a)

If the purpose of the display is to make a comparative check reading between several systems, circular dials are superior. Dials can all be arranged so that their nominal position is, for example, at 12 o'clock. This allows a fast scan of the instruments enabling an emergent feature of the group of displays as a whole to reveal itself, which is not apparent when each display component is considered in isolation. An abnormal value is easily and quickly spotted as the emergent feature of the vertical line formed by all the pointers when in their nominal position is broken. Tape-type instruments have a slight disadvantage in this respect. It is easy to make a swift check reading across a row of tapes but not down a column of tapes. Their emergent features are not as strong as dials and they do not have an implicit nominal position (see also Figure 12.5). Check readings can also be aided by the selective use of colour on the displays around the nominal position. When information from several sources needs to be mentally integrated in some way, there is a benefit in arranging the display elements in close proximity to one another and in such a way so that they form an emergent whole (the proximity compatibility principle – see Barnett and Wickens, 1988).

'Open window' displays of a moving-index, fixed-pointer design have all the disadvantages of both traditional dials and counters, with none of their benefits. It is impossible to make a check reading from these displays, as the pilot has no idea where in the range of potential values the current value lies without actually reading the value. Additionally, they do not possess the accuracy of a counter. They do, however, transmit some trend information, although not as efficiently as a fixed-index, moving-pointer displays.

The Primary Flight Display

The PFD is the evolutionary manifestation of the attitude indicator (AI); air speed indicator (ASI); the altimeter and a section of the horizontal situation indicator (HSI). A generic form of a typical modern PFD is shown in Figure 12.3. The original PFD was a classic example of the compromises a display designer faces. In this case the Human Factors Engineer was not only concerned with the optimal format for the display of information but also with the constraints on the format of these instruments imposed by the space available and the demands of the regulatory authorities.

The AI is the instrument that best exemplifies the problems posed to the display designer. In Western designed and built aircraft, the AI is an 'inside looking out' instrument where the artificial horizon line always remains aligned with the natural horizon. This is prescribed by the regulations in CS/FAR 25. An alternative design is the 'outside looking in' design, where the aircraft symbol moves in relation to a fixed horizon line. Bryan, Stonecipher and Aron (1954) found that the 'inside looking out' (conventional) horizon encouraged a loss of attitude awareness during instrument flight that subsequently resulted in a loss of control. Fitts and Jones (1961) also found that this design was inferior both in terms of speed of interpretation and number of errors of interpretation. However, with familiarity, performance with the 'inside looking out' design improved considerably.

The advantage of this 'outside looking in' accrues for several reasons. The moving component of the display in this type of instrument is compatible with the movement of the primary flight control inceptor (there is control–display compatibility). This format also complies with the principle of the moving part (Roscoe, 1968). This principle requires that the animated portion of a display should move in accordance with the observer's

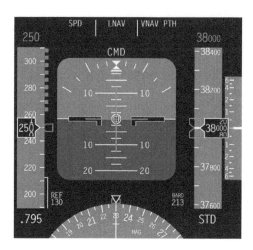

Figure 12.3   Generic Primary Flight Display

mental model of the behaviour of that variable. In the conventional 'Western' AI, the horizon bar (the moving component of the display) apparently moves left when a roll to the right is initiated, that is it *appears* to deflect to the left *relative* to the pilot's immediate surroundings, i.e. the flight deck. Green and Farmer (1988) also observed that human beings maintain their orientation in the environment by virtue of the mechanisms in the peripheral visual system, which utilise strong horizontal and vertical features in the environment, such as the natural horizon. The foveal (central) visual system, used to read the flight deck instrumentation is not normally used to maintain orientation, nevertheless this is the part of the visual system that the pilot is required to use to when interpreting the AI. The foveal visual system is used for detailed, close up work, not as an orientation system and movement detector (see Chapter 11 – Environmental Stressors – concerning pilot disorientation). The 'inside looking out' format does, however, have an advantage when maintaining a turn, when a one-to-one correspondence between the real and the artificial horizon is desirable.

The altimeter has probably attracted almost as much research attention as the AI. In this case the problem faced by the designer is balancing the demands of reading accuracy with the scale length required (in the region of 45,000 feet) in a compact instrument that is quick to read (state information), conveys good rate information (e.g. to aid pilots in tasks such as anticipating when to commence levelling off) and also results in very few reading errors being made. The original three-pointer altimeter was slow to read and generated an unacceptable number of reading errors, even when used by experienced pilots (Grether, 1949). The three-pointer altimeter produced an 11.7 per cent error rate and took a mean time of 7.1 seconds to interpret. During operations there was a tendency for the 10,000-feet hand to get lost behind the other larger hands and in times of high workload or stress pilots occasionally mis-read the altimeter by 10,000 feet. When descending in poor visibility, this could result in a controlled flight into terrain accident. The electro-mechanical counter-pointer altimeter replaced the three-pointer altimeter. This removed the 10,000- and 1,000-feet hands and incorporated a digital counter. This was much faster to read and resulted in few reading errors.

On the PFD altitude is displayed as a windowed, fixed-pointer, moving index display in the form of a vertical tape (see right-hand side of Figure 12.3). This format is poor for conveying rate information and for making a check reading. This window is also typically only 1,000 feet (500 feet on either side of the current altitude). The windowed tape-format altimeter also contravenes the principle of the moving part (Wickens and Hollands, 2000). To convey an increase in altitude, the moving part of the display (the tape) actually moves downwards. Unfortunately, the alternative to contravening the principle of the moving part, which would require the tape to move upwards as altitude increases, would contradict another population stereotype as it would require numbers to increase from bottom to top. AC 25-11 specifies that for both altitude and airspeed the larger numbers should be at the top of the display.

However, despite all the criticisms levelled at it above, Grether (1949) observed that it actually resulted in a slightly lower error rate (expressed in terms of the percentage of errors in excess of 1,000 feet) than the counter-pointer display format (0.3 per cent versus 0.7 per cent). However, it did take longer to read (mean of 2.3 versus 1.7 seconds).

Many of the problems posed by the display of airspeed are the same as those created by the display of altitude. Although the range of values is somewhat smaller in this case the precision required is greater. Many of the criticisms levelled at the altitude tape can equally as well be applied to the indication of airspeed on the PFD, which uses exactly the same format (left-hand side of Figure 12.3). In this case the window is usually capable of accommodating a range of only around 100 knots (50 knots either side of the current speed).

However, the PFD is more than just the sum of its component parts. It must be considered as a whole instrument. When this is done some of the previous criticisms of the individual components need re-evaluating. The development of the PFD also needs to be put into context. With the advent of early CRT displays, the actual area available for the display of primary flight data was actually reduced, hence more compact versions of the altimeter and airspeed indicator were required. The windowed, fixed pointer moving-index format for these instruments serves this requirement well. The format and layout of the instrumentation on the PFD is an adaptation of the US Air Force 'T-line' concept (itself a development of the RAF 'basic T' developed during World War II). The rationale underlying the 'T-line' concept was to display all relevant air data in a horizontal line across the central point of the AI and navigational data along a vertical line running through the centre of the same instrument (hence the inclusion of a section of the HSI). This arrangement promotes an efficient scan in instrument meteorological conditions. Although it could be argued that the section of the HSI is an unnecessary duplication of information also contained on the ND, it considerably reduces the length of the pilot's instrument scan, especially in situations such as performing an instrument landing system approach. Furthermore, the information on this section of the HSI can be correlated directly with the localiser and glideslope deviations displayed around the AI in approach mode. The development of this format for the display of primary flight data was pursued at the behest of the airworthiness authorities during the development of the first CRT-based integrated PFDs. Modern flight decks utilising much larger LCD displays often display a full compass rose below the PFD (in addition to the moving map on the ND). In short, while the criticisms of the components of the PFD are valid when considered in isolation, they are perhaps less important when the PFD is considered as a whole.

The Navigation Display

The ND is the single display that best demonstrates the advantages conferred by electro-optical display technology when combined with high levels of automation on the flight deck. To address such navigational questions as 'where am I?', 'which way am I going?', 'how fast am I going?' and 'when will I get there?' no re-coding of data are required to transform ranges and bearings to navigational beacons (or from GPS data) into a navigation solution. In older 'clockwork' cockpits the re-coding of indicated airspeed, altitudes and VOR/DME data was demanding on the pilot's WM. Monitoring progress against the flight plan was a high workload exercise done periodically, rather than continually. Revising a route or plotting a diversion during flight was demanding and prone to error. Additionally, the opportunities to look ahead were minimal.

Modern NDs provide a 'God's Eye' view of the planned and actual progress of an aircraft's flight (Figure 12.4). The position of airports and beacons are clearly shown. Perhaps more importantly, all the corrections for side winds, altitude, etc. are now performed by the computers in the aircraft and the results are displayed on a screen in a manner compatible with the mental model the pilot has of the aircraft's progress. It is also current practice to overlay the weather radar picture on the ND and TCAS (Traffic Alert and Collision Avoidance System) information, which further aids tactical flight planning allowing the pilots to anticipate poor weather and plan an alternative route around it (in the long term) or avoid conflicting traffic (in the much shorter term). Displaying these critical pieces of information in such a way allows a rapid assimilation of the situation as a result of spatial information being presented in a spatial code.

One of the shortcomings of the initial instantiation of the ND was that while such a display was good at conveying Lateral NAVigation (LNAV) information, it was poor at depicting vertical (VNAV – Vertical NAVigation) flight profiles and showing potential conflicts with the terrain. However, Kelly (2004) notes that providing the equivalent VNAV awareness for climbs and descents is much more of a challenge as parameters such as flight path angle, rate of climb, thrust, drag, speed and acceleration are inextricably linked by the laws of physics. Corresponding autoflight and thrust management modes for VNAV are inherently more complex and more difficult for pilots to understand than those for LNAV. Boeing has supplemented the ND in its latest generation aircraft with a Vertical Situation Display (VSD) at the bottom of the ND map display (see Chen, Jacobsen, Hofer, Turner and Wiedemann, 2000; Kelly, 2004). VSDs can present either a lateral projection or a projection of an 'unwound' FMS route, depending upon which display option is selected (see Figure 12.4). Modern NDs can also provide further additional external hazard information to the crew, such as terrain and predictive windshear alerting.

Future Air Traffic Management (ATM) practices will require aircraft to navigate in a different manner. This concept, known alternatively as Direct Routing, Trajectory Management or 'Free Flight' will significantly affect the pilot's role and responsibilities. Limited Direct Routing has already been implemented in parts of US airspace. In 'Free Flight' responsibility for ATM will be delegated to the flight deck (self-assured separation). Aircraft will fly direct routes and manoeuvre freely at their optimum speed and altitude, spending less time in the air and using less fuel, significantly increasing efficiency. As a result the ND is being further adapted for the Cockpit Display of Traffic Information (CDTI). Changes in the airspace to allow Free Flight cannot be fully exploited if aircraft are not equipped with suitable display technology to allow pilots to manoeuvre to maintain separation from other traffic, avoid weather and to undertake other aspects of

Figure 12.4   Boeing Navigation Display (ND) formats with a Vertical Situation Display (VSD) on lower part. The left figure depicts an encounter with terrain; the right figure shows normal approach to landing (Kelly, 2004)

real-time flight planning. NASA Ames Research Center has engaged in a great deal of effort developing such systems, principally concentrating on the real-time representation of four-dimensional traffic information to aid Situation Awareness and decision making (e.g. Johnson, Battiste, Delzell, Holland, Belcher and Jordan, 1997; Johnson, Battiste and Holland, 1999). The display of all this additional information is only possible, though, with the development of the much larger LCD displays. Implementing all this information on earlier, smaller CRT displays would have led to unacceptable clutter. The more 'cluttered' a display is, the longer it takes to find information and there is a greater potential to mis-interpret it. The ability to 'switch out' information removes clutter and hence decreases search times but anything that is 'switched out' cannot be monitored.

Display of System Information

The display of system information is concerned with enabling pilots to monitor the performance of the aircraft's engines and to permit them to maintain an overview of other aspects of the aircraft's operation, such as the configuration of the high lift devices, the status of the hydraulic, DC and AC electrical systems, and the operation of the environmental conditioning system, etc. The requirements for displaying engine-related data and other systems information are quite different, which has resulted in quite different display formats being developed.

Engine instrumentation does not require continuous monitoring. Emphasis is placed upon periodic check readings of the various parameters to monitor health and performance. However, should the need arise the information on these displays should be capable of conveying precise, diagnostic information. Figure 12.5 shows the primary and secondary engine instrumentation from the Boeing 747-400 and Airbus A320. The instrumentation from the 747 uses a tape format (plus a digital counter) as a result of the need to display parameters for four engines in a limited space. This allows for swift check readings to be made and the digital counters provide a source of accurate information, if required. The A320 uses the slightly superior circular dial format for its engine instrumentation as a result of the need to transmit information from only two engines but while still having roughly the same amount of display area available.

Secondary systems in modern aircraft are managed automatically. However, when a failure occurs in one of these systems it is necessary to display the status and the configuration of that system. Oliver (1990) argues that with the advent of systems that automatically re-configure themselves in the case of a failure there is a need for even more detailed feedback to the pilots.

Most modern commercial aircraft utilise synoptic display formats for the provision of system information for everything other than the engines. The *Oxford English Dictionary* defines 'synoptic' as 'taking or affording a comprehensive mental view'. However, this type of display can only be used to transmit status (qualitative) information to the pilot. Rapidly changing values should not be incorporated into these displays. Typical examples where this format has been used include the configuration of fuel control systems (see Figure 12.6). The advantage of this type of display is that the information is presented in a manner compatible with the pilot's mental model of the aircraft (e.g. the left-hand wing fuel tank is on the left-hand side of the display). Not only is it possible to present fuel contents on such a display, it is also possible to present simultaneously other fuel-related information, such as valve status and pump pressures, thereby providing the pilot with all fuel-related information on a single page. This greatly aids in the identification and diagnosis of problems. However, if the system is complex, perhaps with many pumps, lines, controls and actuators (as in the hydraulic system) it may not be possible to display the schematic of such a system in limited display space. It may also lead to clutter, making the display difficult and slow to read. Selective use of colour can aid in this respect. The next generation of synoptic displays take things one step further. Using cursor-positioning devices the pilots can interact directly with the display in the same manner that a user of a desktop computer navigates around a Windows™ interface. It can be seen that when taking this approach the distinction between what constitutes a control and a display becomes blurred.

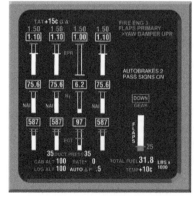 

Figure 12.5   **Primary engine instrumentation from the Boeing 747-400 (left) and secondary engine instrumentation from the Airbus A320 (right)**

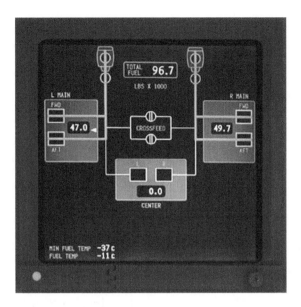

Figure 12.6   Boeing 777 EICAS (Engine Instrument and Crew Alerting System) fuel synoptic

Head-Up Displays

Head-Up Displays are very commonly used in military aviation, particularly in fast jets, but are being increasingly deployed in commercial aircraft. Many modern airliners can be specified with a HUD. The HUD is a device that interposes images on a transparent optical combiner in the pilot's line of sight. The images are focused at infinity, which means that the pilot does not have to re-focus their eyes when transferring their gaze from the symbology on the HUD to the outside visual scene (Figure 12.7). In commercial aircraft it is common for the HUD to be mounted in the overhead panel and deploy downwards when required, a little like a sun visor in a car. This introduces one of the key differences between civil and military applications of HUD technology. In a military aircraft the HUD is used continually for the display of primary flight (and weapons) instrumentation. In a commercial aircraft a HUD is normally only deployed when required, mainly during approach and landing, particularly in low visibility. The remainder of the time it is stowed away and the 'normal' head-down flight instruments are used. Proctor (1997) claims that aircraft with a HUD can eliminate more than 60 per cent of the disruptions in flight schedules as a result of low visibility at their destination airport. HUDs allow manual approaches to be undertaken in much lower visibility than is authorised without a HUD. HUDs certificated for Category IIIB minima enable landings with a decision height lower than 50 feet (15 m) and a runway visual range of less than 656 feet (200 metres) but not less than 246 feet (75 metres).

Figure 12.7   Lockheed Hercules C-130J showing Co-Pilot's Head-Up Display

The content of the information on the HUD is mainly composed of the primary flight information normally also found on the PFD, although in a slightly different format. For example, in contrast to the PFD a pitch ladder is used to indicate attitude and an extended lubber-line to mark the natural horizon. Additional flight path symbology such as the flight path vector can also be displayed (usually in the form of a large 'O' with flattened legs; the centre of the 'O' represents the point in the sky towards which the aircraft is heading). On some HUDs airspeed and altitude is displayed using a type-type format almost identical to that used on the PFD. In Figure 12.7 these are replaced with digital representations of these parameters surrounded by a pointer which provides an indication of rate of change information.

Various studies have shown there is no beneficial effect on performance when using a colour HUD compared to using a monochrome version (Dudfield, 1991). In general, it has been found that the optimum colour for all HUD symbology is a slightly fluorescent yellow-green. This is the part of the spectrum to which the eye is most sensitive and it is a colour that rarely appears in nature. Remember that the HUD symbology is frequently displayed against a background of the real world.

The principal argument in favour of the HUD is that it obviates the need for the pilot to switch their gaze between the instrument display and the operating environment. The main justifications for the use of HUDs relate to three factors (Weintraub and Ensing, 1992; Newman, 1995).

- They reduce the eye movements required to view the aircraft instrumentation and the outside world supporting better the allocation of visual attention. However, the use of small symbology (approximately 28 min of arc in height) on the typical HUD still makes eye movements necessary, albeit smaller ones than would be required to look at the main instrument panel. As the major time cost of an eye movement is in preparation for the movement and not the movement itself (with large movements taking only slightly longer than small movements) the advantage of head-up presentation is not as large as would initially be thought (Zambarbieri, Schmid, Magenes and Prablanc, 1982).

- HUDs reduce the need to re-accommodate the focus of the eyes when transitioning between the display and the outside world. However, there is still the need to switch attention between objects in the far domain (the outside world) and the near domain (the HUD). Although the symbology on the display is collimated (focused at infinity) the Human Information Processing system still regards the information on the HUD and in the outside world as belonging to two different domains (Wickens and Carswell, 1995). It takes time to switch the focus of attention between the two (Wickens, Ververs and Fadden, 2004). HUDs help to promote processing of aircraft information while scanning the external environment but this may not be true parallel cognitive processing, just rapid attentional switching.
- They enable the presentation of symbology that directly overlays (is conformal) with features in the outside world (e.g. the runway during approach and landing). However, the overlay of symbology onto the view of the outside world can obscure or laterally mask objects in the environment and may also disrupt effective scanning (Wickens and Long, 1995; Martin-Emerson and Wickens, 1997).

Stuart, McAnally and Meehan (2001) suggest that HUDs may promote attentional capture (attention to one source of visual attention at the expense of another) to the detriment of performance. The salience/compellingness of HUD symbology captures attention as a result of characteristics such as sharpness, contrast, and colour. Clutter, as a result of a large number of display elements on the HUD affects the speed and accuracy of visual search for relevant information. The expectancy of the user for events also determines the likelihood of attentional capture. For example, it was observed that when events in the environment were unexpected, for instance a runway incursion during a simulated approach, pilots were more likely to miss these actions compared to when they were expected (Fischer, Haines and Price, 1980; Wickens and Long, 1995). A further reason for attentional capture is that pilots may exclusively concentrate on HUD symbology when the display itself contains all the necessary information required (e.g. when using a conformal runway representation on the HUD during approach). Hence, the HUD induces the user to fly by the instruments without reference to the outside world as the information from the far domain is redundant. Finally, Stuart, McAnally and Meehan (2001) identified perceptual load as a factor in promoting attentional capture, particularly when the task for which the HUD was being used was demanding. Attention becomes focused on the primary task reducing the pilot's ability to process other stimuli. This effect is likely to have adverse consequences for their ability to monitor the outside world while performing a high workload task, for example landing in high turbulence.

Although these factors would seem to suggest that many aspects of HUDs are undesirable this is not the case. It is just that HUDs cannot be regarded as a universal panacea for flight path guidance-related display design. They have drawbacks as well as advantages. Fadden, Ververs and Wickens (1998) in a meta-analysis of studies comparing performance when using head-up and head-down display formats, identified three advantages for HUDs. The speed of traffic detection was superior when pilots fly using a HUD (e.g. Beringer and Ball, 2001; Martin-Emerson and Wickens, 1997) however detection rates of unexpected events was poorer. Detection of events on the display was superior when using a HUD, a benefit from its head-up location (Ververs and Wickens,

1998; Martin-Emerson and Wickens, 1997). Flight and ground path tracking were superior when using a HUD (Ververs and Wickens, 1998; Wickens and Long, 1995).

Enhanced and Synthetic Vision Systems

Recent developments in display technology have resulted in Enhanced and Synthetic Vision Systems (EVS/SVSs) which combine database and sensor technologies with display symbology to create enhanced Situational Awareness systems constructing an intuitive representation of such things as the projected flight path of the aircraft or of the required taxi route to the gate. With EVSs the outside scene is augmented with data from on-board databases. When using an SVS the pilot no longer views the outside world directly but views a representation of it derived via sensors collecting real-time data, further augmented by information from databases. EVSs and SVSs may be displayed using either HUDs or head-down displays.

Mulder (2003a; 2003b) developed a 'tunnel-in-the-sky' display to project the required flight path directly onto the sky using a HUD in front of the pilot. This representation was reasonably successful for depicting straight tracks but less so for curving trajectories. Other display systems can now overlay representations of the projected route of the aircraft relative to the terrain with attitude, airspeed and altitude information on a modified, head-down PFD display, also integrating elements of the navigation solution (see Figure 12.8).

Other SVS applications include overlaying the outside visual scene with taxiing aids derived from an onboard database and the aircraft's GPS position (e.g. directions, locations or cautions and warnings). This was designed to aid pilots finding their way around complex airport runway and taxiway systems, particularly in restricted visibility

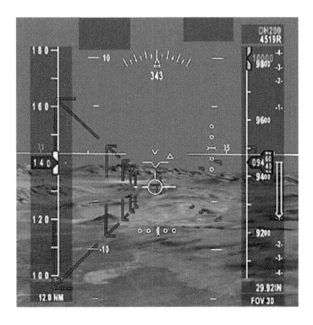

Figure 12.8   NASA Synthetic Vision Display

(Molineros, Behringer and Tam, 2004). Other simpler systems using novel applications of existing sensors on board the aircraft and the available data broadcast by other nearby aircraft (e.g. the Mode S transponder) have been developed to update moving map displays on the flight deck giving advanced warning of conflicting aircraft on the runway both during take-off and the last stages of approach and landing (Szasz, Gauci, Zammit-Mangion, Zammit, Sammut and Harris, 2008).

All of these systems that augment reality with data from sensors and on-board databases aim to enhance the pilot's Situation Awareness by combining these data to provide relevant information in an integrated format that enables the crew to look ahead (i.e. achieve Endsley's Level 3 SA). Information (not data) is presented to the pilot in a naturalistic manner; the amount of interpretation and re-coding of data from these displays (*q.v.* early clockwork cockpits) is vastly reduced.

AUDITORY DISPLAYS

The auditory channel has various advantages and disadvantages over the visual channel when it comes to putting across information to the pilot. Its principal advantage is that it is not dependent on their direction of gaze. Thus, the auditory channel is particularly suitable for the transmission of alerts and warnings. Auditory displays also require no panel area. However, they also have severe drawbacks. The data/information is transient and can be lost against a background of noise. In general, they are best used as 'attention grabbers' to direct the pilot's awareness to a particular situation or visual display which contains the relevant non-transient data. The auditory channel is best used when:

- the message is simple and short;
- when the message will not need to be referred to later;
- when the message deals with events in time (not space);
- when the message calls for immediate action; and
- when the visual system may be overburdened.

Whenever the auditory channel is used to convey data, it is implicit that the phonological loop component of the pilot's WM will be involved, with all its inherent limitations (see Chapter 2 on Human Information Processing in Part One).

All warning systems (be they auditory or visual) have the following functions (Noyes, Starr and Kazem, 2004):

- *Monitoring* – Assessing the situation with regard to deviations from pre-determined fixed limits or with respect to a threshold value.
- *Alerting* – Drawing the pilot's attention to a hazardous or potentially hazardous situation.
- *Informing* – Providing information about the nature and criticality of a problem in order to facilitate a response.
- *Advising* – Aiming to support human decision-making activities in addressing an abnormal situation.

For an auditory display system to be effective the information that it transmits must first be heard by the pilots. There are two basic problems in the detection of an auditory signal:

frequency masking and amplitude masking. Frequency (pitch) masking occurs when the target signal is masked by other sounds in the environment of approximately the same frequency. Amplitude masking occurs when the target signal and the noise environment have different frequency spectra, however the amplitude of the noise components in the environment overpower the signal (Patterson and Milroy, 1979). This requires that a noise spectrum analysis is required of the operating environment across all frequencies in the audible range. This provides data concerning the most appropriate amplitudes and frequencies to use to avoid both types of sound masking. Simply increasing the power of a warning signal is not necessarily an appropriate option as it can have exactly the opposite effect to that desired in a warning system. This can result in evoking the startle reaction in the pilot which can cause a dramatic increase in arousal levels (the fight or flight response). This has the effect of narrowing their attention and simultaneously decreasing their cognitive capacity, which is exactly the wrong thing to do when there is a problem. However, it is not just the level of the sound that is responsible for the startle reaction: it is also the manner in which the warning signal sets off. In some flight decks, the onset of the sound is particularly sudden, going from 0–100 dB in under 10 ms. Simply by increasing the onset period of the sound to about 20–30 ms, the startle reaction can be avoided (Patterson, 1990). Even if warning sounds do not stimulate the startle reaction, loud, continuous alerts can divert attention away from processing task-related information thereby increasing the potential for error (Peryer, Noyes, Pleydell-Pearce and Lieven, 2005). A pilot's first reaction is to deal with the warning system and not the problem that it is signalling.

The correct level for a warning tone should be in a band 15–20 dB above the auditory threshold. The auditory threshold is the sound pressure level (amplitude) at which a signal can be reliably discriminated against the background noise, which will vary with frequency. In certain parts of the background noise spectrum where the sound pressure level is already high it would be unwise to incorporate a warning sound at that frequency as it would need to be excessively, perhaps dangerously, loud. Under no circumstances should the noise levels ever exceed 120 dB. Furthermore, any auditory warning signal should not be a pure tone but should be composed of at least four different tones at different frequencies. In this way even if one or two of the frequencies of the warning sound are masked by components in the background noise there should be enough parts of the signal above the auditory threshold level for the warning to be heard.

Auditory warnings come in two basic types: non-speech warnings (comprised of sounds alone) and speech warnings (usually based on a combination of warning tones coupled with a short, informative, message).

Assuming that the crew have been able to detect an auditory warning the pilots also needs to discriminate it from the other warning sounds (and thence recall its meaning and act upon it correctly). To an extent, the discrimination of the warning sounds and their correct identification are two aspects of the same problem, however acting appropriately once the warning has been identified is the problem of the flight training department! It is often easy to discriminate between several warning sounds when they are played one after another (comparative discrimination) but when played in isolation the task becomes more difficult. The task becomes even harder when it is considered that (hopefully) pilots will not encounter these sounds very often. Edworthy and Patterson (1985) suggest four characteristics that can aid in the discrimination of non-speech auditory warnings. It is possible to alter their:

- Pitch
- Harmonic qualities (if mixing several different frequencies to produce a single chord-like sound)
- Timbre (wave shape/quality)
- Rhythm.

These options may be applied either singly or in combination.

A sound alone also has no implicit meaning associated with it; for example, it must be learnt that a bell in the cockpit means 'fire'; a horn means 'gear not lowered', etc. Even if the warning sounds can be reliably discriminated from one another, their meaning and subsequent actions are still dependent upon the LTM of the pilot.

One way that has been investigated to enhance the meaning of non-speech warnings is by developing auditory icons (Gaver, 1989). Ulfvengren (2003) produced a prototype suite of auditory icons for use as warnings in a helicopter. Auditory icons consist of acoustically complex, environmental sounds which represent the event they are trying to warn about. It was found that the use of auditory icons related to the alert functioned better than other warning sounds. Perry, Stevens and Howell (2006) found similar results for a further set of auditory icons developed as an experimental set for various functions in a fixed wing aircraft. The main drawback of this approach, though, is that these icons are very likely to get lost in the background noise as they are specifically designed to mimic environmental or system-related events.

A problem with non-speech warning systems is that there is no prioritisation in the signal. All warning sounds, even though they may represent system deviations of differing levels of consequence, effectively have the same level of priority. However, by incorporating rhythmic pulses into the warning sound, it has been observed that different patterns of pulses suggest different levels of urgency. Faster pulses denote increased urgency. It was also observed that acoustic warning tones each with a unique rhythmic pattern were more memorable than tones simply using timbre and pitch, and hence, were less open to confusion (Edworthy and Patterson, 1985).

Speech warnings (or more correctly integrated tone/speech warnings) have the advantage that they have meaning attached to them. However, they also have great potential for getting lost in the background noise in the cockpit (a bad case of both frequency and amplitude masking) unless steps are taken to obviate this problem. The best way of overcoming this is to introduce the message with a tone (or tones). These can be also used to convey the priority of the message. For example, in an Airbus top priority warnings are introduced by a single chime; cautions are introduced by two chimes and advisories by three. Furthermore, with prioritisation of warnings, the most pressing warning is presented first, thus freeing the pilots of the secondary task of working out which system failure is the most important. Warnings can also indicate the fundamental cause and not also signal the subsequent cascade of failures related to it. Such a warning system should also be slaved to the relevant display computers to make the relevant information readily available.

However, if a speech warning is to be introduced by a tone it is essential to leave a gap of around 0.5 seconds between it and the speech message, otherwise a phenomenon known as 'forward masking' will occur, obliterating the first part of the speech message. The Human Information Processing system needs time to re-focus the object of its attention and to allow new information through the Sensory Register (the Echo). Speech messages should also be kept as short as possible (another feature of the limited capacity of WM).

The function of the auditory warning system is only to gain the attention of the pilot and orient them to the relevant part of the flight deck where less transient information can be found. Messages that are too long and/or too complex will over burden the limited capacity of WM. This also means that this type of warning will be slower to respond to so is not a good choice in a highly time critical situation.

HAPTIC DISPLAYS

There has been increasing interest in the last decade about the use of haptic display devices (i.e. those that use the mechanoreception senses). Research from cognitive psychology suggests that the haptic channel can provide a further valuable way of providing information to the pilot in addition to the auditory and visual channels. The haptic channel has several potential benefits over the other senses:

- It is omnidirectional and is not blocked by either high levels of background noise or low light levels.
- It can be used to convey spatial information. Haptic localisation is reliable (Tan, Barbagli, Salisbury, Ho and Spence, 2006) and is superior to auditory localisation.
- The haptic sense cannot be 'switched off' making it useful for providing warnings (nevertheless it can be blocked, e.g. if the user wears thick gloves). Warning information delivered through the haptic sense may also be less distracting and not prone to evoking the startle reaction.
- The haptic sense has a longer STM effect than the other major senses (ten times longer than the visual sense and at least three times longer than the auditory sense).
- van Veen and van Erp (2001) observed that tactile feedback was not substantially impaired during high g-load conditions (up to 6 g).

The haptic sense has been used for some time in the design of aircraft warning systems in the form of the stick shaker, used to warn pilots of the impending stall. Similar systems are now being developed for road cars, sending vibrations through the seat, to warn drivers of such things as lane departure or potentially conflicting traffic in their blind spot (Ho, Tan and Spence, 2005; Ho, Reed and Spence, 2006). Pilots also receive haptic feedback from control systems. For example, the feel system in the Boeing 777 increases control weights to the yoke in the event of an excessive bank angle. The inter-linkage of controls can also provide valuable transmission of information between the pilots on the flight deck and provide feedback from the automated systems in the aircraft (see Field and Harris, 1998 in the next chapter). In a study by Sklar and Sarter (1999) it was observed that mode transitions signalled by a combination of visual and tactile alerts produced superior performance in terms of both speed and accuracy of detection, compared to tactile or visual only warnings.

A more complex haptic display device not intended as a warning system has been developed for military applications, where the visual workload is high and there is additionally a strong emphasis on keeping the pilot 'head up and eyeballs out'. The Tactile Situational Awareness System (TSAS) has been developed by the US Naval Aerospace Medical Research Laboratory to convey both navigational information and target discrimination and tracking cues (Cholewiak and McGrath, 2006). It is made up of a vest worn beneath the flying suit, lined with vertical rows of 'tactors' (vibrating pads).

These tactors vibrate to indicate directional drift, target type and target direction. As well as relieving the visual sense, such haptic devices also provide further redundancy, for example, in preventing spatial disorientation. It has been suggested that such systems may be used to aid pilots in complying with unseen airspace restrictions (restricted areas, minimum safe altitudes, etc.) and also to supplement guidance when using SVS systems, such as 'tunnel-in-the-sky' displays. Such factors may be signalled by increasing the intensity or frequency of stimulation as the airspace restrictions approach (van Veen and van Erp, 2001).

Haptic research on the civil flight deck still has a long way to go. It is only in its relative infancy and demonstrable benefits need to be established in the commercial aviation context where the concurrent demands on the pilot are not as high as in the military domain.

FURTHER READING

There are several excellent books on flight deck design available, including display design considerations. For a slightly more engineering perspective with a military emphasis, try:

Jarrett, D.N. (2005). *Cockpit Engineering*. Aldershot: Ashgate.

From a cognitive ergonomics perspective, and dedicated solely to the design of commercial aircraft flight decks there is the absolutely excellent:

Harris, D. (2004b). *Human Factors for Civil Flight Deck Design*. Aldershot: Ashgate.

For a broader look at the design of warning systems, including auditory warnings, a very comprehensive overview can be found in:

Stanton, N.A. and Edworthy, J. (1999). *Human Factors in Auditory Warnings*. Aldershot: Ashgate.

Finally, although not directly related to aviation, the following book is well worth looking at if you have an interest in multi-model interfaces, including haptic aspects:

Ho, C. and Spence, C. (2008). *The Multisensory Driver*. Aldershot: Ashgate.

13
Aircraft Control

In many ways, the manner in which a commercial aircraft is controlled has changed vastly in the first century of aviation. The automation in the *Machine* plays the central role in determining the three-dimensional (even four-dimensional) flight path of the aircraft. Nevertheless, this has not changed the basic skill set required by the hu*Man* in charge of the *Machine*: it has, however, added to it considerably. This chapter looks at the basic problem of aircraft control from the perspective of the pilot but it is impossible to do this without some consideration of various other aspects of the design of the aircraft and its flight deck, specifically issues in automation and interaction with the aircraft computer systems. As a result, there is a lot of overlap between the content of this chapter and the following two chapters. Furthermore, the problem of aircraft control also cannot be separated from training. I think that I mentioned somewhere that it was almost impossible to separate many aspects of human factors in aviation as the interrelationships were manifold and complex!

Good control design translates the pilot's intent into action with a minimum of effort. The design of good flight deck control systems minimises training requirements and the number of opportunities for design-induced error while reducing crew workload. However, in addition to the primary flight controls, the interfaces to the automated flight control systems are equally as important (perhaps more so) in the way that pilots now exert control over the flight path of the aircraft.

At a very basic level, the pilot(s) now exert control over their aircraft in one of three basic control modes:

- *Manual control* – where the pilot actually 'hand flies' the aircraft using the thrust levers, control column/sidestick and rudder pedals (the primary flight control inceptors).
- *Tactical control* – where in-flight modifications are made to the strategic (longer term) plans in the FMS using the autopilot systems (using the pilot interface on the Mode Control Panel – MCP) to deal with real-time events during the flight.
- *Strategic control* – in which detailed level flight plans are entered into the aircraft's FMS using the CDU (Control and Display Unit). These commands are then implemented by the other automated systems in the aircraft (e.g. the autopilot).

Some of the various components of the flight deck used to control the flight path of the aircraft are depicted in Figure 13.1.

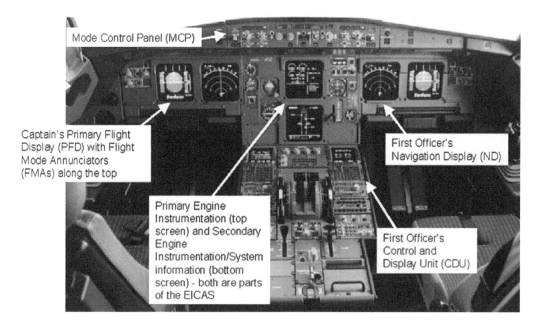

Figure 13.1   Primary flight deck interfaces to the FMS on the Airbus A320

Having said all this, the actual control of the aircraft is rarely as simple as described above, with certain aspects of the flight being controlled by the FMS while other aspects are under the control of the autopilot systems using the MCP. Furthermore, pilots only occasionally fly commercial aircraft manually. For many years now, the pilot has been a supervisory controller of systems and a flight deck manager, rather than a 'hands on throttle and stick' flyer. They are a setter of high-level goals rather than an inner loop controller. This is not something that has happened with the advent of the highly automated 'glass cockpit' aircraft than began to be produced in the 1980s. These aircraft just exacerbated a trend that had been developing since the first autopilot systems were developed in the 1930s.

Göteman (1999) describes a typical company ethos for the utilisation of the various components by which control may be exercised over the aircraft's flight path. The CDU is used for long-term, strategic objectives, such as the input of flight plans for the management of the overall flight. Data input and manipulation using the CDU is undertaken in non-time critical circumstances, for example prior to flight or during the cruise phases. When the aircraft is under the control of the flight plan stored in the FMS the central role of the flight crew is to monitor the progress of the flight against objectives. The MCP is used to execute, minute to minute, tactical inputs to the FMS where short-term deviations to the stored flight plan are required. The MCP has less functionality than the CDU (being restricted to immediate VNAV and LNAV functions, and the control of speed and vertical speed) but is much less complex and faster to use. The main control inceptors are only used where manual control of the aircraft is mandatory (e.g. on take-off) or where immediate decisive control of the aircraft is required, for example during emergency avoidance manoeuvres. Billings (1997) has argued that pilots are certainly becoming more peripheral to the direct aircraft control task as automation increases, as shown in the Figure 13.2. Pilots now interact with devices several steps removed from the direct control of their aircraft.

Applegate and Graeber (2001) described the increasing levels of integration and interdependency of aircraft systems in Boeing twin-engined, jet transport aircraft (see Figure 13.2). Early airliners such as the Boeing 707 and 727 had relatively independent systems managed by a Flight Engineer. The initial Boeing 737s (737-100 and -200) had simplified systems with greater levels of automation for the management of systems to allow two crew operations. Nevertheless, the aircraft still utilised analogue technology. Later Boeing 757/767 models were the first to use digital technology, however, their basic architectures were simply digitised versions of earlier analogue systems with little integration. The Boeing 777 employed new system architectures with greater use of digital technology. The Boeing 777 possesses highly integrated systems with inputs providing data for a variety of aircraft functions, both for control and display functions.

More generic issues surrounding automation and HCI on the flight deck are considered in the next two chapters. This chapter concentrates on the specific issue of flight path control. It doesn't matter if the aircraft is being flown under manual control or using its autoflight systems, some aspects of the control problem remain exactly the same.

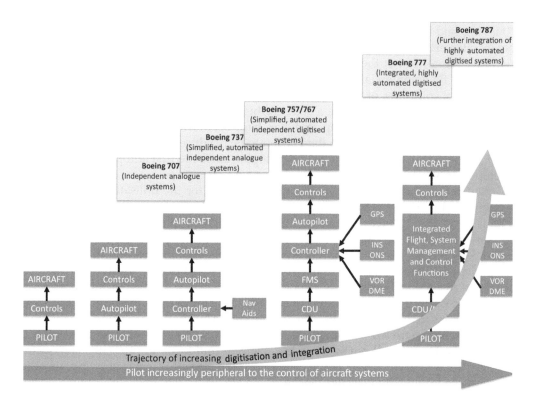

Figure 13.2   An evolutionary perspective on the pilot's relationship with their
aircraft (adapted from Billings, 1997) overlaid with Applegate
and Graeber's (2001) conceptualisation of increasing digitisation
and integration in Boeing Aircraft

THE AIRCRAFT CONTROL PROBLEM

The control of an aircraft is best described as a hierarchical control problem. The basic
nature of a hierarchical control problem is that the parameter which needs to be controlled,
in the case of an aircraft its flight path, can only be controlled indirectly via other lower
order parameters (e.g. pitch rate or roll rate/heading – see Figure 13.3). There is no method
of directly controlling altitude. This is actually accomplished via pitch rate control which
changes pitch attitude, which causes the aircraft to climb. In terms of changing the aircraft's
heading this is done by rolling the aircraft to the appropriate roll attitude, which changes
its rate of change of heading, which changes the heading.

Aircraft primary axis flight controls are cross-coupled. A change in one parameter will
affect other parameters. The most obvious example of this is the cross-coupling between
speed and pitch. In a conventional-technology aircraft it is impossible to change the pitch
attitude of an aircraft without affecting either its vertical position or airspeed. As speed
increases, the aircraft will climb as lift over the wings increases, unless the trimmed angle
of attack is reduced. Similarly, if the aircraft is pitched up without an increase in thrust,
airspeed will decay. In terms of lateral control, rolling the aircraft will also result in a

Altitude control via elevator input

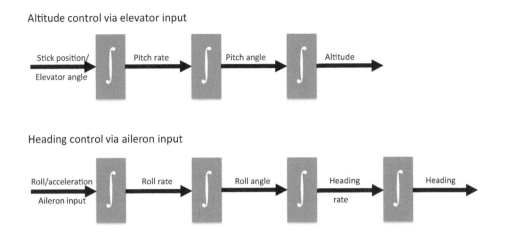

Heading control via aileron input

Figure 13.3   The hierarchical nature of the aircraft control problem (adapted from Baron, 1988)

concomitant downward pitching moment as some lift is lost across the relatively slower-moving wing on the inside of the turn. Yaw inputs also exhibit cross-coupling with both the roll and pitching axes of motion. These aspects of system dynamics all affect operator control strategies and psychomotor skill acquisition.

With further consideration it is also slightly ambiguous which parameter an input to the control column in a conventional technology aircraft actually controls. Consider the time history response to a pitch input, as described in Figure 13.4.

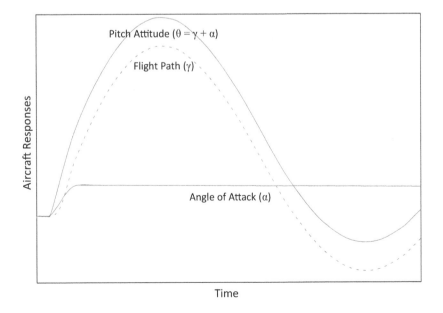

Figure 13.4   Time history responses for a conventional aircraft (Field, 2004)

Figure 13.4 shows the angle-of-attack, pitch attitude and flight path time history responses of a conventional aircraft (i.e. one without a FBW flight control system) to a step control inceptor input (a small pitching input held indefinitely). After the initial dynamics settle, the angle-of-attack response remains approximately constant at the new trimmed angle-of-attack. To maintain this new angle-of-attack the pilot must either hold the constant stick position or trim out the forces. This direct link between stick position and angle-of-attack results in this type of response being labelled an angle-of-attack response-type. However, this takes a longer-term view of the problem and describes it very much from an engineering perspective. What the pilot actually experiences is a pitch rate response: larger stick displacements result in a greater pitch rate. This is the short-term response to a pitch input (seen in Figure 13.4). All the other responses described in this figure exhibit a further longer-term, non-steady-state response, called the phugoid mode. The phugoid mode results in deviations in pitch attitude, flight path and airspeed, emphasising the cross-coupled nature of aircraft control. However, as long as the phugoid mode is well damped these responses all decay to new steady-state values in the longer term but it is unlikely that this will ever occur before the pilot makes further inputs (see Field, 2004).

Most new designs of commercial aircraft are not conventional technology aircraft where there is a direct mechanical link between the flight deck's control inceptors and the control surface effectors (elevator, aileron or rudder). A number of different flight control laws are possible FBW aircraft which further complicate the relationship between the control inceptors and the flight path response. For example, in a pitch rate response-type a constant control inceptor pitch input produces a constant pitch rate in the aircraft: if the input is held, the aircraft will continue pitching indefinitely providing sufficient thrust is available. When the pitch input is released, the aircraft will maintain the pitch attitude at release. However, the flight path response now does not mirror that of pitch attitude as it does in a non-FBW aircraft (as described in Figure 13.4). As will be seen, this has considerable implications for the manual control of the aircraft's flight path. In contrast, with a pitch attitude response flight control system a constant pitching input to the controls results in a new pitch attitude. If the input is held, the pitch attitude will be held at the new value indefinitely provided sufficient thrust is available. When the control inceptor is released, the aircraft will return to its original pitch attitude. Unlike the pitch rate response control law, in this case the initial response in flight path is similar to that of the pitch attitude of the aircraft meaning that pitch attitude can be used by the pilot as a surrogate of flight path. For a fuller description of the various flight control laws and the implications for both the aircraft and the pilot, see Field (2004).

MANUAL CONTROL

The Series Pilot Model

While these descriptions begin to describe the requirements of the control problem they say relatively little about the control strategy employed by the pilot. To this end one of the most common control models is the series pilot model, which describes the nature of the pilot/aircraft system as a whole (McRuer, 1982). An adapted form of this model from Field (2004) is outlined in Figure 13.5. The control of an aircraft can be characterised as a compensatory tracking task, where the objective is to minimise any errors from the desired three-dimensional flight path. However, it has been illustrated that the one thing that the pilot cannot directly control is flight path. This is done through a series of surrogates, for example in the vertical axis the pilot controls pitch rate in the short term and angle of attack in the longer term. But, as has already been alluded to briefly, there is also one further problem. The pilot cannot actually observe either flight path or angle of attack directly; she/he can only observe the aircraft's pitch attitude. In the series pilot model, the flight control problem is broken down into a short-term and a long-term problem. In the vertical axis, the short term problem is one of the control of pitch attitude. Pitch attitude is used here as the control parameter because the pilot cannot directly observe angle of attack, which is actually the parameter that has the effect on flight path and is also the parameter that the pilot has direct control over. As a 'rule of thumb', you cannot control a parameter that you cannot observe. The longer-term control problem is one of flight path or altitude control.

The short-term (pitch attitude) control problem can be considered to be 'nested' within the longer-term (flight path angle or ultimately altitude) control problem. What the pilot is doing is attempting to 'close' the inner control loop (i.e. the attitude control problem) in an attempt to control the 'real' problem, which is that of altitude/flight path angle control. In the horizontal plane, the pilot is dealing with a similar control problem, only in this case aircraft roll attitude is being used as a first approximation (inner loop parameter) in an attempt to control the aircraft's heading, and subsequently its track across the face of the Earth. As in the vertical control problem, the hapless pilot again has no direct control over the aircraft's heading; she/he only has control over roll attitude.

To a certain degree, pilots of larger commercial aircraft actually have slightly more direct control over the flight path problem via their autopilot or FMS. They can command an altitude (or flight level) and/or a desired track, and the aircraft then selects the correct thrust, pitch attitude and heading to attain the desired flight path parameters. However, the way in which the aircraft tackles the control problem is similar to the way the pilot does it, and the same laws of physics apply!

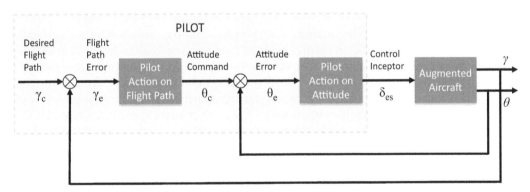

Figure 13.5   The series pilot model illustrating a response in pitch. γ is the flight path angle; θ represents pitch attitude; γc is the commanded flight path angle; γe is the error between commanded and actual flight path; θc is the commanded pitch attitude; θe is pitch attitude error; δes represents a pitching input to the stick made by the pilot (Field, 2004)

Other Factors Affecting Control

There are (at least) four other factors that are products of the machine that that serve to complicate the aircraft control problem and impair human performance.

System lags considerably impair performance. These come in two basic types:

- *Response (or transport) lag* which is the time between a control input being made and the system responding; and
- *Display lag* which is the time between the aircraft responding and feedback to the pilot being provided by the display systems.

Large aircraft at low airspeed respond slowly to control inputs. You can begin to appreciate the problems when you consider the quite trivial problem of controlling the temperature in a shower (the response lag in this case is a product of the distance between the shower head and the mixer valve regulating the ratio of hot to cold water). Accurate control requires anticipation and human operators are poor at anticipating the future state of a system based on its present condition. They are unable to perform the many and complex calculations required to predict its future state and hence make the required input(s). Predictor displays (e.g. those found on a HUD) can help in this respect.

Display lags increase the time required before a further control input can be made and furthermore, may not reflect the current state of the system. Most trainee pilots during their first few flights will at some stage have been told off for trying to fly straight-and-level by attempting to zero the rate of climb indicator. There is a considerable lag in this display and the one thing you can guarantee is that if you try to level the aircraft using this method the only state the aircraft will not be in is level!

Not surprisingly, minimisation of both these lags improves performance of the pilot and controllability of the aircraft.

Control gain is also referred to as sensitivity or control:response ratio. In a high gain flight control, a small input by the pilot will result in a large system output. Different types of aircraft require different control gains. This will become more evident in a following section on the psychomotor skill of flying but, to pre-empt this slightly, as the novice pilot flies in a compensatory tracking manner with emphasis on the cancellation of flight path errors, they require a lower gain control system. Furthermore, the novice pilot also does not undertake gross flight manoeuvres, concentrating just on the basics. In contrast, the military fast-jet pilot will have amassed many more flight hours and therefore will employ a more open-loop (or 'pre-cognitive') control strategy, partly as a result of experience and partly as a result of the necessities imposed by the gross manoeuvres necessitated by the air-combat environment. As a result, she/he will require flight control optimised for large, open loop control movements (i.e. a higher gain control).

However, for both types of pilot, when it comes to the approach and landing task where precise control is required, a lower gain flight control is required. This is not a problem in a flight training aircraft but it does provide a challenge for highly agile combat aircraft. This conundrum is usually overcome in one of two ways, by either having a non-linear gain on the stick (gain increases as displacement increases) or by switching to a different, lower gain control law when flaps and/or gear are selected for landing.

Bandwidth refers to the frequency with which the pilot can make meaningful control inputs. This is usually limited by the design of the hu*Man* limb, the central, information-processing rate and the design of the *Machine's* control interface. In the case of a commercial flight deck with a large control column, the typical bandwidth achievable is unlikely to be in excess of 1.0–1.5 Hz (Baron, 1988). However, with sidestick controllers, such as those found in Airbus aircraft, this can be higher.

Stability mis-matches between system lags (particularly response lag), flight control gain and operator bandwidth can cause stability problems (Wickens, Lee, Liu and Gordon Becker, 2004). These are in the form of Pilot Induced Oscillations (PIOs) which occur when the pilot unintentionally commands an increasing series of control corrections in opposite directions (usually in the pitch axis) each one in an attempt to counteract the aircraft's response to the previous input. This results in an overcorrection in the opposite direction. These result from a mis-match between the frequency of the pilot's inputs and the response frequency of the aircraft. PIOs are a product of a large response lag in the system associated with a high gain control, and the control bandwidth of the pilot being either too high or of the same natural frequency as that of the response of the aircraft but being 180° out of phase (see McRuer, 1980; Wickens and Hollands, 2000). Such instability can usually be avoided by reducing the control gain of the aircraft and/or through pilot training, encouraging them not to 'over control' the aircraft.

CONTROL INCEPTORS

The most common control inceptors are the wheel column (or yoke), centrestick and sidestick. An aircraft with conventional responses to a control input always requires trimming to a new angle-of-attack as the airspeed or aircraft configuration is altered. Initially, any out of trim force will be held by the pilot during the manoeuvring portion of the task. The trim system will then be used to reduce control forces in the longer term. Conventional response-type aircraft therefore require control inceptors that allow reasonable displacement to permit accurate control around the trimmed control position, away from the trimmed position and to allow re-datumming of the trim position. The inceptor also needs to be large enough to accommodate the trim control itself. In comparison, rate command response-types do not require trimming. Unlike in conventional technology aircraft the pilot uses pulse type inputs to change the attitude of the aircraft and then the inceptor returns to its neutral position (usually via sprung assistance like a games joystick). Hence, inceptor displacements for rate command response-types which do not require trimming need not be as great as for conventional response-type aircraft. However, it is still necessary to provide sufficient displacement to allow accurate control for both small amplitude (precise) and large amplitude flight path control tasks.

The wheel column inceptor arrangement is suitable for both conventional and non-conventional (FBW) response-types. It can be designed to allow sufficient displacement to permit precise control around the datummed trim position while also accommodating larger control displacements. It allows re-datumming of the trim position for different flight conditions and aircraft configurations. The main disadvantage of the wheel column arrangement is its size. It takes up appreciable space in the flight deck and can obscure the displays. It is also the heaviest of the inceptors.

The centrestick shares many of the features of the wheel column arrangement but is used in very few large aircraft. In recent years the only large aircraft to have used this type of control inceptor are the McDonnell Douglas (Boeing) C-17 and (slightly earlier) the Avro Vulcan! It is well suited to both conventional and FBW control laws and has the advantages over the wheel column of being smaller; not obscuring the displays to the same extent; nor taking up as much space and it is also lighter.

Many sidesticks employ only a limited displacement (none at all in the case of force sticks) and are operated through wrist action only. These are best suited to FBW rate response control laws which only require occasional inputs from the pilot and no trimming (there is little space available for stick-mounted controls). The sidestick does have the advantage, though, in that as a result of its smaller size and outboard mounting position it does not obscure the PFD. However, space must be allowed aft of the sidestick control for a pilot's wrist support which means that the outboard areas of the flight deck that are used in some aircraft for ancillary systems or even storage are now unavailable.

For a full description of matching control inceptor types to the various FBW control laws, see Field (2004). Photographs of the three main control inceptor types can be found in Figure 13.6.

There are also anthropometric considerations for the design of the major controls on the flight deck. The reach and accessibility of controls is determined from the ERP and should include the major flight controls, landing gear, thrust levers (including reversers), rudder pedals and the tiller for the nose wheel steering. The seat and rudder pedal adjustments must allow appropriate accessibility for pilots ranging in stature from 153 cm (5'2") to 183.5 cm (6'3") (see FAR/CS 25.777). However, as Lovesey (2004) notes, humans are slowly but surely getting bigger. Turner and Birch (1996) reported that UK male aircrew have increased steadily in most anthropometric dimensions (e.g. stature, thigh, waist and chest circumference) since World War II. Important dimensions for the design and layout of flight deck controls, such as functional reach, have increased by 0.5mm annually (equivalent to 0.06 per cent). The critical anthropometric dimension is when the controls are at their maximum deflection. They should also be sufficiently adjustable to accommodate ingress and egress without interference with the flight controls.

Figure 13.6   Three basic types of primary flight control inceptor. Left to right: Airbus A320 sidestick; McDonnell Douglas C-17 centrestick; and Boeing 777 wheel column (yoke)

However, the primary flight control inceptors should not be thought of as just controls: they are also sources of information for the pilots. In modern, FBW commercial aircraft the requirement for the primary axis flight controls and thrust levers to be physically linked to the aerodynamic control surfaces or engine fuel control units has been removed. Control deflections instigated by the crew are now electronically signalled via transducers, interpreted by a suite of flight computers and implemented by actuators at the control surfaces or in the engine. Moreover FBW systems have also removed the requirement for the controls on the flight deck to have a physical interconnection, as in the Airbus family of aircraft where the sidesticks are completely independent. Thus, when the flight controls are moved on one side of the cockpit, their deflections are not now necessarily mirrored in the controls of the other pilot. In addition, the flight controls and thrust levers do not mimic any autopilot inputs.

The removal of these linkages takes away one of the lines of communication between the two pilots and also between the aircraft's automated systems and the pilots. The position of the stick conveys information between the pilots concerning the status of the aircraft and the flying pilot's intentions without the need for either verbal or visual information transfer. Pilots have also suggested that when in a conventional aircraft that is being flown under the command of the FMS, the first cues for a descent or a turn being initiated are often from the back-driven movement of the control columns or thrust levers. Thus, in some implementations of FBW technology where there is no interconnection of flight controls across the cockpit or there is no backdrive of the flight controls from the autoflight systems, the pilot is now heavily dependent upon information gained via their central visual system (i.e. the pilots have to consciously search for information). This can be compared to the multiplicity of information channels that pilots have available to them in conventional aircraft (see Figure 13.7). Crossfeed between the pilots' control inceptors has been found to provide useful anticipatory information to the pilots (Field and Harris, 1998). Feedback from the autoflight system to the control inceptors was found to be slightly less important than the interlinkage between the pilot's controls, however backfeed from the autothrottle to the thrust levers was considered to be very important.

The implications for the use of non-backdriven throttles are exemplified in the following extract of the accident report concerning a China Southern Airlines Boeing 737-300:

> The aircraft banked dramatically right and hit a mountainside. FDR data indicate that, after the aircraft levelled at 7,200 ft during a descent towards the airport, with autopilot and autothrottle engaged, the left throttle lever advanced while the right did not.

> The initial report says that there was no indication as to where in the throttle system the problem was, but the FDR indicates that occurred earlier in the same flight and had been corrected manually. (Learmount, 1994)

Unlinked, non-backfed primary flight controls not only overburdens the pilot's central visual system and removes a line of communication between the crew, it also makes the monitoring of the automatic systems much more difficult. The automation becomes more 'opaque'.

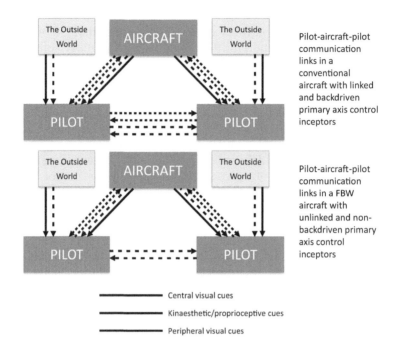

Figure 13.7   Matrix of control cues between the pilots and the aircraft in a conventional technology aircraft and a FBW aircraft without either interlinked or backfed controls (Field and Harris, 1998)

THE PSYCHOMOTOR SKILL OF FLYING

Learning how to control the flight path of an aircraft using the stick, rudder and throttle is an example of psychomotor skill acquisition. It has been suggested that one way of characterising flying is as a compensatory, closed-loop control task in which the pilot is continually aiming to minimise the error component by attending to feedback about their performance and acting accordingly. This can be contrasted with an open-loop task, in which the operator is not (and cannot) be dependent upon continuous feedback about their performance. A typical example of this is striking a golf ball. No matter how hard a golfer tries to influence the flight of the ball after it has been struck, they will usually find that their efforts are to no avail (I have tried…). Furthermore, a golfer selects a certain type of swing to play a shot. This may be a slight variation on their normal swing but the essential characteristic is that it is pre-selected prior to executing the shot. Once the swing is commenced it becomes 'fully automatic' and the motor program selected is executed.

Various stages have been proposed for the acquisition of a skill. Fitts and Posner (1967) suggested that trainees progressed through three stages when acquiring a skill: the cognitive, associative and autonomous phases. In the cognitive phase trainees are required to attend consciously to the cues which guide them concerning when the task should be performed, the individual components of the task and their inter-relationships, and the feedback about their performance. These cues may be from components of the task or task environment, or they may be in the form of guidance from their instructor or training system. In the associative phase, the trainee's actions begin to be guided directly by environmental cues but still some cognitive attention is required. In the final phase, the autonomous phase, task performance becomes automatic and no longer requires conscious attention. Anderson (1982) suggested that there were two stages to skill development: declarative and procedural. These roughly correspond to Fitts and Posner's cognitive and autonomous phases. Jaeger, Agarwal and Gottlieb (1980) proposed a three-stage developmental model of human-skilled performance. They suggested that the student initially learns the directional relationships, followed by their timings and finally by their spatial relationships. Wightman and Lintern (1985) suggested that in the development of three-dimensional tracking skills, such as those required by a pilot, a fourth factor, that of control co-ordination, was the final aspect to develop.

Up to this point it has been convenient, especially for the discussion of aircraft dynamics, to regard flying as a closed-loop, error-cancelling, compensatory tracking task. While this is what the task requires and it is also the manner in which an autopilot system tackles the problem (i.e. minimising the flight path error until it tends towards zero) it is not necessarily the way in which a skilled human pilot undertakes the manual flight task. McRuer and Jex (1967) in their 'successive organisation of perception' model of tracking behaviour proposed that skilled pilots could potentially operate in three modes, the highest of which could only be achieved with extensive practice. They suggested that novices operated at the lowest level, which was a purely compensatory control strategy. In this mode, the pilot simply minimises any flight path error. However, this mode results in relatively poor control performance, as the response times of the human operator, control lags and display lags all play a part. With increasing practice it becomes possible to operate in a pursuit mode where the pilot responds directly to their control input, thereby improving performance by avoiding many of the lags inherent in the system (and these can be considerable in a large commercial aircraft being flown at low speed). At the highest level, the pilot is responding in a pre-cognitive mode. This is an entirely open-loop behaviour that operates independently of any feedback. A highly skilled pilot can exhibit all of these modes of control behaviour. For example, as flight path always shows an appreciable time lag behind the aircraft's angle of attack (and hence also the pilot's control inputs) operating in a pursuit mode allows them to minimise or avoid many of the lags inherent in the flight control system. However, the pilot's reactions to the disturbing effects of turbulence must be compensatory, because such effects cannot be predicted. Rotating an aircraft to the appropriate pitch attitude on take-off or in the flare may be considered as examples of the pre-cognitive mode of control behaviour.

As a further example, when performing an Instrument Landing System (ILS) approach, in the initial stages of training the pilot will be operating in a compensatory fashion, attempting to minimise the glideslope and localiser error with reference to their displays. With experience, upon observing a flight path error, the pilot will put in an input of 'about the right size' (an initial open-loop, pre-cognitive control input) which will bring the aircraft (approximately) back on the desired flight path. This initial (large) input will then be modified with additional smaller inputs until the ILS error is zeroed. Similarly, when the pilot wishes to execute a left-hand turn, she/he makes a large, open-loop left deflection of the stick to initiate the aircraft rolling in the desired direction at about the right rate of roll. When the rate of roll has resulted in approximately the desired bank angle, the stick is centred. This is then followed by a series of smaller closed-loop adjustments to obtain and maintain exactly the required bank angle. A similar process will be undertaken to roll out of the turn onto the desired heading (the pilot's surrogate control parameter for track).

As long ago as 1899, Woodworth noted that as skill levels increase, the initial movement becomes larger and more accurate and the smaller control movements decrease both in size and number. Furthermore, the time between the phases of movement also decreases. Because the initial movement is open-loop, it is not dependent upon extrinsic feedback. Feedback is required, however, if the operator is to develop this initial 'ballistic' open-loop skill component.

When evaluating performance on any tracking task, such as flying an aircraft, it is common to measure the end product of performance (e.g. errors between the tracked parameter and a target value). Metrics such as the arithmetic mean error and standard deviation of error have strong validity when applied to parameters such as flight path or airspeed deviation especially when associated with a well-prescribed flight task that demands a high level of performance, such as flying an ILS approach. The arithmetic mean error gives an indication of the overall flight path error (on a particular axis) and its associated standard deviation gives a measure of the 'smoothness' of the pilot's performance (see Hubbard, 1987). These two parameters are often used in preference to the Root Mean Square Error (RMSE). Taken in combination, the arithmetic mean error and the standard deviation of error completely define the RMSE. Furthermore, the RMSE also has the additional disadvantage that it produces identical values for quite disparate performances. For example, being consistently high, consistently low, or at the correct mean height but with great variations in height-keeping can all result in the same RMSE value.

However, there is a certain disassociation between the control inputs of the pilot and the flight path response of the aircraft. The series model of pilot control (McRuer, 1982) illustrates this. The pilot cannot directly observe the aircraft's flight path and so instead exerts control through use of the primary flight controls via a lower order surrogate (i.e. the aircraft's attitude). The pilot is essentially trying to close an inner loop (related to aircraft attitude) as a surrogate for controlling flight path (see Figure 13.5). In a large transport aircraft, the relationship between control input, aircraft attitude and flight path variation is mediated by factors such as inertia, control response lags, control power and the relatively high stability of the machine. As a result there is a significant delay between control input and the aircraft's response. Consequently further control inputs after the initial input may serve to cancel it out or reinforce it before the initial input has taken effect (assuming they don't induce a PIO)! As a result, significant pilot control input activity may not be reflected in changes in either the aircraft's attitude or its flight path. It has been observed that these basic flight path measures alone (measures of product) do not have the sensitivity required to discriminate between fine variations in manual flying skills.

However, a pilot's control strategy can be assessed directly by measurement of their inputs to the primary flight controls as an adjunct to flight path performance metrics (Ebbatson, Harris, Huddlestone and Sears, 2008, 2010). Appraisal of the pilot's control inputs to the primary flight controls provides a direct measure of performance associated with closing the inner (attitude) control loop, whereas evaluation of flight path performance is a measure of success in closing the outer (flight path) control loop (see Figure 13.5). McDowell (1978) used control input power spectra-based measures to evaluate pilot performance from the distribution and weighting of pilots' control input frequencies. It was observed that in an agile combat aircraft, more experienced pilots generally used higher frequency control inputs, particularly in the roll axis. McDowell concluded that there were changes in a pilot's control movement power spectra as a function of skill level. Measures of this property could be used effectively to discriminate pilot skill/experience level. However, when flying a large jet transport aircraft a different control strategy is required. More experienced airline pilots use a smaller range of lower frequency inputs in both pitch (control column) and yaw (rudder pedals) (Ebbatson, Harris, Huddlestone and Sears, 2008). This was also observed with commercial pilots flying a large jet aircraft in an asymmetric condition. In this case there is even less flight control and engine power available which makes the aircraft even more sluggish in its control responses. The aircraft's yaw response is compromised as a certain degree of rudder is constantly required to offset the effect of the asymmetric thrust condition, hence less it has little control power in yaw is available when turning away from the dead engine. Pitch response vertically is compromised simply as a result of having 50 per cent less engine power available. As a result, high frequency inputs are likely to be of little benefit.

FLIGHT PATH CONTROL USING AUTOFLIGHT SYSTEMS

Almost all of the time post take-off and prior to the approach and landing phase, commercial aircraft are flown under some degree of automatic control. In many cases, even landings are now performed automatically ('autoland'). As described at the beginning of his chapter, the aircraft is flown under the control of the FMS. The FMS takes inputs mainly from two places in the flight deck, the CDU and the MCP. Feedback to the pilots comes from a variety of sources but principally the CDU itself, displays on the MCP, and also the PFD and ND. In general, the CDU interface to the FMS is used for longer-term strategic planning of the route (i.e. the entry and editing of flight plans, etc.). The MCP tends to be used for shorter-term tactical manoeuvring, (e.g. VNAV; LNAV; heading select; level change; altitude hold, etc. – Göteman, 1999). These interfaces have been illustrated in Figure 13.1. However, as will be seen this is somewhat of an over-simplification. And this is where the problems begin to occur.

The autopilot systems have more than a single mode within them. The number of modes varies from aircraft to aircraft but typically the VNAV system will encompass: a vertical speed mode; a mode where the rate of climb is dictated by a target airspeed; a cruise (economy) climb mode; and a flight path angle mode, etc. LNAV modes may encompass steering modes commanding either track or heading. The Airbus A320 has nine autothrottle modes, ten VNAV modes and seven lateral navigation modes.

Selection of the wrong VNAV mode using the MCP interface can result in accidents (see the description of the Air Inter Strasbourg Airbus A320 accident in Chapter 6 on Error in Part One). However, transitions between some of these autoflight modes are automatic, for example if the vertical speed originally commanded is too great for the current altitude, the aircraft may automatically revert to level change. This change occurs without any instruction/input from the crew and often without any adequate annunciation on the flight deck to make them aware that this change has occurred (a mode change is usually signalled by small elements on the PFDs). This can result in the flight path angle of the aircraft shallowing in the climb and the aircraft flying through an intersecting airway (one that it would have climbed above had it continued at its commanded rate of climb). There is no display on the flight deck that shows the implications of the mode change. Over one half of all the incident reports in the NASA ASRS (Aviation Safety Reporting System) database are concerned with altitude 'busts' (see Hardy, 1990). Uncommanded mode transitions are one of the main categories of automation 'surprise' identified by Sarter and Woods (1992).

Such an automatic mode transition was implicated in the McDonnell Douglas MD 82 accident which crashed in fog on approach to Urumqi airport. In the MD 82, if the ILS signal is lost during approach the aircraft will transition from approach mode to vertical speed mode at the same descent rate. This should allow it to be in the correct position when the ILS signal is reacquired. However, vertical speed mode causes the aircraft to maintain a constant rate of descent until the crew intervenes but will not guarantee that the aircraft is on the correct ILS glideslope profile. During the ILS approach to Urumqi the autopilot disconnected resulting in a mode transition to vertical speed mode at 800 feet/ per minute. The crew did not notice the mode transition resulting in the aircraft crashing 2 km short of the runway.

Sarter and Woods (1992) identified several other issues relating to aircraft control, particularly during VNAV. It was suggested earlier that the strategic control of the aircraft was undertaken via the CDU and tactical control was exerted via the MCP, but as stated, this is somewhat of an over-simplification. In some cases the control of the aircraft can be 'split' between the CDU and the MCP, with, for example, LNAV and speed being executed under the strategic control of the flight plan but with the VNAV component being controlled from the MCP. A typical result of this 'split' in control is as follows. During the climb the aircraft is assigned a cruise flight level 2,000 feet higher than the initially allocated cruise altitude which had been entered in the CDU. This tactical change in flight level is commanded via the MCP. Sometime later, at the top of descent the airspeed comes back to 190 knots but the automation fails to initiate the descent because the VNAV aspect of the flight is not under the control of the CDU. This results in the top of descent point being passed and an excessive rate of descent subsequently being used to regain the desired flight profile (if this is possible). Another way that this issue may manifest itself is in the problems the crew may have monitoring the active target values. This can again be associated with conflicts between CDU values and those on the MCP. For example, VNAV climbs and descents are often constrained by MCP targets. As a result, at the top of descent an altitude lower than the cruise altitude must be entered on the MCP for the descent to be initiated. Failure to do this can also result in the aircraft slowing but not descending. Pilots have commented that the failure to initiate descents is often attributable to the MCP targets not being constantly available for inspection (e.g. the next target altitude may be entered in the altitude window but not initiated). FMS targets can also be hidden in the CDU page architecture.

Studies have indicated that although pilots are generally proficient in the use of autoflight systems during normal operations there were gaps in their knowledge concerning the functional structure of the automation which became more apparent in non-normal situations. For example, Sarter and Woods (1994) in a later simulator study reported that when they asked 20 pilots all current on the Boeing 737-300 how to disengage the approach mode after localiser and glideslope capture, only three pilots could remember all three ways of doing it. Three pilots knew two of the three methods and seven knew just one. Seven pilots did not know any valid method. Many pilots also suggested at least one method that *would not* disengage this mode. The same pilots were also asked how to abort a take-off at a ground speed of 40 knots or less. Sixteen pilots said close thrust levers, apply thrust reversers and brakes. However, simply retarding the thrust levers below 64 knots will have no effect; they will simply motor forward again. The autothrottles will only disconnect in this manner above 64 knots when the automatic system shifts to 'throttle hold' mode. Below this speed the autothrottles must be disconnected manually (only four pilots knew this). These are not inadequacies in the pilots, their training or the design of the aircraft. This emphasises the interaction between all of these components and how a shortcoming in one aspect of the system causes problems in other components. This is also another excellent example of 'opaque' automation (see the next chapter on Automation). The 'on button' on the MCP for engaging approach mode doesn't switch it off (but this is also a safeguard to avoid accidental de-activation) and the autothrottle's automation decides that below 64 knots the pilot isn't going fast enough to stop …

A report by Wood (2004, Chapter 3, p. 1) for the UK CAA concluded:

> The research indicated that there was much evidence to support the concern that crews were becoming dependent on flight deck automation. Several MOR [Mandatory Occurrence Reports] incidents revealed that crews do respond inappropriately having made an incorrect diagnosis of their situation in which the automation fails. For example, disconnecting the autopilot following an overspeed in turbulence then resulted in level busts. If pilots had a better understanding of the automation then it is likely that the need for manual flying could have been avoided and thus the subsequent level bust. During the course of this research two more fundamental observations were made: First, pilots lack the right type of knowledge to deal with control of the flight path using automation in normal and non-normal situations. This may be due to operators making an incorrect interpretation of existing requirements and/or a lack of emphasis within the current requirements to highlight the particular challenges of the use of automation for flight path control. Second, there appears to be a loop-hole in the introduction of the requirements for CRM training.

As a result of observations such as this, Wood and Huddlestone (2006) suggested a new approach is required to flight crews' training for using the aircraft automation for flight path management. They argue that licence syllabus requirements detail the knowledge required of a pilot to operate the automation but make little comment about the knowledge required concerning how the automation actually operates. In training, the autopilot and FMS are treated in a similar fashion to other aircraft 'systems' (such as fuel, hydraulics or electrics) and as a result students simply learn the basic declarative and simple procedural knowledge required. During later practical aspects of their training the student pilots become aware *that* the automation can perform a particular function but not *how* it actually does it.

The autopilot does not change the manner in which an aircraft flies but it can effectively 'disguise' the effects on speed or flight path that a pilot would usually be aware of were they flying the aircraft manually. Wood and Huddlestone (2006) illustrate this point with reference to the following situation. In a highly automated aircraft fuel-efficient descents are achieved by programming the FMS with the forecast winds for the descent and all of the relevant altitude constraints and navigation waypoints. In this 'profile mode' the system calculates the optimum three-dimensional flight path and manages the energy of the aircraft by allowing it to vary its target descent speed of 300 knots (e.g.) by up to ±20 knots. However, only rarely will an aircraft be allowed to complete such a descent without further restrictions and intervention from ATC. If a hard speed constraint of 300 knots is requested by ATC then the aircraft will no longer be able to remain in the 'profile mode' as programmed. The crew must change the mode of descent to one that manages speed to 300 knots (or less). If the speed of the aircraft is constrained then the aircraft will deviate above or below the required flight path depending on the wind, thus potentially failing to adhere to the ATC altitude constraints. To ensure these constraints are met the autothrust system will apply power if the aircraft drifts below the flight path. However, the crew has to observe and intervene manually to apply speed brake if the aircraft drifts above the flight path. Therefore, a simple ATC requirement to maintain a nominal descent speed can require significant interventions from the crew, for example by selecting a different automation mode and/or intervening manually through the application of speed brake. Wood and Huddlestone (2006) suggested that simply knowing *what* the various automation modes do in these circumstances was not enough. They also needed to know *how* they work so an appropriate mode could be selected and the pilots were also aware of its limitations and the implications of these for the control of flight path against targets. As a result it was proposed that automation management training should be integrated with manual flying skills training, flying the various different types of speed/altitude constraint manually and then implementing the same flight profiles using various aspects of functionality in the automation to demonstrate *how* the automation achieves the flight path goals, what its limitations are and what needs to be monitored.

A subsequent study by Wood (2009) used such a 'whole task' approach. Initially students were instructed how to fly an automated aircraft manually. Once this fundamental skill (and knowledge) was acquired the complexity of the task was gradually increased (i.e. more layers of the automation were introduced) and the pilot learned how to make inputs into the automation to get it to perform the flying task (see Reigeluth, 1999). The student was encouraged to verify that the automation was controlling the aircraft in the same way that he/she had manually flown the aircraft the day before. From the very start trainees were required to link the process of command input to performance output. Monitoring of the automation was also built in to this process to make the pilot aware of what the automation should be doing to control the aircraft. Only once this was mastered was the pilot instructed how to manage the systems. Finally, operational aspects were introduced once they were skilled enough to fly the aircraft and manage its systems. System management skills built upon basic flying skills; they were not regarded or taught as a separate, discrete piece of knowledge.

The 'Double Whammy' of Flight Deck Automation and Recurrent Licensing Requirements

In the previously cited study of flight crews' dependency on automation by Wood (2004) a great deal of anecdotal evidence was found to exist suggesting that pilots of highly automated aircraft experienced manual flying skills decay as a result of the lack of opportunity to practise hand-flying during line operations. Other authors have suggested the same thing over a period of years (see also Curry, 1985; Veillette, 1995; Owen and Funk, 1997). The ability of a pilot to revert to basic manual control is essential, for example in cases where the aircraft's automatic capability is diminished or when reconfiguring the automatics is an ineffective use of crew capacity. Every licensed air transport pilot must demonstrate their ability to operate their aircraft manually during licensing proficiency checks in a full flight simulator. However, there are only infrequent opportunities to exercise manual flying skills in modern flight operations. The threat of skill fade is a concern shared by pilots, operators, regulators, manufacturers and researchers alike (e.g. Childs and Spears, 1986; Parasuraman, Molloy and Singh, 1993; Tenney, Rogers and Pew, 1998).

Ebbatson (2006) describes a number of incident and accident reports in which manual handling deficiencies of pilots flying highly automated aircraft were cited as either a contributory or causal factor. For example, in March 2002 the crew of an Airbus A321 returning to East Midlands Airport in the UK performed a manual approach, disengaging both the autopilot and the autothrust systems. The aircraft became slightly high on the approach and the handling pilot failed to recover the glideslope until late in the approach, deviating substantially both above and below it. The aircraft also deviated significantly laterally from the optimum approach path. During the final stages of the approach the profile was recovered but the airspeed decayed well below the target. As a result, when flaring the aircraft an abnormally large angle of attack was required to reduce the descent rate due to the low airspeed and consequently the tail struck the runway causing damage to the fuselage skin and supporting structure. The handling pilot had been operating the aircraft type for just over a year and previously had extensive experience operating single-seat military fast jets. He later commented that he could not recall having ever operated this aircraft without the use of autothrust before the incident (Air Accident Investigation Branch, 2002). The UK Air Accident Investigation Branch (AAIB) also gives accounts of two similar incidents where on both occasions the autopilot, autothrust and flight director systems were disengaged at an early stage of the approach. Again, there were significant deviations on both the localiser and glideslope coupled with poor airspeed management resulting in high flare angles and tail strike damage. However, there was another common factor linking these three occurrences. In each of the events the handling pilot was about to undergo a licence proficiency assessment and had deliberately disengaged the autoflight systems to practice handling the aircraft manually prior to this evaluation. The AAIB remarked that 'the wish to practice manual flying skills is understandable but degrading the aircraft's systems in order to do so is something that should be approached with great caution' (Air Accident Investigation Branch, 2002).

No one in their right mind would wish to go back to hand-flying large aircraft all day, however something is clearly amiss with the integration of the design of the aircraft's automation, pilot training and normal practices during flight operations. This really begins to emphasise the importance of the HFI approach. Fixing only one of these components in isolation would have relatively little effect. However, coordinating the response would have major benefits, not only in safety but also in reducing through life costs.

FURTHER READING

For a deeper consideration of issues surrounding the Human Factors of aircraft control, try the following very readable chapters:

Baron, S. (1988). Pilot Control. In: E.L. Weiner and D.C. Nagel (eds) *Human Factors in Aviation* (pp. 347–85). San Diego, CA: Academic Press.

Or:

Field, E. (2004). Handling Qualities and Their Implications for Flight Deck Design. In: D. Harris (ed.) *Human Factors for Civil Flight Deck Design* (pp. 157–81). Ashgate: Aldershot.

For those who enjoy a more challenging read and are mathematically enthusiastic, it may be a little old but it is still full of useful concepts, try:

McRuer, D.T. and Krendel, E.S. (1974). *Mathematical Models of Human Pilot Behavior* (AGARDograph 188). Neuilly-sur-Seine: AGARD/NATO.

14
Automation

Automation is all pervasive on the flight deck. It isn't just about the control of the aircraft. Automation is also implicated in almost everything else, from cabin pressurisation, to engine start up, to dimming the displays at night time. Parasuraman, Sheridan and Wickens (2000) defined automation as the full or partial replacement of a function carried out by a human operator. Furthermore, the aircraft flight deck can be thought of as one huge flying computer interface. It is a relatively crude example of ubiquitous computing, where there is computer functionality everywhere but is 'hidden' behind user interfaces that look nothing like a PC. Automation falls very firmly within the realm of the *Machine* however it has profound aspects on the hu*Man* component. Not only does it change the way people do their jobs (hence the education and training requirement) it has a profound affective effect on the user. Furthermore, it is also very difficult to draw a hard and fast line between 'automation' and 'human–computer interaction' (covered in the following chapter). So I won't.

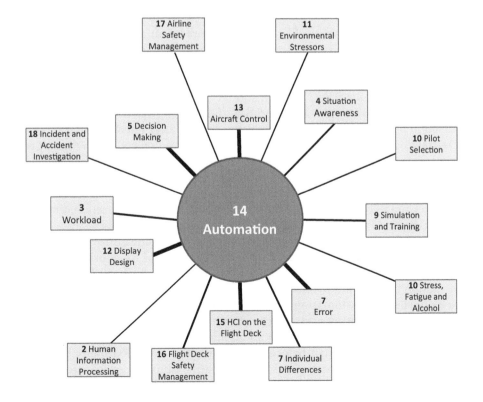

Aircraft are mainly operated using the on-board automation but automated systems still require displays by which to monitor them and control interfaces (either traditional or computerised) via which to exert control. This is critical to avoid design induced errors. However, modern highly automated systems place considerable demands on flight deck decision making.

The trend in flight deck design over the past half century has been one of progressive 'de-crewing' and increased digitisation and system integration (see Applegate and Graeber, 2001). The common flight deck complement is now that of two pilots (for operational and procedural purposes usually designated 'pilot flying' or PF and 'pilot-not-flying' or PNF). Fifty years ago, it was not uncommon for there to be five crew on the flight deck of a civil airliner (two pilots; Flight Engineer; Navigator and Radio Operator). Now just two pilots, with much increased levels of assistance from the aircraft, accomplish the same tasks once undertaken by five. There are even proposals to reduce the crew of short-range airliners to just a single pilot, with increased on-board automated assistance and surveillance from the ground (Harris, 2007). More reliable systems with increasing levels of automation resulted initially in the demise of the Radio Operator, followed by the Navigator and finally the Flight Engineer. Many of the functions once performed by flight crew are now wholly (or partially) performed by the automation in the aircraft. The emphasis in the role of the pilot has changed from one of being a 'flyer' to one of being a systems (or flight deck) manager. The aircraft and its systems are now more usually under supervisory control, rather than manual control. The manner in which the pilots exercise control is now one of being an outer-loop controller (a setter of high-level goals) and monitor of systems, rather than that of an inner loop ('hands on', minute to minute) controller. This has changed dramatically the nature of the pilot's control task in the cockpit. Emphasis is now much more on crew and automation management rather than flight path control *per se*.

The President's Task Force on Flight Crew Complement (McLucas, Drinkwater and Leaf, 1981) analysed the safety implications of the reduction in the number of flight crew from three to two in the 'new generation' (at the time) of medium-range airliners (e.g. Boeing 757/767 series). It was concluded that from an analysis of data, such as the relative accident and incident rates of the two-crew shorter-range airliners operating at the time (e.g. Boeing 737, McDonnell Douglas DC9 and BAC 1-11), there was no threat posed to safety by the deletion of the Flight Engineer's position and the associated automation of his/her tasks. De-crewing has not posed a threat to flight safety when coupled with appropriate developments in the technology available on the flight deck. Despite increases in air traffic density, current generation, two-crew commercial jet aircraft exhibit an accident rate approximately 15 times lower than that of the four-crew, first generation commercial (e.g. de Havilland Comet) or ten times lower than that of the second generation, three-crew jet airliners (e.g. Boeing 707/720) (Boeing Commercial Airplanes, 2009).

When writing a chapter like this one it is tempting to become quite negative about the whole topic and concept of automation. The vast majority of the work published tends to concentrate on the negative effects of automation on the human, the 'inappropriate' implementation of automation and the accidents 'caused' by automation. It is very easy to overlook the benefits automation has reaped. These are many and varied. For example, no one in their right mind would wish to start hand-flying airliners across the Atlantic or Pacific again; developing navigation solutions is now faster, more accurate and has a lower likelihood of error than before. Automation has provided many benefits but it has changed the nature of human work a great deal. A failure to appreciate the nature of these changes (and hence the design of the pilot's job and the nature of their training) in addition to inappropriate implementation of automation, a poor understanding of what automation actually is (and what it can offer), and unrealistic expectations are often at the root cause of a great deal of dissatisfaction in this area. This is not to say that there are not problems with the design of automated systems and the way that the human operator has been incorporated into their development and use. However, a degree of perspective is required. As a sweeping statement to finish off this short introduction it can be suggested that the first decades of the implementation of automated systems have been dominated by engineers developing systems and solutions about what they *could* automate. As a result, the human has had to adapt to the automation. We are now entering an era of human-centred automation. It is up to the Human Factors communities and Systems Engineering to identify what *should* be automated and provide the engineers with these specifications. Emphasis should now firmly be placed upon adapting the automation to the human, not vice versa.

WHY AUTOMATE?

There are many lists describing the benefits and disbenefits of automation. Many items on such lists remain constant; others change with the application area. Advantages typically cited (e.g. Wiener and Curry, 1980) include:

- Increased capacity and productivity.
- Reduction of manual workload and fatigue.
- Relief from routine operations.
- Relief from small errors.
- Precise handling of routine operations.
- Economical utilisation of machines (energy management).

In particular, in the aviation domain several specific reasons have been identified for the introduction of automation (see Harris, 2003):

- *Availability of technology* – Aviation is a complex business and pilots need help to manage their aircraft owing to the complexity of both the machines and the airspace that they operate within. However, the other side of this philosophy is that we will automate what we can automate (as opposed to what we should automate).
- *Safety* – Without a doubt, high levels of automated assistance can aid the pilot and have contributed to safety, for example CAT III autoland capability, but it is essential to avoid the fallacious argument that safety can be improved by removing the operator from the system entirely, thereby avoiding human error. There are two problems with this approach. Firstly, the automated devices are themselves designed and built by human beings (it's just the nature of the error that changes). Secondly, the devices themselves are not perfect and have the potential for generating errors, largely as a result of the fact that they are 'context unaware'. The pilot is the essential link between the context and the appropriate use of the aircraft's automated systems.

- *Economy and reliability* – Automatic devices, such as autothrottle and autopilot, can fly the aircraft much more smoothly, accurately *and* economically than the human pilot. They can adapt to environmental disturbances faster and can fly more complex thrust management schedules. As a result, under normal circumstances, the aircraft can be operated more economically using the autoflight control system, and as it can also be operated more smoothly producing less 'wear and tear' hence reducing maintenance costs.
- *Reduction in workload* – Autoflight systems certainly reduce the physical workload associated with flying an aircraft. Computerised display systems have also reduced the mental workload associated with routine mental computations associated with in-flight navigation (and have also considerably increased navigational accuracy and reduced the number of errors). However, it is wrong to say that automation reduces workload. It simply changes its nature. Wiener (1989) has called the automation in most modern aircraft 'clumsy' automation. It has reduced crew workload where it was already low (e.g. in the cruise) but has increased it dramatically where it was already high, for example in terminal manoeuvring areas. In the latter case the workload has also changed in nature to become almost exclusively mental workload associated with the management of the aircraft's automation.
- *More precise control and navigation* – In addition to the economy and reliability issues, automation also allows for precise navigation (which may involve shorter flying times, and hence increases in efficiency) as well as increased safety. With airspace becoming increasingly crowded vertical separations are being reduced in some areas to 1,000 feet and aircraft are being required to self assure in-trail separations in certain 'free-flight' areas. As a result, accuracy of navigation is becoming an increasing safety issue.
- *Display flexibility* – While the display systems (and the navigational computations underlying the displays) can now relieve the pilots of a great deal of the mundane navigational workload there is a tendency to overload the flexible display areas with system-related items in the belief that this makes the automated systems more transparent and easier to monitor. This has been called the 'lots of data but no information' approach!
- *Economy of flight deck space* – There are more systems in the modern aircraft than ever before but conversely fewer crew and less display space than 20 years ago. All system information now appears on one of six flexible electro-optical display areas. More information is now hidden than is displayed. As a result, the pilots do not have to opportunity to continuously monitor most systems, hence the need for automated monitoring.

LEVELS OF AUTOMATION

Automation is not an 'all or nothing' thing: there are degrees of automation. There are various ways in which the degrees and dimensions of automation can be described. The following is just one, taken from the British Aerospace EAP (Experimental Aircraft Programme) for describing cockpit functions.

- *Human Functions* – Functions that can only be performed by the human operator with no support or augmentation from the aircraft, for example, verbal communication, map reading or keeping a visual lookout.
- *Cognitive, Supported Functions* – Basically decision-making activities on the part of the pilot that are aided by the information provided by the aircraft (e.g. map reading on a head-down, moving map display or a take-off configuration go/no-go light).
- *Human, Augmented Functions* – Human, continuous control operations that are augmented by aircraft systems, for example, attitude and airspeed control via the aircraft's FBW systems.
- *Human, Augmented, Automatically-limited Functions* – As in the previous bullet point, however, the aircraft is protected from potential 'out-of-limits' inputs by monitoring of the pilot's actions, (e.g. FBW alpha limit systems or checking waypoint entries against fuel load).
- *Automatic, Limited, Continuous Functions, with Human Override* – Automated functions that can be overridden by human control inputs, for example control being taken over from the autopilot by continuous inputs to the throttle and stick.
- *Automatic, Limited, Discrete Functions, with Human Override* – Automatic functions that can be overridden with a discrete human input, for example a sanction/cancel button.
- *Automatic, Autonomous Functions* – Tasks which are continuously being undertaken and are also continuously being monitored independently of the pilot. Examples of this include automatic systems checks and status monitoring. The pilot is only informed if there is a malfunction or encounters an out-of-limit situation.

Sheridan and Verplank (1978) described the hierarchy of automated control in a slightly different way (see Table 14.1). Many other descriptions of the degrees and types of automation are also available (e.g. Endsley and Kaber, 1999).

The degree of task delegation to the machine itself has several stages, from simple *management by delegation* (a boss/slave type of relationship – 'I say, you do'), to *management by consent,* where the machine suggests various options but the pilot has ultimate control of which to select (all future courses of action must be sanctioned), to finally *management by exception* where the machine is operating almost autonomously (the operator is only informed of major deviations from normal operations, otherwise the system remains silent) (see Dekker and Hollnagel, 1999). As should be obvious, though, the higher the degree of automation the lower the operator's workload when the system is operating within parameters. However, as a result the operator is concomitantly less aware of system configuration and any pertinent environmental parameters which may be problematic if they are required to intervene for any reason. At this point workload may increase quite considerably. Additionally, as you progress through 'higher' degrees of automation, i.e. the amount of authority and autonomy delegated to the aircraft increases, then the potential for problems to develop also increases (an issue that will be picked up later).

What is also evident in any of these taxonomies is that there is both a cognitive and a control element to automation. Parasuraman, Sheridan and Wickens (2000) identify four generic types/functions of automation: information acquisition; information analysis; decision and action selection; and action implementation. They suggest that these are roughly parallel to the human processes/structures of sensory processing; perception/ WM, decision making and response selection, respectively.

Table 14.1   Sheridan and Verplank's (1978) levels of automation of decision and action selection overlaid with Dekker and Hollnagel's (1999) concepts of task delegation

Level	Description
1	Automation offers no assistance: pilot must take all decision and actions. **MANAGEMENT BY DELEGATION**
2	Automation offers a complete set of decision/action alternatives, or
3	Narrows down the selection of options, or
4	Suggests one alternative, and DECREASING
5	Executes that suggestion if the pilot sanctions it, or LEVELS OF CONSENT
6	Allows the pilot a restricted time to veto before automatic execution, or
7	Executes automatically, then necessarily informs the pilot, and
8	Informs the pilot only if asked, or INCREASING
9	Informs the pilot only if the automated system decides to. LEVELS OF
10	The automation decides everything and acts autonomously, ignoring the pilot. AUTONOMY

As mentioned many times by now, the role of the pilot has changed from one of being a 'flyer' to one of being a supervisory controller of automated systems. They are responsible for the management of the aircraft's automation almost in the same way that they would manage another crew member. Sheridan (1992, p. 1) defines supervisory control as a situation in which:

> ... one or more human operators are intermittently programming and continually receiving information from a computer that itself closes an autonomous control loop through artificial effectors to the controlled process or task environment.

However Sheridan also notes that there is no single model of supervisory control but there are six basic supervisory functions:

- Planning;
- programming the automation;
- monitoring the automation;
- diagnosing problems;
- intervening if necessary; and
- learning from experience.

PROBLEMS WITH AUTOMATION

Studies of the implementation of automation in the cockpit have suggested that even after over a year of experience on type 55 per cent of pilots indicated that occasionally the FMS did things that surprised them and 20 per cent of pilots did not understand all the modes or features available to them (Sarter and Woods, 1994). James, Birch, McClumpha and Belyavin (1993) observed that the basic problems identified with the control of an aircraft using an FMS were characterised by three common questions asked by the flight crew:

- 'What is it doing?'
- 'Why did it do that?'
- 'What will it do next?'

Slightly more recently, a fourth question was added by Courteney (1999):

- 'I wonder if it will do that again?'

These questions begin to identify some of the symptoms of the information shortcomings on the flight deck which are associated with giving the pilots inadequate feedback or not making it apparent what the current system goals are; or not making the pilots aware of the process by which the FMS is flying the aircraft; and/or not making them aware of what is the next system goal. These questions also demonstrate that pilots have difficulty understanding the logic in automated systems and, as a result, cannot anticipate the automation's next move. As the pilot is now an outer loop controller (she/he sets high level goals and monitors the aircraft's progress) this type of feedback is essential. Error can only be assessed by comparing what you have got with what you wanted.

The problem is basically that automated systems are 'opaque'. Unlike a bicycle, where with a little imagination and deduction, function can be deduced from form; this cannot be done with computers and automation. Starting at the box or the pilot interface will not usually help. Function cannot be deduced from form. Norman (1988) argues that automated (computerised) systems are opaque largely due to two characteristics: the Gulf of Execution and the Gulf of Evaluation.

- *The Gulf of Execution.* Does the system allow actions that correspond to the intentions of the pilot?
- *The Gulf of Evaluation.* Does the system clearly provide feedback that is easily interpretable in terms of the difference (if any) between what was wanted and what is actually happening? Can the pilot monitor the state of the system?

Norman adds two further dimensions: the degree of system Autonomy and Authority:

- *The degree of Autonomy* of an automated system describes the number of sequences of actions a system may perform without further instructions from the pilot once she/he has given the high level command to initiate them.
- *The degree of Authority* of a system describes the power that a system has to take command of various parameters and/or override the control of the pilot.

In a very complex, highly automated system the combination of Autonomy and Authority may combine to produce the illusion of perceived Animacy. Put another way, the automatic systems appear to develop a mind of their own. All automated systems are deterministic, however, they can appear to lead an independent existence at times. Looking back over an incident which may have happened, with a full appreciation of the operating logic of the system, it can usually be established why the system did what it did. But more often than not crews do not have the luxury of a retrospective analysis of what has happened. What crews need to know is what the aircraft is doing and why it is doing it, and they need to know it as it is happening.

A good example of the role of high levels of Authority and Autonomy in a complex system can be found in the China Airlines Airbus A300-600R accident at Nagoya in Japan (Aircraft Accident Investigation Commission: Ministry of Transport, Japan, 1994). The aircraft was being flown manually on the approach by the First Officer (under flight director guidance) with the autothrottles engaged. For some reason, TOGA (take-off/go-around) power was accidentally applied but then the throttles were retarded slightly and TOGA mode was disengaged. The autopilot was re-engaged at this time. However, as TOGA power was being delivered and the flight director system was linked to the application of go-around power, when the autopilot re-engaged it assumed a go-around was taking place and went into that mode and started to command a nose-up attitude. To counteract the nose-up pitch the pilot applied down elevator to regain the glide slope. The A300 system uses stabiliser trim if the elevator does not have enough authority, thus as the pilots applied down elevator, the aircraft compensated and tried to continue flying the go-around profile by applying opposite stabiliser trim (which used the all-moving tailplane). Applying a force on the control column would not (at that point in time) disconnect the autopilot in the A300 when in go-around or landing mode, or when it was below 1,500 ft. The pilots reduced the power and the autopilot was disconnected. The aircraft halted its climb and started to descend back to the glideslope again. At this point the autothrottle was again re-engaged. Thrust increased and the aircraft started to climb once more, despite full down elevator being applied. As the nose passed through about 40° the slats and flaps were retracted: airspeed finally fell to about 70 knots and the aircraft stalled and crashed killing all on board. It can be argued that the problems in this case lay with the Authority and Autonomy of the system. The autopilot had more Authority over the control of the aircraft than did the pilots. The pilots also could not understand the mode transition in the autopilot system. The mode transition was a product of the Autonomy granted to the autopilot.

Another way in which the Authority of autoflight systems can be illustrated is by comparing the differences in philosophy of the two major airliner manufacturers. The Airbus approach in its FBW aircraft is to endow them with 'hard' envelope protections, which limit the aircraft to pulling fewer than 2.5 g and does not allow speed to decay to the point of the aircraft stalling. The aircraft has total Authority in these cases. Boeing, on the 777 has installed 'soft' envelope protections which require the pilot to apply a significantly higher force on the yoke once bank angles exceed 36° but the aircraft does not limit the pilot in the amount of 'g' or bank angle that can be applied in an emergency situation. Put another way, the Authority of the aircraft's automation is reduced considerably in favour of the pilot.

In what is now considered to be a seminal paper, Bainbridge (1987) described many of the ironies in the implementation of advanced automation. A basic irony lies in the view that the designer may regard the pilot to be unreliable and inefficient, so as a result considers that she/he should be replaced. There are two ironies of this attitude: designer errors can be a major source of operating problems, and the designer who tries to eliminate the pilot still leaves him/her to do the tasks that the designer cannot automate. This in itself causes problems as the pilot can be left with an arbitrary collection of tasks with little coherent support from the machine to accomplish them.

As a result, the pilot flying a highly automated aircraft is expected to monitor the automatic systems to verify that they are operating correctly and if not they are then expected to takeover. This results in the irony of manual takeover – the pilot still needs to be able to fly the aeroplane (but has less chance to practise these skills – see the previous chapter on the Aircraft Control) but also needs further cognitive skills to identify and diagnose a fault and adopt an appropriate strategy for recovery of the situation. Basically, the operator of any automated system needs more training, not less.

A more serious irony of automation is that the automatic system has been installed because it can (supposedly) do the job better than the pilot, but the pilot is still required to monitor that it is working effectively. There are several problems with this. The first problem is that in complex modes of operation the system monitor (pilot) needs to know what the correct behaviour of the aircraft should be, for example in flight path control. Such knowledge requires training and/or displays. The second problem is that if all the decisions can be fully specified, then a computer can make them more quickly, taking into account more dimensions and using more accurately specified criteria than the pilot. As a result there is therefore no way that the pilot can check that the aircraft's computers are following their rules correctly. The pilot can only monitor the automation's decisions at some meta-level to decide whether they are 'acceptable'. If the automation is being used to make the decisions because human judgement and reasoning are deemed not adequate in this context, then which of the decisions should the pilot accept? The human monitor has been given an impossible task. Finally, human beings are very poor monitors of systems, so automatic systems monitor the aircraft systems and the pilot monitors the monitoring system … Vigilance is difficult to sustain beyond about 40 minutes, especially if the system being monitored is highly reliable and the likelihood of an event is very rare.

THE EFFECTS OF AUTOMATION ON THE PILOT

Three decades ago, Wiener and Curry (1980) recognised that there were many potential negative effects of increasing the level of automation on the flight deck on the pilots. This was predicated on the assumption that pilots were no longer either 'hands on' controllers of the aircraft or active system monitors; the automation in the aircraft assumed almost all control and monitoring functions. Despite reducing the fatigue and workload associated with long periods of manual control, the downside of much increased automation included greater levels of boredom and complacency. Other factors identified were erosion of competence (in terms of manual control skills) which has already been addressed in the previous chapter, a lack of Situational Awareness and the negative effects of underload, topics that have also been covered elsewhere. In many ways their observations were quite prescient, however this simple two-factor model of aircrew functions ignored the increased management load on the pilots (both in terms of CRM and management of the automation itself – this is not necessarily a passive process) and it also implied a simplistic 'all or nothing' approach to the implementation of automation.

Picking up on several of the issues Wiener and Curry (1980) identified, automation complacency is closely linked to two related aspects of automation: overtrust and distrust. Automation is often endowed with many human characteristics, the most usual one of which is 'trust'. Muir (1994) observed that trust in automation was shaped by the reliability and predictability of the machines. Higher levels of reliability were related to higher levels of trust but in the event of a subsequent failure in what was perceived to be a highly reliable system it took a great deal of time for trust to recover. Lee and Moray (1992) subsequently found similar results. It takes a while to establish trust in an automated system, which may initially lead to over-monitoring of that system. Once operators have come to rely on what they perceive to be a highly reliable system, though, the opposite effect tends to occur; there is a failure to monitor the automation (assuming that it will perform as advertised) especially if the cost of performing the operation manually or the act of engaging or disengaging the automatic system is high. Lee and Moray (1994) further demonstrated that when humans have little confidence in their own abilities but they perceive the automated system to be highly reliable, they are more likely to use it.

Bailey and Scerbo (2007) further reported that as operator trust increased in an automated system, monitoring performance decreased as a function of increasing system reliability. However, their results also indicated that the complexity of the monitoring task heavily influenced operator monitoring performance; poorer performance was observed with more cognitively demanding tasks. This was attributed to what has been labelled 'automation complacency'. This phenomenon has been observed by earlier researchers and has been attributed as a causal factor in several incidents and accidents (e.g. the Eastern Airlines L-1011 near Miami, Florida, 1972 – NTSB-AAR-73-14; the Northwest Airlines DC-9-82 near Detroit, 1987 – NTSB-AAR-88-05; and the incident to American Airlines Airbus A300-600 near West Palm, Florida, in 1997 – see Dismukes, Berman and Loukopoulos, 2007). Wiener (1981) defined it as 'a psychological state characterised by a low index of suspicion' (p. 117); Billings, Lauber, Funkhouser, Lyman and Huff (1976) suggested that automation complacency is 'self-satisfaction which may result in non-vigilance based on an unjustified assumption of satisfactory system state' (p. 23).

Another View

Dekker and Woods (2002), however, criticise many authors who have made the suggestion that the principal problem with automation is that it removes the pilot (or operator) from 'the loop' and hence, causes accidents as a result. They argue that there is no empirical evidence which suggests such notions as 'automation complacency' or a failure to monitor automated systems have resulted in any accident involving a highly automated aircraft (but other authors would disagree with this – Parasuraman, Sheridan and Wickens, 2008). Furthermore, in many accidents the automation was performing exactly as advertised – there was no failure in any system. Dekker and Woods (2002) suggest that the aircraft were 'managed' into their accidents. The pilots were not passive, 'out of the loop' monitors, too complacent or too far 'out of the loop' to intervene effectively. In all cases the pilots were actively engaging with the aircraft's systems in the role of 'active' managers, searching for information, programming the FMS, planning for a coming phase of flight, responding to system demands and communicating both across the flight deck and with other agencies. While doing all this, the aircraft were actively managed into an accident, be it a stall or flown into the ground. The consistent factor underlying these accidents was a breakdown between the pilots and the automated flight deck agents. The manner in which the automation was implemented was difficult to use and obtuse. Dekker (2004a) suggests that typical automation problems contributing to such 'management' accidents included:

- *Mode errors* – where the pilot thought the aircraft's computer was in one mode, and they did what they thought was the right thing but the computer was actually in another mode (so they actually did the wrong thing).
- *Getting lost in the software* – usually as a result of having relatively few displays compared to the number of active processes being undertaken by the computer. It is often difficult to find the right page or data set.
- *Not co-ordinating computer entries* – especially when many operators are interfacing with the same system.
- *Becoming overload* – computers are supposed to off-load people in their work but often demand interaction during peak workload times ('clumsy' automation). Furthermore, operators are often required to monitor large amounts of data and can become unable to locate vital pieces of data.
- *Not noticing changes* – usually as a result of poor display design or the operator being in another display mode, effectively 'hiding' potentially critical pieces of data. In a highly computerised system, more data are usually hidden than are displayed as a result of a limited amount of display area available.

ENHANCING INTERACTIONS WITH AUTOMATION

There are two basic ways in which pilot's interactions with their automation can be improved. The short-term solution involves improving their training. This is the manner by which Wood and Huddlestone (2006) suggest that pilots can achieve a better understanding of what the aircraft automation is trying to do when controlling the aircraft (see Chapter 13 on Aircraft Control). Another manner by which pilots' interactions with automation can be improved is by specifically considering the manner by which automation will be managed on the flight deck as part of their CRM programme. The CAA (2006), in their publication CAP 737 (*Crew Resource Management Training: Guidance for Flight Crew, CRM Instructors and CRM Instructor-Examiners*) raised various concerns about CRM training in highly automated aircraft. These are discussed in Part Four concerning *Management* and specifically, in Chapter 16, management on the flight deck.

User-friendly Automation

The emphasis in this chapter is on making the automation itself more 'user friendly' or, as several authors suggest, developing the anthropomorphic aspects attributed to automation further, making it into a 'team player'. Dekker (2004a) suggested that the question for successful automation should not be 'who has control' but 'how do we get along together' (p. 194). Flight deck designers and engineers require guidance to support co-ordination between pilots and automation. Billings (1997) developed a philosophy for the implementation of automation on the flight deck which demanded that the aircraft systems worked co-operatively with the pilots in the pursuit of their flight goals. To support this philosophy he suggested that:

- The pilot must be in command.
- To command effectively, the pilot must be involved.
- To be involved, the pilot must be informed.
- The pilot must be able to monitor the automated aircraft systems.
- The automated systems must be predictable.
- The automated systems must also be able to cross monitor the pilot, and
- Each element of the system must have knowledge of the other's intent.

Boeing adopted a human-centred design philosophy in the development of the Boeing 777 (Kelly, Graeber and Fadden, 1992). This philosophy is outlined in Table 14.2. It will be noted that in terms of the dimension Authority, the automation philosophy applied places the pilot firmly in control, making them the ultimate arbiter of all action.

Table 14.2   Human-centred automation design principles for the Boeing 777 flight deck (from Kelly, Graeber and Fadden, 1992)

Pilot's role and responsibility
The pilot is the final authority for the operation of the aircraft
Both crew members are ultimately responsible for the safe conduct of the flight
Decision making on the flight deck is based on a goal hierarchy

Pilot's limitations
Expected pilot performance must recognise fundamental human performance limitations
Individual differences in pilot performance capabilities must be accommodated
Flight deck design must apply error tolerance and avoidance techniques to enhance safety
Flight decks should be designed for crew operations and training based on past practices and intuitive operations
Workload should be balanced appropriately to avoid overload and underload

Pilot's needs
When used, automation should aid the pilot
Flight deck automation should be managed to support the pilots' goal hierarchy

Automation can make a system 'opaque'. This has previously been attributed to factors such as the Gulf of Execution and the Gulf of Evaluation (Norman, 1988). Closing these 'gulfs' enhances a pilot's interactions with the automation. Christoffersen and Woods (2000) developed these ideas to suggest ways in which to optimise human–automation interaction to turn it into a 'good team player'. A 'good team player' makes their activities observable for their fellows and they are easy to direct. To do this they suggested that system information should be event-based (not state-based displays conveying the status of the machine and its mission goals). The displays must be future-oriented, allowing the operator to enhance their Situation Awareness by being able to project ahead. The display format should also be form-based, which enhances the operator's ability to detect patterns in data rather than them having to engage in arduous calculations, integrations and extrapolations of disparate pieces of data. This theme has already been described in the discussion of the design of flight deck displays. Dekker and Hollnagel (1999) have described highly automated aircraft cockpits as being rich in data but deficient in information. As noted earlier, it should be the objective of display designers to convey information (and even knowledge) not simply to provide data (see Ackoff, 1989). These solutions begin to address the Gulf of Evaluation imposed by automation.

Addressing the Gulf of Execution problem is a little more difficult. The root of this problem is that an aircraft is a flexible tool that can perform many and varied missions. Great flexibility in its automation requires complexity in the interface. However, if you simplify the interface and hence you decrease the aircraft's flexibility, it is very possible that the effect will be to widen the Gulf of Execution (you can't make it do what you want it to). Incorporate all of the many functions that *may* be required, even in the most remote circumstances and this will serve to increase the system complexity but narrow this particular Gulf. Human Factors is no different to any other engineering or design discipline – compromises are always required. It's just that more people tend to notice these compromises!

Adaptive Automation

An alternative method for implementing automation, avoiding some of the shortcomings of traditional automated systems, is by using adaptive automation. Adaptive automation changes the level or number of systems operating under automatic control with regard to aspects of the situation. The level of automation can be modified in real time and changes can be initiated by either the pilot or the machine, enabling the level and type of automation to be matched to the pilots' requirements for that phase of flight (Parasuraman, Bahri, Deaton, Morrison and Barnes, 1992). It can be argued that the flight deck is an adaptable systems interface as allocation of functions to either the pilots or the aircraft are already initiated by the users. In contrast, in adaptive systems either the user and/or the system can initiate changes in automation allocation. The machine may allocate more functions to itself under certain circumstances, for example if it detects an emergency situation, if it detects shortcomings in the real-time measurement of the pilot's performance, or if it detects that the pilot's workload is becoming unacceptably high. This may be done on the basis of the measurement of various physiological parameters (Hancock and Chignell, 1987; Scerbo, Freeman, Mikulka, Parasuraman, Di Nocero and Prinzel, 2001).

At the moment, though, there is only very limited implementation of adaptive automation. Some civil applications have been developed in research aircraft (e.g. the Cockpit ASsistant SYstem (CASSY) – Onken, 1994, 1997). These systems monitor the actions of the pilot and compare them against data derived from the position of the aircraft, status of the onboard systems and other external environmental factors. Algorithms are then employed to determine if there is any difference between the expected and actual states. In the advent of a significant difference a warning may be given to the pilot, one (or more) potential solutions to resolve the problem may be suggested or, in extreme circumstances, the solution may be automatically implemented. Bonner, Taylor, Fletcher and Miller (2000) describe an approach whereby adaptive automation automatically adjusts the amount of assistance offered to the pilot on the basis of contextual factors associated with pilot workload. In low workload conditions the pilot has manual control and total Authority over aircraft control parameters. In extreme workload conditions the computer has complete control and Authority over flight path control, allowing the pilot to undertake other strategic tasks. Various levels of automated assistance are available between these two extremes (see Table 14.3).

Table 14.3   Pilot Authorisation of Control of Tasks (PACT) adaptive automation hierarchy (Bonner, Taylor, Fletcher and Miller, 2000)

Automation Level	Operational Relationship	Computer Autonomy	Pilot Authority	Adaptation
5	*Automatic*	Full	Interrupt	Computer monitored by pilot
4	*Direct Support*	Action unless revoked	Revoking action	Computer backed up by pilot
3	*In Support*	Advice, and if authorised, action	Acceptance of advice and authorising action	Pilot backed up by computer
2	*Advisory*	Advice	Acceptance of advice	Pilot assisted by computer
1	*At Call*	Advice only if requested	Full	Pilot assisted by computer only when requested
0	*Under Command*	None	Full	Pilot

The only operational systems which can be described as being adaptive automation are to be found in military aircraft. For example in the General Dynamics F-16D adaptive automation performs a specific, safety-related function. The Ground Collision-Avoidance System continuously predicts the time until the aircraft breaks a pilot determined minimum altitude. At a certain critical point, the automation issues a warning to the pilot. If no action is taken a further auditory warning is issued and the automation takes control of the aircraft to manoeuvre it away from the ground before returning control of the aircraft to the pilot. The system can recover the aircraft quicker than any human pilot and is essential in a highly agile aircraft which can induce G-LOC (G-Induced Loss of Consciousness) in the event of extreme manoeuvring. An earlier, and slightly more alarming version of adaptive automation was implemented in the Russian-built Yakovlev (YAK) 38. This was a Soviet version of the British Aerospace Sea Harrier V/STOL aircraft. One of its more advanced features was that it could complete a hands-free landing onto the aircraft carrier. Telemetry allowed it to be guided onto the deck with no interaction from the pilot. In this landing mode, ejection seat activation was automatic. The YAK 38 was not a popular aircraft … Adaptive automation still has some way to go before it will be commonplace on the civil flight deck.

FURTHER READING

Anyone interested in flight deck automation must read the original, seminal work on the topic:

Billings, C.E. (1997). *Flight Deck Automation*. Mahwah, NJ: Lawrence Erlbaum Associates.

The following report is also essential reading. It was the starting point for the development of the first Human Factors certification criteria for aircraft flight decks (now EASA Certification Specification CS 25.1302) and as such its influence cannot be underestimated. It is also highly readable.

Federal Aviation Administration (1996b). *Report on the Interfaces between Flightcrews and Modern Flight Deck Systems*. Washington, DC: US Department of Transportation. Freely downloadable from: http://www.faa.gov/education_research/training/aqp/library/media/interfac.pdf

15

Human–Computer Interaction (HCI) on the Flight Deck

It was a very close run thing between combining this chapter with the previous one on automation or giving it a chapter on its own. In the end, it was given a chapter on its own on the basis that not all automation involves a computer. Also, there are issues worth discussing within the context of computing and HCI which may get overlooked in a more generic treatise on automation. Hence, as the modern flight deck is one large, non-traditional computer interface these issues received a chapter of its own.

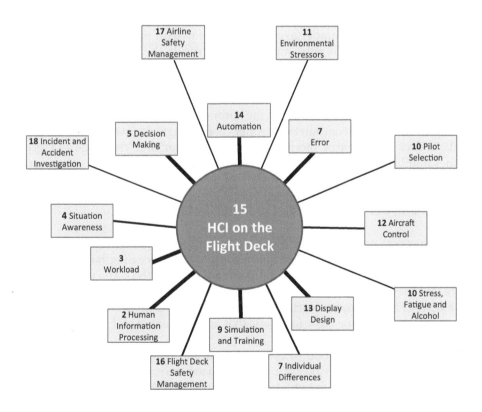

The flight deck is now just one large computer interface – an example of ubiquitous computing. Flight deck displays are interfaces to the flight deck computing. Good design of the interfaces to the aircraft's computers reduces training requirements and the opportunities for design-induced error. Simultaneously they can minimise workload and support the crews' decision-making processes. There is also an intimate link to human memory structures, especially LTM.

HCI has an intimate relationship between the hu*Man* and the *Machine*. HCI is unlike human–*Machine* interaction (HMI) as unlike HMI which is essentially a 'boss–slave' relationship ('I say, you do') HCI often involves the user in a dialogue with the computer ('I say, maybe you do when all things are considered'). Computing on the flight deck certainly does not represent cutting-edge, 'state of the art' technology. In fact, when it comes to the pilots' interface with the computers on the flight deck, on initial acquaintance some of the interface technology is more representative of the early 1980s rather than the present day! However, as will be discussed, this is often for very good reasons; but this is also not to say that there is not vast scope for dramatic improvements in the pilot's interface with the flight deck computing. It poses unique problems for the HCI designer.

Computing on the civil flight deck differs from that on the desktop in several key ways. Many of the reasons for these differences lie not in the availability of suitable advanced technology but in operational and regulatory drivers external to the aircraft: other reasons for these differences rest in the nature of the task. The control of an aircraft is a dynamic, time critical task undertaken by a crew of two (or more) pilots interacting with the *Machine* via a suite of computer interfaces. As a result, the FMS also needs to support a low level form of computer-supported collaborative working. Operating philosophies differ between airlines but in general when the aircraft is relatively close to the ground or is under manual control, the PNF is responsible for interacting with the flight computers. When the aircraft is under the control of the automated systems, the PF is responsible for interacting with the flight computers. In either case, though, the general principle applied on the flight deck is still one of 'monitor and cross-monitor' and 'challenge and response', techniques familiar to all pilots who have ever operated in a multi-crew environment.

COMPUTING ON THE FLIGHT DECK

Modern flight decks are an instance of ubiquitous computing, which has been described as a post-desktop model of HCI where the computer's processing has been integrated simultaneously among many devices and systems (Weiser, 1991). The user may not even be aware that they are interacting with a computer.

The FMS (or FMC) is at the heart of all automation in the modern commercial aircraft. Billings (1997) describes the basic functions that are inherent in all FMSs. These are:

- *Navigation* – determination of current speed and position, and identification and utilisation of navigation aids (e.g. ground-based transmitters and satellites).
- *Performance* – management of lateral and vertical navigation of the aircraft against targets specified in the flight plan. Determination of time remaining and fuel used.
- *Guidance* – generation of steering commands and determination of error against targets.

- *Display management* – generation of symbolic code depicting navigation and performance status on the flight deck displays.
- *Data management* – flight plan construction, maintenance of navigation and aircraft performance data base.
- *Input/output* – processing of received and transmitted data (e.g. via datalink) and management of radio communication and radio navigation aids.
- *Built-in test* – self-test and automatic logging of faults.
- *Operating system* – executive control of the FMS.

In the terminology of Sheridan (1987) the FMS is the Human Interactive System (HIS) that forms the interface (or in this case the plural, interfaces – CDU and MCP) with the aircraft's many Task Interactive Systems (TIS). A generic form of this computer architecture is described in Figure 15.1.

The CDU is the principal device for making strategic interactions with the FMS. It is often composed of a small, text-based monochrome screen (although at the time of writing coloured units are being introduced) and is of the order of 70 mm × 90 mm (although sizes vary slightly) and has an alphabetically arranged keyboard with several keys (hard keys) dedicated to specific flight-related functions. Down the side of the screen are programmable keys (soft keys) that relate to on-screen menu options. In many cases these lead to form fill-in screens into which the various performance and navigational parameters are entered, although many interactions with the FMS are made using MS DOS-like, text-based commands. A typical example of the CDU can be found in Figure 15.2. Most civil airliners have at least two CDUs (one for each pilot) although it is not uncommon for a third CDU to be found on some flight decks to the rear of the pedestal.

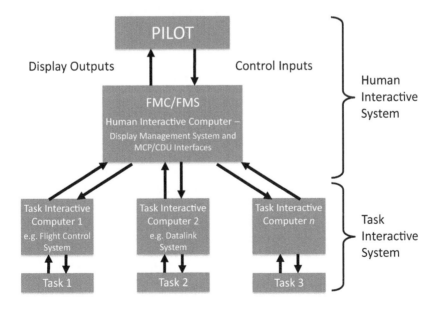

Figure 15.1   Hierarchical nature of task management using the FMS. The FMS/FMC forms the basis of the Human Interactive System in a supervisory control task (adapted from Sheridan, 1987)

In older aircraft, although both pilots could have different information displayed on their CDU, only one pilot at a time could interact with the FMS. However, more modern systems allow both pilots to interact simultaneously with the system. Only recently, with the introduction of the Airbus A380 has the basic design of this interface changed to any significant extent. The A380 uses much larger (A4) sized screens (now possible with advanced LCD technology) with separate, larger keyboards (now in a QWERTY format, more akin to that found on a desk-top computer) and is equipped with graphical pointing devices.

Figure 15.2   Typical flight management computer control and display unit (from a Boeing 777-200)

The CDU is not, however, the only interface to the FMS but it is the one that looks most like a computer, being a self-contained unit with data input and information output components. The PFD (containing essential attitude, airspeed and altitude information), the ND (showing the aircraft's track and heading against a simplified moving map), the EICAS (Engine Indicating and Crew Alerting System, containing information about engine and system functioning and which can also present checklists to support normal and abnormal functions), and the FMA (Flight Mode Annunciator, which shows the status of the automated systems engaged) are also information output sources from the aircraft's computer systems. The MCP and even the stick, rudder and thrust levers (in an FBW aircraft) can be regarded as non-conventional input interfaces to the aircraft's computers.

There are several constraints to the implementation of computing on the flight deck. It is the flexibility of computers (i.e. they are not dedicated to one task as a 'traditional' control and display system is) which result in compromises being made in their screen design and their input interfaces. Computers (in general) are dedicated information handlers, their purpose being to store, transmit, receive and manipulate information. However computers, even in an aircraft, are expected to perform many disparate tasks (see the list of functions of the FMS from Billings, 1997) and it is their multi-purpose nature that leads to their complexity. While computers can perform many tasks simultaneously, they have only a limited number of displays upon which to depict the output or progress of their 'work'. It is this multi-functionality, system complexity and relative lack of display space that leads to opaqueness of operation. Dekker (2004a) has highlighted this, suggesting that:

- Automation can make things 'invisible': it can hide changes, events or anomalies. Displays tend only to show system status (the mode it is in) rather than system behaviour. The interfaces hide the complexity of the underlying computing processes.
- Because the FMS has only a few interfaces (the 'keyhole problem') it can force people to search through many pages of data and hence, if data require integration, to rely heavily on calculation and/or their memory.
- Pilots can be forced into managing the FMS interface rather than managing the flight. These extra interface management burdens often coincide with periods of high workload (e.g. approach and landing).

AN ELEMENTARY MODEL OF HCI ON THE FLIGHT DECK

For the sake of argument and simplicity, a two-component model of computer dialogue design is proposed, the basic constituents being the cognitive and the physical aspects of the flight deck computer systems. Both of these components impose constraints on the design of computer interfaces on the flight deck. These domains are the:

- *Physical domain* – mostly connected with input/output devices that direct information exchange between the pilot and the computer and hence are largely concerned with the human operator's sensory-motor system; and the
- *Cognitive domains* – the knowledge requirements that the pilot must learn and remember to use the FMS: these are the actions and procedures that must be performed and the pilot's (cognitive) representation of components in the FMS.

This elementary model is described in Figures 15.3 and 15.4 to illustrate how these domains impose themselves upon the seen and unseen flight deck systems and processes.

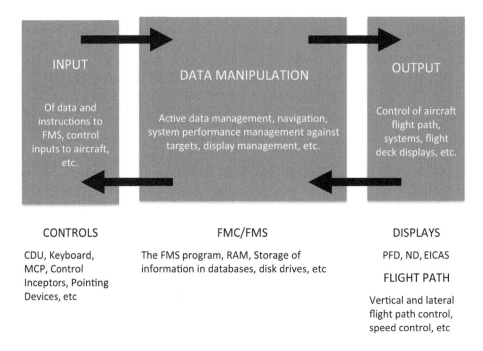

Figure 15.3   An elementary model of HCI on the flight deck

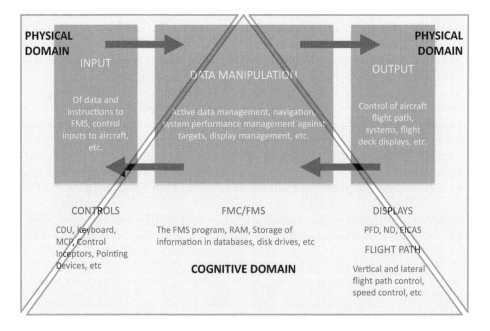

Figure 15.4   Model of HCI components on the flight deck extended to encompass the areas of influence of the physical and cognitive domains

The control of any computer system has an intimate relationship with the information on screen: input choices may be limited by the computer's program which displays what is (and is not) a legitimate input at that point. Interacting with a computer is not so much the 'I say – you do' approach that characterises a lot of 'traditional' HMI – it is more of a dialogue between the pilot and the computer. The program (which is resident in the box which lies between the input and the output devices) dictates what the pilot can and cannot do at a particular point in time. However, the user can only 'see' (in a metaphorical sense) a small part of what the computer can do. There is no relationship between form and function, as in a mechanical device, and so the user can 'get lost' in the system. They often do not know 'where' they are in the program, what they can do, where they should 'go' to do what they want to do and/or how they should get there. Computing of any type abounds with metaphors. To aid this process, the user needs a 'mental model' of the system. The user has to develop this to use it effectively. The 'structure' of the computer program itself is invisible to the user and only manifests itself via the user interface. Because the pilot cannot see the 'structure' of the computer program, the user is very dependent upon their LTM and hence to aid interaction they need guiding and prompting to navigate their way through the system. Harking back to the earlier description of Human Information Processing (Chapter 2), it is appropriate to start considering HCI in terms of scripts, schema and recall memory (for the structure of the program) and recognition memory as a way of helping the user navigate through the user interface.

There are four basic types of interaction style found on modern flight decks. The advantages and disadvantages of these are summarised in Table 15.1.

Table 15.1   Advantages and disadvantages of the four basic interaction styles used on the modern flight deck (adapted from Shneiderman, 1992)

Advantages	Disadvantages
Menu selection	
Shortens learning	May involve many menus
Reduces keystrokes	Getting 'lost' in software structures
Structures decision making	Consumes screen space on CDU
Permits use of dialogue management tools	
Avoids many errors	
Form fill-in	
Simplifies and structures data entry	Consumes screen space on CDU
Requires little training	
Easy data management	
Avoids/traps many errors	
Text-based scratchpad input	
Flexible	Slow to input
	Error prone
	Poor error handling
	Requires substantial training
	Very LTM intensive

Table 15.1   *Concluded*

Advantages	Disadvantages
Graphical User Interfaces	
Flight functions presented visually	Hard to program and quality assure to aerospace
Easy to learn	standards
Many errors can be trapped/avoided	Requires interactive graphics display and pointing
Encourages exploration of FMS functions	devices
	Graphics device cannot be only control interface
	(single point of failure for many systems)

The Physical Domain

On the data input side of the FMS (in the physical domain) the majority of interaction devices commonly found on desktop machines cannot be used because of constraints imposed by the operating environment on the flight deck. The primary constraint is simply one of space, hence the common use of the small alphabetical keyboard integrated into the CDU, as on the flight deck of many aircraft there is no space for a full size QWERTY keyboard. That was until the Airbus A380 arrived with its outsized flight deck! Even when the space was available (as on the slightly earlier generation of Airbus aircraft which use a sidestick control inceptor rather than a full size control column) the use of a full-size keyboard was eschewed in favour of providing a large workspace in front of the pilots. Again though, with the introduction of the A380 a full sized keyboard is now an option.

Modern flight deck computer interfaces also require a pointing device (such as a track ball or touch pad). Pointing devices serve several generic interaction functions, however these tend to be limited on the flight deck to 'selecting' functions (e.g. choosing a menu item on the screen) and 'editing' functions (positioning a cursor for insertion of text, etc.). Direct pointing devices interact directly with the screen to perform these tasks. The two most common forms of this type of device are light pens and touch screens (fingers). Indirect pointing devices do not interact directly with the screen (e.g. the aforementioned trackballs and touch pads, or mice on a desk top machine).

The requirement for a pointing device on the flight deck (either direct or indirect) is partly mitigated by the fact that most interactions with the FMS are via the CDU (which has a text-based interface) or via the MCP, which has dedicated, semi hard-wired controls for each function. Furthermore, the operating environment on the flight deck precludes the use of many varieties of pointing device. Mice and light pens are obviously unsatisfactory due to stowage problems and the effects of turbulence which would make them difficult to use. Trackballs consume less space and are suitable for use on the flight deck if space is available around them to support the wrist of the pilot and if enough resistance is incorporated into their action to counteract the effects of inadvertent activation. However, trackballs tend to attract dirt and moisture that compromise their action. The flight deck is a workplace in which food and drink is consumed! As a result trackballs on the flight deck tend to be mounted vertically in mounts on the central pedestal (as in the Airbus A380) to avoid many of the problems of the ingestion of foreign material. In some cases (as in the stillborn Fairchild-Dornier 728) the track ball was mounted on the underside of the armrest of the pilots' seats.

Touch screens would seem to be a potentially satisfactory solution but they actually have several major drawbacks. In many cases the primary displays are actually beyond the comfortable reach of the pilots, especially shorter ones. The screen on the CDU is too small to comfortably accommodate the pilot's fingers *and* maintain a clear view of the screen item to be selected. Finally, the pilot's fingers leave greasy marks across the display surface. Not only do they obscure the display slightly but after prolonged use, when sunlight falls across the screen from an oblique angle, these marks can totally obscure the display. Acid from sweat can also erode display coatings.

Touch sensitive pads are well-established on notebook computers and are now installed on the Boeing 777 as a means of interaction with the FMS. Touch pads tend to be much smaller than the display area with which they are associated. As a result, the desirable 1:1 interaction device-to-cursor movement ratio cannot be achieved, which results in a higher gain control than is ultimately desirable for the use on a flight deck. Precise cursor control, even after a period of familiarity, can be difficult. However, as touch pads operate by sensing changes in electrical capacitance caused by the presence of a finger on or near the pad they are unaffected by inanimate objects, such as pens and general flight deck grime. Furthermore, as they have no mechanical parts, they are extremely durable. However, as they operate by sensing changes in electrical capacitance they can be prone to erratic operation with moisture on the fingertips or in humid conditions and require software to stabilise their action.

As a result of the above factors it is of little surprise that cursor movement keys on the CDU keypad and the line-select 'soft keys' found around the CDU screen remain the preferred means of interaction with the FMS for operations that require the selection of an item from a menu. If there is a great deal of travel space between items to be selected on screen (as in the new, large A4-sized displays on the Airbus A380) pointing devices are better. However, if there is only a short space, cursor keys provide faster performance.

On the output side of the elementary HCI model described in Figures 15.3 and 15.4, the principal sources of information output are the PFD, ND, EICAS and the screens on the CDUs. While there may seem to be a great deal of display area available this is not the case. The PFD and primary EICAS screen are usually filled with primary flight and engine performance information. Although the FMS is simultaneously controlling many aircraft functions there is only limited space to display a fraction of the information available, hence the 'keyhole' analogy. Much of the information output from the FMS is displayed via the small CDU screen.

As has already noted a computer screen has a limited amount of display space available on it, some of which is taken up with virtual controls which are used in association with the interaction devices (e.g. icons used in association with the pointing devices). The more icons that are on a screen, the less the user's LTM is burdened and the more guidance is available to the pilot, however less space is available to do other things. In general, one of the biggest problems in HCI screen design is not what is on the screens on the flight deck but what *is not* on the screens. More information is hidden in the modern aircraft than is displayed. If you want information, you often have to ask for it.

The Cognitive Domain

The FMS has great functionality (Utility) for the management of the aircraft and the flight, however, its greatest drawback is its Usability. This is notoriously poor. Utility is the ability of a system to perform a task or tasks – the more tasks the system is designed to perform, the more Utility it has. Trying to define Usability is a good game but it is generally taken to be an attribute describing how easy the interface is to use. Shakel (1986, 1991) has suggested four aspects to Usability, but note that these are not necessarily mutually exclusive:

- *Effectiveness* – refers to levels of user performance in terms of speed/accuracy error rates, etc. Note that this will depend heavily on the user group. For example, if you set a performance criterion for a new system of '95 per cent of pilots should complete FMS programming tasks A and B within two minutes and not make more than three errors', the results will depend upon who the pilots are. Pilots new to the system will be unlikely to achieve these goals using a complex system; this might be an acceptable target for experts, though.
- *Flexibility* – How many different ways are there to achieve a goal, for example menus and keyboard short-cuts? Menus can be irritating and time consuming for the expert but can help the novice user know what is possible at any point in time. Many different ways to achieve the same goal can help to accommodate a wide range of users of different experience levels but can also increase complexity, error rates and confusion, and can also serve to increase training requirements. Also remember that the FMS/FMC is a ubiquitous computing system. The MCP is also a system interface.
- *Learnability* – This is essentially a dimension for new and/or infrequent users of systems. How long does it take to learn a set task, or how many errors are made? It is, however, a very good indication of the general quality of the user interface.
- *Attitude* – This is an affective response to the system, in terms of frustration, workload and/or satisfaction. People are less likely to put in the time and effort required to learn a system that is unpleasant to use.

The MS-DOS operating system from the 1980s had a wide variety of powerful system management features but required users to learn and remember dozens of arcane keystrokes to type inputs into the computer and using the grammar and syntax of a text-based command language. It had high Utility (it provided the necessary functionality) but very low Usability (users had to expend a great deal of time and effort to learn and use the system and there was a great deal of frustration when inputs were not accepted due to a simple typographical error). The Windows™ operating system, however, has the same Utility but a much more usable system interface. Many of exactly the same criticisms may be made of the text-based interfaces used on many FMSs. Harking back to Chapter 2 on the basic principles of Human Information Processing, it can be seen that text-based command languages (like MS-DOS) are heavily dependent upon recall memory (all you are presented with to cue recall is a blank screen and a flashing cursor). The user must recall precisely the grammar and syntax required to perform a particular command. Graphical user interfaces and menu-driven systems rely on recognition memory (you only need to recognise all the options that are open at a particular point in time). Recognition-based LTM recall paradigms always outperform pure memory recall tasks.

HCI PROBLEMS ON THE FLIGHT DECK

Estimates vary, but it is typically suggested that most pilots can exploit only 20–30 per cent of the functionality of the FMS on their aircraft. Given that the FMS is the key system for the operation of the modern airliner this is, on the face of it, a somewhat alarming figure. However, it only takes one minute's thought to recall that every day most pilots can successfully find their way from airport to airport anywhere in the world both safely and efficiently. This would seem to imply that there is a great deal of functional redundancy in the modern FMS. This is partially true. However, other parts of the FMS are dedicated to abnormal operations (e.g. single engine climb performance, maximum cruising altitude) and these functions are required only in the extremely rare event of an emergency. In an analysis of incidents involving highly automated 'glass cockpit' aircraft (Sarter and Woods, 1992) it was observed that the use of such infrequently used FMS modes was one of the key contributing factors. The manner in which the automation has been implemented in many modern hi-technology airliners also came in for a great deal of criticism in the FAA report on the interfaces on the modern flight deck (FAA, 1996) (see also the previous chapter). The fundamental causes of the interaction problems with the modern FMS lie in the cognitive domain.

As previously described, the principal user interface to the FMS for the input of instructions and data is a text-based interface with limited implementation of menu and form fill-in interaction styles to help guide the users through the system. Outputs from the FMS, in the form of feedback of system operation to the pilots, come in the form both of text and graphical information types. Several graphical user interfaces to the FMS have been developed and their performance assessed in representative flight environments (e.g. Romahn and Schaefer, 1997). In all cases, the graphical interfaces to the FMS outperformed the 'traditional' FMS interface in terms of measures such as ease of use, speed of use, error-inducing properties, etc. However, despite these advantages graphical user interfaces to the FMS are only just being implemented. The Airbus A380 (and at the time of writing the forthcoming A350) and the Boeing 787 Dreamliner use large screen, full colour interfaces to the FMS (in place of the small, monochrome text-based systems) with cursor control devices and options selectable from menu-driven systems. Not only has the physical size of the flight deck contributed to these advances, the development of cheap, light, low power consumption, large screen display technologies during the twenty-first century has also played a considerable part.

There are many aspects of the FMS that are less than optimal, but it is worth emphasising that most FMSs found in many commercial aircraft have never really been designed to undertake all the tasks that they currently perform. The FMS has evolved over the past three decades to encompass more and more tasks. The FMS was initially conceived and implemented as an aid to navigation and flight control, particularly during the cruise phases of flight. This functionality was subsequently increased to encompass control during climb and descent (VNAV). The FMS further evolved to undertake other flight planning and navigation tasks, fuel calculations, weight and balance management, communication management and many other functions. It is no longer solely a navigation aid. It has developed into the aircraft's central information management system. Unfortunately, as the FMS's functions have increased in number, the control and display interfaces have remained largely unchanged, one consequence of which is that the FMS menu structure and organisation is compromised. This has meant that the demands on the cognitive domain have increased many-fold but this complexity is largely hidden from the pilot as it cannot be easily represented on the small, multifunction displays that are available. Shortcomings in the physical domain of the FMS have a knock-on effect of causing representational problems in the cognitive domain.

As a result of the increased number of functions that the FMS must perform, there are a vast number of menus and pages that need to be available for display and completion on the CDU. In many cases, however, they do not use an optimal menuing structure. Not all menus can be accessed from all pages. FMSs tend to use a cyclic or acyclic menu (network) structure, rather than a tree-type structure (see Figure 15.5). Although this has the benefits of allowing shortcuts between functions, this option also has the disadvantage of disorientating the user if the structure of the network is not obvious (Shneiderman, 1992). Due to the text-based nature of the CDU display, its small size and the use of the other

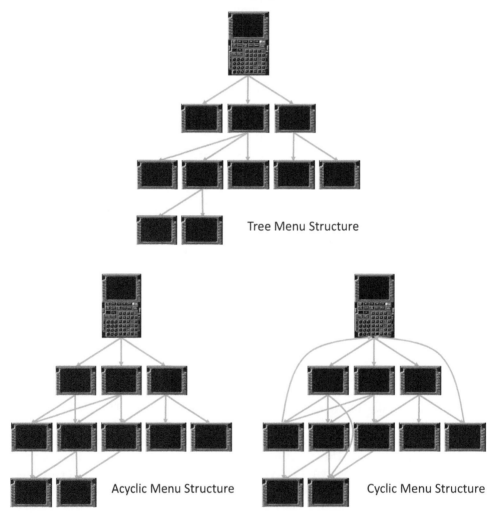

Figure 15.5   Types of menu structure implemented on the CDU interface to the FMS using Shneiderman's (1992) typology

displays on the flight deck for other flight-critical information, there is no opportunity to depict such a notional structure of the organisation of the FMS page architecture.

There is a further shortcoming commonly found in the FMS user interface that is again a direct result of the size of the CDU screen. The menu structures employed tend to be 'narrow and deep' (i.e. many items to select from in just a few menus) rather than 'broad and shallow' (i.e. a few items in many menus). Consensus within the research literature tends to suggest that breadth is superior to depth in terms of speed and accuracy of menu item selection, and also in user preference (Kiger, 1984; Landauer and Nachbar, 1985).

The structure of the task of managing an aircraft's flight between airports and the software structure of the FMS is not always compatible. It has already been suggested that it is difficult for the user to ascertain the organisation of the pages and sub-pages of the FMS as this is part of the cognitive domain and hence is not visible to the user. Finding out this organisation is made more difficult as the tasks are often arranged in a different manner to the way in which the pilot conducts the flight. The pilot is required to develop a different mental model of the FMS's software organisation and functionality to the one that they already possess for the normal management of the flight of the aircraft between airports. The FMS's structure is not 'user centred'.

The organisation of data on the FMS pages is a product of the requirement that the software programmer has to maintain neat, tight data flows (especially when system memory is limited) rather than reflecting the order in which the pilots receive these data. This results in many of 'workarounds' being improvised by the pilots to make their systems function in a manner commensurate with their task. For example, on certain FMSs the pilots are required to enter their departure weight as one of the first steps in setting up the flight plan. The system will not allow the pilots to proceed until this is entered. However, the final passenger and cargo load (and hence fuel) is not one of the first pieces of information that the pilots obtain when entering the flight deck before a departure, but they still need to initialise the navigation systems, enter the flight plan, etc. This leads to them entering a fictitious value into the system to allow them to continue with the business of setting up the rest of the flight plan. Unfortunately, a great deal of the performance data (e.g. ceilings and ranges) is based upon these figures. These need re-calculating when (and if) the pilots remember to enter the correct take-off weight when it is supplied to them by the dispatchers!

Data entry conventions for the FMS are also not necessarily consistent, largely as a result of software modules being added as the system has evolved over a period of years. For example, on certain FMSs the latitude and longitude information on some pages may require a leading '0' if the lat/long has only a single digit. On other pages this will not be required. As it is a text-based system no prompt is given to the user about which of the two formats is actually required! To page between FMS pages, on some occasions the 'Next Page' key is required, on other occasions a prompt appears by one of the Line Select Keys to progress between pages.

EXTERNAL DRIVERS AFFECTING PROGRESS ON THE PILOT-COMPUTER INTERFACE

While it may seem slightly strange that the computing in a modern airliner remains relatively 'primitive' compared to what can be found in the modern home or office (even in the newer Airbus and Boeing products) it is necessary to understand the constraints which dictate that aviation computing progresses in a slow, evolutionary manner rather than in the revolutionary manner that has been seen in desktop computing over the last three decades. These constraints directly affect the pilot–computer interface.

The primary factor that has militated against faster development of the user interface is the certification requirements. No component on the flight deck associated with the control of an aircraft can be installed and operated without the approval of the airworthiness authorities. FAR/CS part 25 describes the design and construction regulations for large, passenger carrying commercial aircraft. Systems that are critical for the safe conduct of the flight are required to show extremely high levels of reliability. For example, FAA AC 25.1309 requires that systems such as the FMS/FMC are required to show a level of reliability (in terms of system failure) in excess of 1×10^{-7} per flight hour. Attaining and demonstrating this level of reliability in a joint software/hardware system is no small matter and it certainly isn't cheap. However, like the vast majority of the certification regulations only 'machine' issues associated with system reliability are addressed. As many incidents, accidents and much research has demonstrated, the major source of unreliability in the joint cognitive system composed of the pilot and the aircraft lies on the human side of the equation. This is not to say that the pilot is to blame; far from it. The difficulties the pilots experience are as a result of the poor design of the human–computer interface. Courteney (1999) describes some of these shortcomings, such as being incompatible with the pilots' working environment; unclear system logic and having to workaround shortcomings in the design of the system. Singer (1999) describes the (very few) sections of the airworthiness regulations that pertain to the pilot–computer interface on the flight deck and these provide very little guidance to the designer.

A further factor to consider is the longevity of commercial aircraft. It is not uncommon for a basic design to be in production for 30 years (the first variant of the Boeing 737 flew commercially in 1967). Furthermore, the service life of a commercial aircraft can itself be over 30 years. So when it is also taken into account that the design freeze for the flight deck can occur about five years before entry into service, it is possible that the design of the flight deck computers and their interface will have to survive for well over half a century.

Updates are possible to both the software and hardware during the service life of the aircraft but these can cause logistical problems and hence are not undertaken with great frequency. Take the hypothetical example of the requirements imposed on the airline operator when performing a mid-life update on a commercial aircraft's FMS from the current text-based interface to a more 'modern' graphical user interface.

The licensing requirements for holders of an Air Transport Pilot's Licence (ATPL) dictate that a commercial pilot can only hold one type rating at any given time (although variants of this basic type rating are allowed). If there are significant variations between variants of the same basic aircraft type the airworthiness authorities may require one of several things. They may demand either a supplemental endorsement on the pilot's licence or they may suggest that the two aircraft variants are so significantly different that a pilot may not hold simultaneously a qualification to fly the two aircraft. Both of these options will involve the airlines in great expense and cause major logistical problems.

With regard to the training requirements the airworthiness authorities must approve all training courses. As a result the airlines will need to make investments in equipment to train the pilots, for example developing computer-based training programs for introducing the new GUI (Graphical User Interface) and investing in updating part-task flight deck simulators and full-flight simulators. The re-equipment of approved simulation facilities will also require approval by the airworthiness authorities. Furthermore, there is also the expense of removing pilots from line flying and placing them on training courses to qualify to fly the aircraft equipped with the new interface! These factors only address the flight-crew related issues and do not even touch on maintenance and spares-holding considerations. This is why it is essential to take a system-wide perspective when acquiring such a new piece of equipment (and why an all-new aircraft type is a golden opportunity to introduce major flight deck updates).

A final factor to consider is the size of the market for FMC/FMSs. The cost of developing and certificating a system to the standards required by the airworthiness authorities is considerable, however the market is relatively small. The development of a radical interface to the FMS would vastly increase the unit price to the end purchaser (this being the airline, via the aircraft manufacturer).

As can be seen, there are considerable regulatory, financial, training and organisational problems associated with even modest updates to the pilot–aircraft user interface. Even a modest update will be very expensive for an airline when the organisational costs are considered. The cost associated with the development and implementation of an improved user interface to the FMS is probably relatively small when considered in the wider scheme of things. These external factors have considerably hindered the development of the pilot–computer interface on the flight deck in comparison to desk-top computing.

The FMS/FMC in the modern commercial aircraft is a classic example of how a computer system has evolved over the years, to expand its functions way beyond what it was originally intended to do. However, as a result of this its Usability has suffered enormously and is only now being radically addressed in new commercial aircraft types. However, pilots expect twenty-first-century Usability standards on what is basically a 1970s interface. With the advent of new ATM strategies such as ('free flight') the FMS will become an even more important piece of equipment. As a result, in the short term training will have to compensate for deficiencies in the user interface. Furthermore, HCI can never be independent from training as a result of the amount of functionality that cannot be 'seen'. On a desk-top computer user exploration may be an acceptable way of finding the hidden functionality in a computer program. This is not really desirable on the flight deck … As a result, training is essential.

FURTHER READING

Even though it is a little old now, there is only one book that deals exclusively with computers on the flight deck:

Dekker, S.W.A. and Hollnagel, E. (eds) (1999). *Coping with Computers in the Cockpit.* Aldershot: Ashgate.

For a general book on Human–Computer Interaction, try:

Shneiderman, B. and Plaisant, C. (2009). *Designing the User Interface* (5th edition). Reading, MA: Addison-Wesley.

PART FOUR

The Management

The function of *Management* is central to all aspects of Human Factors in aviation. In the case of the following chapters in this Part of the book the emphasis is firmly on Safety Management. However the role of airline *Management* is also to ensure a return on investment for shareholders (by providing an efficient and expedient operation for the paying passengers) and to ensure that all operations fall within the legislative requirements of the Societal *Medium*. The *Management* function ensures 'fit' between the HuMan and the *Machine* to undertake the *Mission* safely and expediently within these safety requirements imposed by the wider, international society.

The *Management* function in the aircraft, from a safety perspective, is complementary to the wider Safety Management function in the airline. The role of the pilot was originally to undertake the three basic tasks of Aviate, Navigate and Communicate. To these functions the broader role of *Management* must now be added: this is management of the aircraft systems and, in the case of the Captain, the management of the crew in the aircraft. In many ways, flight deck management is a microcosm of the wider Safety Management function within the airline. All management is dependent upon leadership, clear communication and teamwork. These also form the basic building blocks of CRM: it is just a question of scale.

The development of safety initiatives of any size will require the identification of risks and hazards, followed by the implementation of the safety programmes and the monitoring of their effectiveness. These are all essential audit functions. Line Operations Safety Audits are now an essential part of the flight deck Safety Management processes. Their results feed both into CRM development and provide a basis for the improvement of all the other training in the airline (part of the Training Needs Analysis process described in Part Two concerned with the huMan component). It is a moot point whether LOSA fits into Chapter 16 concerned with Flight Deck Safety Management but Chapter 17 on the wider Safety Management of airlines was already getting too big! It would be wrong, though, to overly restrict LOSA activities to just the flight deck. It is convenient to put it in Chapter 16, though, simply for structural reasons in this book. But as noted in the Preface, all boundaries and divisions created by man are (to some extent) artificial, and subject matter boundaries in the complex world of aviation operations can be both artificial and arbitrary …

The Captain is the hands-on, minute-to-minute manager but she/he is also the essential link to the airline-wide Safety Management function. However, the dividing line between the flight deck and airline functions is decidedly blurred … Perhaps they are best thought of as living on a continuum defined only by the size and complexity of the activity. Or maybe one is nested within the other. Draw the diagram any way that you want!

The airline-wide Safety Management function should take a wider, more systemic approach than that on the flight deck. From a Human Factors perspective, Safety Management requires knowledge of a bit of everything. There is an intimate link with the organisational roots of human error, training, workload, physical and psycho-social stressors, and the design and operation of procedures. However, it has to be acknowledged that commercial aviation is a very 'open' system (if that makes any sense). You cannot divorce if from the wider cultural context, be that the Safety Culture that has developed within the airline, the national culture or the differing cultures of the nations across the world. Aviation is an international business: you cannot ignore cultural variations in the way that people go about their business. Furthermore, these are more than just superficial differences. In addition to different values and beliefs there are fundamental differences in the way that people from different cultures construct, process and interpret information (Nisbett, 2003). It is a cliché, but culture is more than just 'skin deep'. You cannot change it, just accommodate to it.

Finally, if the worst comes to the worst the incident and accident investigation process is an essential component in closing the safety loop. There is a human component in nearly nine out of ten accidents (CAA, 2008) but in the same way that a structural failure is a failure in the design and engineering process, any human failure in an accident can be regarded as a failure in Safety Management. A great deal of time and money is spent on accident investigation, but accidents are thankfully rare events and are usually the result of a unique set of circumstances. Incidents are more common, can be investigated faster, cheaper and produce better quality data. Aggregated incident data can be indicative of failing safety trends long before an accident finally happens. Perhaps it would be better if more time and money was devoted to incident investigation and analysis and less on accident investigation. However, this would require a shift in the policy emanating from the international regulatory *Medium* – the International Civil Aviation Organisation (ICAO) .

The book concludes with a consideration about the way ahead for Human Factors in civil aviation. The majority of this book still implicitly reflects the emphasis that has been placed in the first half century of the discipline, namely that of enhancing safety. It is in this area where the majority of the science has developed. However, with Human Factors maturing as a discipline the careful application of its principles can enhance performance and reduce operational and through-life costs. The emphasis in the final chapter is on improving efficiency as a result of taking an integrated, through-life approach. The Human Factors discipline can now 'add value'.

16
Flight Deck Safety Management: Crew Resource Management and Line Operations Safety Audits

Failures in CRM still remain the fourth most common causal factor in fatal aircraft accidents and the third most frequently cited contributory factor (CAA, 2008). The evolution of the abbreviation of CRM itself exemplifies the change in culture in commercial aviation in the last 20 years. When CRM was first introduced, the abbreviation stood for Cockpit Resource Management and applied only to flight deck crew. Subsequently, the concept evolved to encompass all flight crew (i.e. including the cabin crew) and became known as Crew Resource Management. Further evolution of the approach has seen CRM extend beyond the aircraft to ramp operations and maintenance, and even beyond the airline (e.g. to ATC).

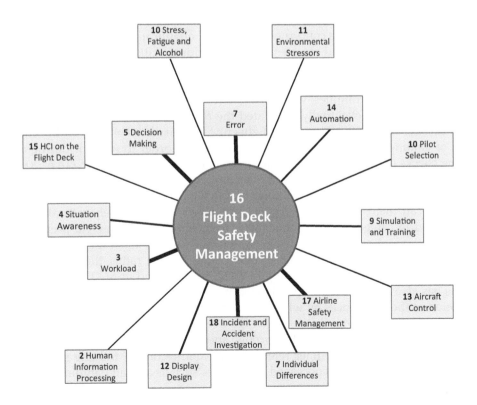

Flight Deck Safety Management is a microcosm of the wider topic of Airline Safety Management. The emphasis is upon the optimal utilisation of all the human resources on the flight deck (CRM), enhancing crew decision making while avoiding (or managing) error. Revisions in Flight Deck Safety Management processes are often driven as a result of the investigation of accidents and incidents.

Up until relatively recently, pilot training and licensing concentrated on flight and technical skills (manoeuvring the aircraft, navigation, system management and fault diagnosis, etc.). However, with increasing technical reliability, it became evident that the major cause of air accidents was human error. The failure of the flight deck crew to act in a well co-ordinated manner further contributed to this end on many occasions. Perhaps the most famous of these accidents which stimulated the development and adoption of CRM involved a Lockheed L-1011 (Tristar) in the Florida Everglades in 1972 (National Transportation Safety Board, 1973). In this case an aircraft with a minor technical failure (a blown light bulb on the gear status lights) crashed because none of the crew was actually flying it. All the crew were 'head down' trying to fix the problem. Other accidents highlighted instances of such things as the Captain trying to do all the work while other flight crew were almost unoccupied or, as in the Staines Trident Accident (UK Department of Trade and Industry, 1973) a lack of crew cooperation as a result of an overbearing and autocratic Captain. In this case, the Captain effectively discouraged communication and questioning of his actions, which eliminated any cross-checking for errors. The aircraft subsequently crashed after a series of errors were made, which seemed to go uncorrected.

Such instances resulted in a series of intra-cockpit flight crew management programmes being instigated. In parallel with these developments during the 1980s, there was also a vast increase in the degree of flight deck automation that also required a change in operating philosophy and hence resulted in further new (at the time) training requirements emerging. While regulations concerned with pilots' qualifications still almost exclusively deal with the acquisition and maintenance of individual skills, there is now a mandated requirement for crew concept training as part of the ATPL (Airline Transport Pilot Licence) syllabus.

Initially CRM programmes were instigated voluntarily by airlines, although the regulatory authorities eventually mandated the requirement for crew concept training. Within Europe, the JAA/EASA require flight crew to be assessed on their CRM skills both as part of gaining a professional pilot's licence and in order to retain it during annual licensing checks (*Joint Airworthiness Requirement – Operations (JAR – OPS)* JAA, 2000/EU-OPS 1, EASA 2009 and *Joint Airworthiness Requirement – Flight Crew Licensing (JAR – FCL)* JAA, 1999). The US FAA adopts a different approach. CRM is part of the AQP in which the airlines can themselves set the proficiency objectives. CRM behavioural objectives are identified in categories that either relate to 'enabling objectives' (to prepare crews for further training) or 'currency items' (activities that crew must remain proficient in, which can be demonstrated through frequent performance rather than regular formal evaluation). Details of the AQP relating to CRM can be found in AC 120-54 (FAA, 1991) and this has also been touched upon in Chapter 9 on Training and Simulation. Further description of the differences by which the two approaches to CRM approval differ can be found in Maurino (1999).

It is difficult to separate CRM concepts and processes from CRM training: so I won't. CRM training is sometimes referred to as 'non-technical' training (in contrast to 'technical' training, which is concerned with issues such as aircraft handling and system management). But it is acknowledged that this is a non-perfect distinction to draw, as will be highlighted at the end of this chapter. However, CRM is a skill developed in the hu*Man* through training. Appropriate and approved training must be provided by the airline's *Management* component to this end, and as CRM is mandated by the airworthiness authorities (part of the social *Medium*) this leads to another issue that must be addressed. How should CRM programmes be evaluated to establish their effectiveness and how should crew competence be demonstrated to the satisfaction of the regulatory authorities?

CRM is the management process to encourage the optimal use of human resources on the flight deck, however this does not provide the whole answer. It needs to be established if the CRM processes in which the flight crew are engaged are both appropriate and effective: does what is taught in the simulator and classroom actually transfer to the line? Over the last decade (or so) LOSA has become an essential part in closing this loop. The LOSA process is predicated upon a model of the management of human error on the flight deck, the Threat and Error Management model – see Helmreich, Klinect and Wilhelm (1999) and Figure 6.2 in Chapter 6 on Error. FAA Advisory Circular AC 120-90 (2006) likens a LOSA to an annual physical examination; an annual 'health check' which provides a 'diagnostic snapshot of strengths and weaknesses that an airline can use to bolster the "health" of its safety margins and prevent degradation' (AC 120-90, p. 2). Audits are undertaken by trained observers on the flight deck during scheduled flights. These observers collect data on environmental conditions, operational complexity and crew performance which may represent a threat to safe operations. These data are then used to inform crew training and flight deck practices, among other things.

These developments can be thought of as Safety Management for the flight deck (as opposed to airline-wide Safety Management, discussed in the following chapter). In many ways, the CRM, TEM and LOSA processes are a microcosm of the wider Safety Management processes but in no way should be considered as being divorced from the airline-wide Safety Management activity (in fact quite the reverse).

CREW (COCKPIT) RESOURCE MANAGEMENT

If this chapter had been written ten years ago it would probably have been much longer and it would also have been thought of as being much more radical in its scope and content. CRM, however, is now simply a 'way of life'; its concepts and processes permeate every aspect of operations in an airline.

The JAA (1998) defined CRM as 'the effective utilisation of all resources (e.g. crewmembers, aeroplane systems and supporting facilities) to achieve safe and efficient operation'. CRM evolved as an operating concept after a series of accidents in which the aircraft involved had no major technical failures (if any at all). The principal causes of these accidents were a failure to utilise all the human resources available on the flight deck in an appropriate manner.

UK CAA Civil Aviation Publication 737 (CAA, 2006) suggests that a CRM syllabus for flight crew should comprise:

- Human error and reliability, error chain, error prevention and detection.
- Company Safety Culture, SOPs, organisational factors.
- Stress, stress management, fatigue and vigilance.

- Information acquisition and processing, Situation Awareness and workload management.
- Decision making.
- Communication and co-ordination inside and outside the cockpit.
- Leadership and team behaviour synergy.
- Automation, philosophy of the use of automation.

The NOTECHS framework (van Avermaete, 1998), which is essentially an evaluation framework developed for the JAA to assess CRM skills, contains four dimensions each comprised of several elements:

- *Co-operation* – Team building and maintaining, considering others, supporting others and conflict solving.
- *Leadership and managerial skills* – Use of authority/assertiveness, providing and maintaining standards, planning and coordination, and workload management.
- *Situation Awareness* – System awareness, environmental awareness and anticipation.
- *Decision making* – Problem definition/diagnosis, option generation, risk assessment/ option choice and outcome review.

These topics are covered in some detail throughout the various chapters in this volume. This chapter looks at the integrated whole and the operationalisation of these concepts that has become CRM.

Both Helmreich (1994) in the USA and Pariés and Amalberti (1995) in Europe have suggested that CRM has progressed through four distinct eras. First generation CRM focused on improving management style and interpersonal skills on the flight deck. Emphasis was placed upon improving communication, attitudes and leadership to enhance teamwork. Although first generation CRM training placed emphasis on teamwork, in many airlines only Captains underwent CRM training!

Second generation CRM built upon first generation concepts but also included instruction in stress management, human error, decision making and group dynamics. However, CRM also began to extend beyond the flight deck door. In third generation CRM, cabin crew also became very much part of the team. Training was extended to include whole crews together, rather than training the flight deck and aircraft cabin separately. The CRM concept also began to extend into the organisation as a whole, embracing further concepts such as national and Safety Culture.

By fourth generation CRM, training *per se* was beginning to disappear as the concepts were being absorbed into all aspects of flight training and the development of flight deck procedures. Helmreich, Merritt and Wilhelm (1999) suggested that fifth generation CRM would extend throughout the organisation and basically involve a culture change. Early CRM approaches were based upon *avoiding* error. Fifth generation CRM, if it can still be considered to exist as a distinct concept, assumes that whenever human beings are involved, error *will be* pervasive. Emphasis is on using the error management troika: avoid errors; trap errors; and/or mitigate the consequences of errors. Fifth generation CRM recognises that humans are fundamentally fallible, especially under stress. Error is part of the human condition, and in common with airline-wide Safety Management programmes, if reported accurately and in a timely manner (as part of a non-jeopardy reporting programme) are there to be learned from.

Early studies on flight deck leadership used 'management style' tools such as the least preferred co-worker approach (Fiedler, 1967) and the managerial grid (Blake and Mouton, 1978).

Fiedler (1967) posited that leaders (Captains) prioritise between task-focus and people-focus. People-orientated leaders were regarded as tending to have positive relationships with their co-workers and act in a supportive way often putting personal relationships first, prioritising them over task performance. Task-orientated leaders gave the task priority and focused on interpersonal relationships only when the job was progressing well. Fiedler suggested that there were three factors relating to the leader, group members and the task:

- *Leader–member relations* – the extent to which the leader has the support of team members and the tone of team inter-relationships (friendly and cooperative).
- *Task structure* – the degree to which tasks are standardised, documented and controlled.
- *Leader's position–power* – the degree to which the leader has authority.

Blake and Mouton (1978) simplified the complex inter-relationships between task and leadership style hypothesised by Fiedler and produced the managerial grid. The managerial grid has two dimensions:

- *Concern for People* – the degree that a leader considers the needs and interests of team members when deciding how best to accomplish a task, and
- *Concern for Production* – the degree that a leader pursues objectives efficiently when undertaking a task.

Initially, it was thought that task-oriented Captains would produce better crew performance, however group-oriented leaders have also been found to be effective on the flight deck (see Foushee and Helmreich, 1988). By the second generation of CRM, though, the emphasis had changed to training pilots in leadership techniques and group dynamics, rather than trying to describe what makes a good leader and team player. Simultaneously, changes in pilot selection criteria began to take place. As discussed in Chapter 8 on Pilot Selection, new hire pilots began to be selected on their social and management skills in addition to their technical proficiency (e.g. Stead, 1995; Bartram and Baxter, 1996; Hörmann and Maschke, 1996). CRM skills tests are being developed to further aid in this process (Hedge, Bruskiewicz, Borman, Hanson, Logan and Siem, 2000).

Helmreich and Foushee (1993) described CRM in terms of a generic Input–Process–Outcome (IPO) model (see Figure 16.1). Features on the input side of the model include aspects of the hu*Man* (personality, attitudes, abilities, etc.) the *Medium* (regulations, national culture, operating environment), and other aspects of the company *Management*. Process factors include intra-crew communication, team formation and management, Situation Assessment/Awareness and decision making. Outcome factors evaluate the performance of both the *Mission* and the crew.

Members of the flight deck crew gather data. These data need to be communicated to enhance everyone's Situation Awareness, which itself is necessary to promote good decision making, awareness of the decisions made and of the actions subsequently taken. These processes are promoted by good management, leadership and teamwork.

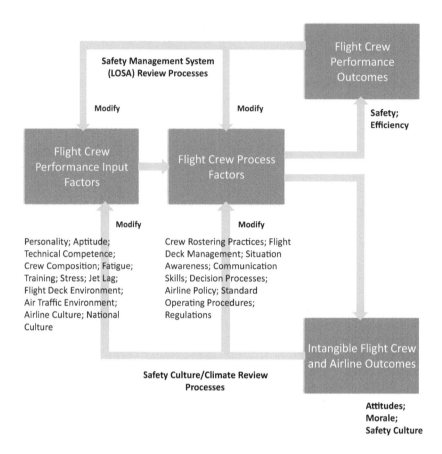

Figure 16.1   Input–Process–Outcome model of CRM (adapted from Helmreich and Schaefer, 1994)

The basic building block of CRM is intra-flight deck communication. In a review of aircraft accidents (Murphy, 1980) it was found that errors were more likely to be the product of failures in communication and coordination than deficiencies in technical proficiency. Communication on the flight deck should be a two-way process. There is a sender who encodes the message and transmits it to the receiver who decodes it. The receiver should then become the sender and re-encode the message and transmit it once again back to the receiver who decodes it once more and checks that the meaning of what was sent was understood properly. Kanki and Palmer (1993) describe five ways in which communication facilitates CRM performance. Communication:

- provides information;
- establishes interpersonal relationships;
- establishes predictable behaviour patterns;
- maintains attention to task and monitoring; and
- is a management tool.

It has generally been found that crew performance improves with an increase in task-related communication (e.g. Foushee and Manos, 1981; Foushee, Lauber, Baetge and Acomb, 1986; Kanki and Palmer, 1993). However, during abnormal situations the relationship between communication and performance is less clear. Kanki, Folk and Irwin (1991) found that the best performing crews in an emergency exhibited homogeneity in their communication patterns: communicating in a highly standardised manner enhanced task performance in these circumstances through increased predictability. During abnormal situations, though, the over standardisation of communication protocols was detrimental (Kanki and Palmer, 1993). Mjøs (2001) observed that in high-pressure situations, informal (but task related) communication was beneficial in improving crew performance, since this approach contributed to the pilots developing a shared mental model of the situation.

Communication is the basis for flight deck-wide decision making as a crew. Huddlestone and Harris (2003) observed that communication was a necessary pre-cursor to promoting team Situation Awareness which is itself the precursor for all decision making (Lipshitz, 1993; Nobel, 1993; Prince and Salas, 1997). Studies in flight simulators have observed that high performing crews discussed in-flight problems in greater depth than did crews that performed less well. These crews also used the low workload flight phases to plan ahead and talk about options (Orasanu and Fisher, 1992). Furthermore, they talked more about strategy and planning but issued fewer commands during an emergency.

The theoretical basis for decision making has been described in an earlier chapter. As far as decision making on the flight deck is concerned within the context of CRM, though, most emphasis has been placed on the development of easily trained decision-making processes. Aviation abounds with mnemonics and aids to help aircrew adopt an appropriate decision-making strategy (e.g. DECIDE – Detect, Estimate, Choose, Identify, Do, Evaluate – Benner, 1975a; QPIDR – Questioning, Promoting Ideas, Decide, Review – Prince and Salas, 1993; FOR-DEC – Facts, Options, Risks and benefits, Decision, Execution, Check – Hörmann, 1995). Empirical evaluation of these guidelines is sparse, although there is some evidence that their training is beneficial to performance. During a series of assessment flights in which they were faced with a series of multiple problems, Jensen (1995) observed that pilots trained in the DECIDE technique performed better than those without training. Murray (1997), using a variation on the DECIDE decision-making approach (DESIDE – Detect change, Estimate the significance, Set safe objectives, Identify options, Do best option, Evaluate) observed a positive attitude amongst pilots to a propaganda campaign promoting its use to structure in-flight decision making. When assessing performance of commercial pilots during line checks, Goeters (2002) specifically regarded the appropriate use of the FOR–DEC decision-making process as evidence of 'good' CRM.

Effective communication and good decision-making skills are of little benefit if the crew do not act together as a team under the direction of a leader (i.e. the Captain). Unlike military aviation, where aircrew fly together as members of a fixed team, in most airlines Captains and First Officers rarely fly together as members of a regular crew. The assumption is that this approach avoids 'corner cutting' and complacency that is occasionally evident in long established teams, hence every flight has to be conducted exactly to the prescribed standard operating procedures. As a result, quick and effective team building is essential for flight safety in commercial aviation.

The Captain in an airliner sets the 'tone' and is crucial in making the crew into a team. Ginnett (1993) found that the Captains particularly good at establishing effective teams held extensive crew briefings and debriefings before and after each flight. During these briefings they established their credibility as a leader, spelling out the required goals for performance and leading by example, yet at the same time encouraged all crew to be active participants in the management of the flight. As was noted earlier, effective communication is the entire basis for effective team management on the flight deck but communication (in terms of the transmission side of the equation) is easier to achieve for the Captain than for the First Officer! The First Officer must be both assertive and a subordinate, which is a very fine balance to achieve if proper communication and coordination is to occur. The trans-cockpit authority gradient was subject to a great deal of research in the earlier days of CRM (e.g. Edwards, 1975; Wheale, 1983). It has been suggested that there is an optimum authority gradient to promote effective communication and coordination between crew members. If the gradient is too 'flat' (e.g. two Captains or a junior Captain and a very senior First Officer) then the dividing line of who is in charge can be unclear. If the gradient is too 'steep' with a domineering senior Captain and an unassertive junior First Officer, this can inhibit communication, coordination and the cross checking of errors. One of the roles of the 'other' pilot on the flight deck (whatever their designation or role is, be it First Officer of Pilot-Not-Flying) is as an error checker and early warning system. This can only happen if open communication is established. As will be discussed, this is a bigger issue in some cultures compared to others.

CRM and LOFT (Line Oriented Flight Training)

In addition to classroom-based training for CRM concepts a great deal of emphasis in airlines is placed upon the implementation and practise of these skills, usually during regular LOFT sessions. LOFT places emphasis on training as a crew and acting as a crew member. During a LOFT session (which usually takes place in a full-flight simulator) crews will fly a complete trip (or part thereof) just as they would in normal operations. However in this case there will be a series of in-flight problems and emergencies that require the pilots to act as a team to resolve them satisfactorily. During these simulated flights the instructors will not intervene but crews' actions will be recorded for later analysis. After the LOFT session, the crews' performance will be reviewed with respect to how they handled flying the aircraft, the technical aspects of the problem and perhaps most importantly, how the human capital was employed to address the issue (see Foushee and Helmreich, 1988).

The effectiveness of the LOFT approach depends upon two key factors: the development and implementation of appropriate flight scenarios within which the training takes place, and the adequacy with which crews are de-briefed by the instructors after the exercise. The FAA publishes guidelines for the development of LOFT scenarios (Advisory Circular 120-35B – FAA, 1990). Prince, Oser, Salas and Woodruff (1993) expanded upon these principles, recommending that to ensure the maximum effectiveness of any LOFT session, trainees should undergo two briefings prior to their session. The first briefing ought to outline the objectives of the training and how the session will progress. The second briefing should be a regular briefing (route, weather, aircraft servicability, etc) to put the flight into a normal operational context. Prince, Oser, Salas and Woodruff (1993) emphasised that scenarios should be designed to address specific training objectives, and that they should be realistic, including mundane items such as flight deck paperwork, ATC communication and periods of inactivity that characterise many flights. All these factors should enhance the reality of the simulation and help to ensure flight crew 'buy in' to the training. Hamman, Seamster, Smith and Lofaro (1993) recommended that the problems faced by the crew require that the pilots should be able to exercise both their technical and inter-personal skills. They should comprise of an initiating event; irrelevant occurrences to distract and/or mislead the crew, and other supporting factors. Furthermore, there should also not be just a single solution to the problem.

The role of the facilitator during the de-briefing following a LOFT exercise is central to its training effectiveness. The FAA (1990) put forward that the de-brief should include a discussion of both the positive and negative aspects of the crews' performance but the facilitator in these sessions should not be judgmental. However, Dismukes, McDonnell and Jobe (2000) observed great variations in facilitator performance. After a two-hour simulator ride it was noted that the mean duration of a post-LOFT simulator debriefing was only 31 minutes. One-third of this time was often spent reviewing incidents on the video and during the remaining time, the Training Captain (facilitator) overseeing the session frequently spent more time talking than did the crew undergoing the training.

While LOFT and CRM programmes have been thought to provide a great many safety and economic benefits to airlines, this has been difficult to establish empirically (Edkins, 2002). O'Connor, Flin and Fletcher (2002) assessed the effectiveness of CRM programmes using the four-level framework for training evaluation described by Kirkpatrick (1976, 1998), as described in Chapter 9 on Training and Simulation.

Almost half the studies utilised a Reactions measure to evaluate CRM training and in all cases bar one, crews were found to receive it positively. Scenario-based instruction and case studies were well-received, however instruction based solely upon lectures was not so well regarded. Studies of attitudes toward CRM conducted pre- and post-training (often using the CMAQ – Cockpit Management Attitudes Questionnaire – Helmreich, 1984) consistently reported a positive change in attitudes immediately after training (e.g. Gregorich, Helmreich and Wilhelm, 1990; Helmreich and Wilhelm, 1991). However, Irwin (1991) reported that in a longitudinal study these positive attitudes decayed over time and only recovered after subsequent recurrent training.

Only 15 per cent of the studies reviewed reported a Learning measure at the end of the course and in the vast majority of cases it was not formalised in any way (e.g. assessments were made informally by instructors during subsequent LOFT sessions). The main reason given for not using formal evaluations was that the demonstration and application of CRM skills was regarded as being more important than theoretical knowledge of the underlying concepts (O'Connor, Flin, Fletcher and Hemsley, 2002). The most widely reported level at which CRM was evaluated was at the Behavioural level. A number of Behavioural rating systems have been developed specifically for such purposes (e.g. the Line/LOS Checklist developed by Helmreich, Wilhelm, Kello, Taggart and Butler, 1990; NOTECHS – van Avermaete, 1998). Behavioural observations were commonly conducted by a Training Captain during the course of normal operations or in the simulator as part of a LOFT session (O'Connor, Flin, Fletcher and Hemsley, 2002). CRM behaviours generally improved after training (Clothier, 1991; Ikomi, Boehm-Davis, Holt and Incalcaterra, 1999). Self-report measures suggested that improvements were maintained six months after training (Naef, 1995).

Few studies have been reported evaluating CRM at the Organisational level. O'Connor, Flin and Fletcher (2002) report only two in the open literature although O'Connor, Flin, Fletcher and Hemsley (2002) claim that 33 per cent of UK operators do evaluate their programmes internally at this level. The two available reports, though, shed little light on the overall effectiveness of CRM training. In contrast, military data where the accident and incident rate is much higher than in commercial operations show considerable benefits from the instigation of CRM programmes. Diehl (1991a) reports an 81 per cent drop in the accident rate in US military aviation after the introduction of CRM training.

The JAA (now EASA) requires the training and assessment of CRM skills as part of the operational regulations. JAR-OPS 1.965 (now EU-OPS 1.965) requires that:

> ... the flight crew must be assessed on their CRM skills in accordance with a methodology acceptable to the Authority and published in the Operations Manual. The purpose of such an assessment is to: provide feedback to the individual and serve to identify retraining; and be used to improve the CRM training system.

The NOTECHS methodology is approved as one means of compliance with this requirement. The NOTECHS system has been briefly described at a high level earlier but each of the elements within each category are further defined by observable, quantifiable behaviours for the purposes of assessment. For example, in the element of 'Team Building and Maintaining' within the sub-category of 'Cooperation' this aspect of CRM is evaluated by observers assessing *observable* (note emphasis) behaviours such as: 'establishing an atmosphere for open communication and participation'; 'Encouraging inputs and feedback from others'; and 'not competing with others'. In the element of 'Providing and maintaining standards' within the sub-category of 'Leadership and management skills' this aspect is evaluated by evaluating behaviours such as: 'Ensuring SOP compliance'; 'Intervening if task completion deviates from standards'; and 'Consulting with crew when deviating from standards if the situation requires'. Full details of the NOTECHS system can be found in van Avermaete (1998) and its process of operation and method of validation can be found in Flin, Martin, Goeters, Hörmann, Amalberti, Valot and Nijhuis (2003).

CRM and Automation

Automation has had a considerable effect on CRM practices. Civil Aviation Authority publication CAP 737 (*Crew Resource Management (CRM) Training*) notes that 'CRM in highly automated aircraft presents special challenges, in particular in terms of situation awareness of the status of the aircraft' (Appendix 8, p. 1). Although the primary role of the pilot has not changed with the introduction of high levels of automation (it is still to complete the planned flight safely and efficiently) the general operating philosophy is often that the automation takes the role of maintaining basic stability and control and higher-level functions (e.g. flight planning, system management and decision-making) are the preserves of the crew. CAP 737 advises that guidelines on the use of automation should be provided indicating when and when not to use automation (and to what extent). CRM procedures should explicitly take into account the status and role of the automated systems.

In the same document it is also observed that several areas have been identified where automation has impinged on CRM procedures and pilot roles. Co-ordination and crew management is required in the assignment of tasks and the standardisation of their performance when using automated systems, because:

- Automation makes it difficult for one crew member to see what the other pilot is doing. Proper procedures are required to ensure both pilots are aware of any inputs made to the automation. Clear communication is required.
- It is more difficult for the Captain to monitor the work of the First Officer (and vice versa) on a highly automated flight deck. Often, only the output (not the process) can be verified. Procedures and clear cross-cockpit communication are again needed.
- Automation has been observed to break down the role of flying and non-flying pilot: there is less clear demarcation of who does what. Procedural standardisation is a foundation of safety on the flight deck hence clear standard operating procedures are required for interacting with highly automated systems.
- Automated flight decks can unintentionally redistribute authority from the Captain to the First Officer if the latter is more proficient in the use of the automation (particularly in times of high workload). However, judicious use of superior automation-related skills in other crew may, by surrendering limited authority to them, be regarded as a sign of good CRM practice.
- There is a tendency of the crew to help each other with programming duties when workload increases which can dissolve the clear demarcation of duties and may detract from the primary task of flying the aircraft.

Dekker (2004a) suggests that the automation must almost be treated (conceptually) as another crew member on the flight deck. As noted in Chapter 14 on Automation, Christoffersen and Woods (2000) describe a 'good team player' as one who makes their activities observable for their fellows and they are easy to direct. The same applies to CRM processes, be they undertaken by a human pilot or a machine. The greater the degree of autonomy and process complexity the more feedback is required to make the relevant behaviours observable and to avoid misunderstanding and miscommunications. Expanding upon the NOTECHS CRM framework described earlier (van Avermaete, 1998) automation can be incorporated into the CRM processes in the following ways:

- *Co-operation*. Automation outputs must be observable and crew inputs must be observed. The design of the automation should support the flight task not dictate it. Furthermore, the pilot can easily and efficiently instruct them what to do.
- *Leadership and managerial skills*. The human operator should be able to delegate authority to the automation but retain ultimate authority: the automation should clearly communicate when it is approaching the limits of its authority (not when it has passed them). Usually, for example when the autopilot is in control of the aircraft, the pilot will have much greater authority over the size of control inputs possible.
- *Situation Awareness* is dynamic and is distributed across human and machine components (see Stanton, Stewart, Baber, Harris, Houghton, McMaster, Salmon, Hoyle, Walker, Young, Linsell and Dymott, 2006, in Chapter 4): multiple views of the same flight situation are held by all the different agents (pilot or aircraft) but different components in the system represent different levels of Situational Awareness. Clear, coordinated human–machine communication is required to enhance Situation Awareness right across the flight deck. It has been argued previously that Situation Awareness does not reside simply in the human, flight crew components: it is systemic.

- *Decision making.* Automation should support decision-making (e.g. through problem diagnosis or option generation) not dictate it or constrain it. Situation Awareness is a precursor to effective decision making (see Chapter 5).

CRM and Culture

In Chapter 6 on Error it was noted that there were national differences in aircraft accident rates. In an analysis of worldwide fatal accidents (CAA, 2008) CRM is not listed at all as a circumstantial (contributing) factor in any North American accidents during the period of analysis, however it is the second most frequent circumstantial factor in accidents involving Asian and Middle Eastern carriers (to be fair it still remains the second most significant causal factor in the US, though). It can also be suggested that most facets of the aviation system have been constructed from a Western (North American/Western European) perspective. However, as a result, the causal factors underlying accidents and prevention strategies that seem reasonable to Westerners might present problems for East Asian and African people. In the context of this chapter the manner of implementation of CRM practices with the First Officer being actively encouraged to question the Captain and be pro-active runs counter to many East Asian norms (Hofstede, 1984; Merritt and Maurino, 2004) where there is a high Power–Distance culture. People from low Power–Distance countries are less inhibited in speaking out when their opinions may contradict those of their superiors. In high Power–Distance cultures confrontation is generally circumvented and questioning or contradiction of superiors is avoided at all costs (Hofstede, 1984; Helmreich and Merritt, 1998). What is more, Westerners may not even be aware of such a problem (Johnston, 1993; Jing, Lu and Peng, 2001).

High Power–Distance has been implicated in several accidents where CRM was identified as a causal factor (e.g. Columbian Avianca flight 052, as described by Helmreich, 1994). In this case the junior flight crew members repeatedly failed to communicate to the Captain the criticality of the worsening fuel situation. The Boeing 707 subsequently ran out of fuel and crashed. CRM issues attributed to high Power-Distance were also implicated in the China Airlines Airbus A300 accident at Nagoya, Japan (described in Chapter 14 on Automation). In this case the Captain intervened after the First Officer accidentally activated the take-off/go-around mode during approach. The Captain (unsuccessfully and inappropriately) fought the automation that was automatically executing a go-around, resulting in the aircraft adopting an excessive nose-up attitude. It finally stalled, killing all on board. At no point did the First Officer question his Captain's actions. Various recommendations were made in the accident report, including a requirement to review procedures to ensure that CRM became more effective (Aircraft Accident Investigation Commission, 1994). In many similar cases the CRM solutions proffered are often predicated upon changing attitudes and behaviours on the flight deck to overcome the reticence of junior members of flight crew to question the Captain. However, it has also been recognised that 'culture sensitive' CRM training is required: 'Western style' solutions will not work in Asian cultures. For example, assertiveness training for First Officers in high Power–Distance cultures is unlikely to be effective as it runs against highly ingrained cultural norms (Helmreich and Merritt, 1998). A five-day training course cannot overcome the ingrained effects of several thousand years of culture.

However, cultural issues on the flight deck run deeper than simply issues in CRM. The basic operating philosophy in commercial aircraft centres on two pilots cross-monitoring each other's actions. Checklists and standard operating procedures are performed on a 'challenge and response' basis. The system of 'monitor and cross-monitor' and 'challenge and response' is predicated upon the belief that crew members will alert each other to irregularities and errors, but implicit in this is an assumption of low Power–Distance. This is an operating philosophy that is based upon a Western European/North American cultural assumption (Harris and Li, 2008).

Johnston (1993) also observed that there were national differences in the conceptualisation of CRM which have an impact on its manner of implementation and crew's performance. Furthermore, he observed that there was a marked difference in how CRM training is perceived outside the USA. In the USA, Johnston suggested that normally CRM is seen as the primary vehicle through which to address human performance issues. Other countries, notably those in Europe, see human performance issues and CRM as overlapping, viewing them as close but distinct relatives. CRM is regarded merely as one facet of human performance in the aeronautical context.

Personnel Selection to Promote Good CRM

Meta-analyses of pilot selection studies suggest that personality is a weak predictor of performance in general (e.g. Hunter and Burke, 1994 and Martinussen, 1996) however when the criterion measures are closely related to team performance on the flight deck the results are different (see Chapter 8 on Pilot Selection). In 'early generation' CRM (where the 'C' often still stood for 'cockpit') Chappelow and Churchill (1988) argued that the 'right' personality types were required to promote appropriate social interactions and co-ordination on the flight deck. However, Besco (1994) contended that breakdowns in CRM were more likely to be due to problems in 'personnel policy, disrespect for "upward communication," non-existent leadership training and exclusive "solo-Captain" policies' (p. 25) rather than personality clashes. Empirical evidence, though, suggests that this view is debatable. Hörmann and Maschke (1996) analysed personality data (collected on their application for employment) from *circa* 300 pilots. This was in addition to data from a simulator check-ride and other biographical data (e.g. age, flight experience and command experience). After three years of service it was observed that pilots graded as 'below standard' had significantly lower scores on the interpersonal scales and higher scores on the emotional scales of the TSS (Temperament Structure Scales – Maschke, 1987) a personality instrument specifically developed for assessing aircrew. These results supported earlier findings by Chidester, Helmreich, Gregorich and Geis (1991) who found that better performing airline pilots scored higher on personality traits such as mastery and expressivity, and lower on dimensions such as hostility and aggression.

Psychometric selection processes have been developed by the US Air Force aimed specifically at identifying pilots who will make effective crew members in multi-crew transport aircraft (e.g. Hedge, Bruskiewicz, Borman, Hanson, Logan and Siem, 2000). These are based on situational judgement tests rather than personality measures. Such tests require the respondent to answer with what they regard to be the most appropriate course of action in a series of job-relevant scenarios. The authors claim that this approach provides a better prediction of CRM performance than personality tests.

Evolution of CRM and LOSA

CRM (and LOSA) concepts are not static – they continue to evolve. Early in the 1990s there was increasing concern that CRM and technical training were diverging. There was oversight of the fact that flying a modern airliner required both types of skill and that the two were complementary and should be integrated. Lofaro (1992) developed the Mission Performance Model (MPM). This was an attempt to integrate CRM and technical performance into an integrated whole. It was based upon various premises, including:

* Flying is an integrated, mission-oriented activity which must be evaluated as such.

* The crew's performance is not adequately captured by totalling the sum of the component tasks/sub-tasks/elements. The focus must be on crew function – usually at the task and critical sub-task levels.

* Flight proficiency skills/knowledge are interwoven, interdependent, and necessarily interact with the CRM skills/knowledge differentially across tasks and conditions.

(Lofaro, 1992,. p. A-5).

In the MPM flight crew generic functions are identified, irrespective of them lying within either the technical or CRM domains. These functions were:

* Communications process and decision behaviour.
* Workload management and Situation Awareness.
* Team building and maintenance.
* Operational integrity.
* Flight manoeuvres and attitude control.
* Propulsion lift/drag control.
* System operation.
* Malfunction warnings and reconfiguration.

Each of these functions are applied and evaluated with respect to specific phases of flight or flight scenarios, developed from a task analysis of operations. Both the non-technical and the technical aspects of each function are evaluated using a set of behavioural markers similar to the approach used in NOTECHS (see Lofaro, 1992). This approach failed to gain much favour, perhaps as a result of the development of the FAA AQP programme. However, its basic approach is entirely consistent with the AQP.

LINE OPERATIONS SAFETY AUDITS

Relatively recently there have been some questions raised about the basis of LOFT training. The *raison d'être* of LOFT is encouraging effective flight deck management practices through team-based training and a subsequent de-briefing of performance within these abnormal and emergency flight scenarios. However, the threats to flight safety faced in a LOFT session may not be the key threats to operational safety faced by aircrew on a day-to-day basis. LOSA began to be introduced into airlines (initially US carriers) in the late 1990s. LOSA data are collected on a non-jeopardy basis by trained observers during regular line operations. The initial idea was that these data formed the basis of an audit process to check the everyday safety health of an airline at an organisational level.

In a LOSA it is assumed that errors are spawned as a result of threats to safety (Helmreich, 2000). These may be either latent threats (threats that are not always easily identifiable but which predispose the commission of errors) or overt threats (factors that can more easily be identified as having the potential to increase the likelihood of an error). Latent threats include such things as organisational (safety) culture, scheduling or vague policies. Overt threats to safety encompass issues such as the operating environment (PSFs), individual pilot factors (e.g. attitude, training or fatigue) and crew factors (teamwork, leadership, etc). These threats need to be actively managed.

Data collected during everyday operations on the flight deck are encoded into three broad categories based upon the model of Threat and Error Management (described earlier in Figure 6.2 in Chapter 6 on Error). These are: external threats to safety (e.g. ATC problems, adverse weather or system malfunctions); errors and responses to the errors committed by flight crew (further broken down into the category of the error, response to the error and the outcome of the situation); and non-technical skills evaluation, which evaluates CRM behaviours using a set of behavioural ratings. Helmreich, Klinect and Wilhelm (1999) categorised errors into:

- *Communication errors*.
- *Procedural errors* (knowing what to do but doing it wrong).
- *Proficiency errors* (not knowing what to do).
- *Decision errors*.
- *Violations* in policies or procedures.

FAA Advisory Circular AC 120-90 (2006) provides a description of the benefits to an airline of undertaking a LOSA and provides an outline approach concerning the manner in which to implement it. It is suggested that the results of a LOSA can be utilised to:

- *Identify threats in the airline's external operating environment* (e.g. adverse weather, traffic congestion or airport conditions). Frequently occurring threats or those that are often mismanaged by crews can be prioritised for further investigation or training. LOSA can also identify positive behaviours which are beneficial to safety which can be used to benefit operations through developing procedures or advisories.
- *Identify threats from within the airline's operations* (e.g. operational time pressures, dispatch errors or problems with ramp personnel). Operational threats regularly arising from certain departments suggest that these areas should be targeted for improvement.
- *Assess the degree of transference of training to the line.* Data provided by Advanced Qualification Programmes, Line Operation Evaluations and LOFT programmes provide insight into the effectiveness of training and if it transfers to line operations (cf Kirkpatrick, 1976, 1998).
- *Check the quality and usability of procedures.* A LOSA may provide insight concerning problems with SOPs. If several different crews make the same error when applying an SOP it can be an indication that the procedure is ill-timed, over-long, confusing and/or competes with other concurrent flight deck activities.

- *Identify design problems in the flight deck interface.* A LOSA can capture aircraft handling and automation management errors which may be indicative of systemic flaws in design and which can also help in developing SOPs to help circumvent these design defects.
- *Understand pilots' shortcuts and workarounds.* These are rarely observed during line checks as pilots normally operate 'by the book' in such circumstances. LOSA can provide data concerning such non-standard practices, some of which may be superior ways of working to airline SOPs. The adoption of unsafe shortcuts and workarounds can also be identified and remedied.
- *Assess safety margins.* A LOSA can provide information about the prevalence of threats and errors that are mismanaged. These can be regarded as precursors to incidents and accidents. In this way it can be established how close to the edge of the safety envelope an airline is operating.
- *Provide a baseline for organisational change.* LOSA data can be used to provide baseline data against which organisational interventions can be assessed to establish if they have been effective in reducing the instance of threats, errors or undesired aircraft states.
- *Provide a rationale for allocation of resources.* LOSA data provide an insight into both the safety strengths and weaknesses in the airline and hence provide a data-driven rationale for the allocation and prioritisation of organisational resources.

The FAA emphasise the importance of obtaining organisation-wide buy-in before commencing the LOSA process. Like all Safety Management initiatives it is essential that the audit process has the complete backing from the highest levels of management and that the cooperation of all personnel is gained. LOSAs must be completed on a non-jeopardy basis. All data collected must be kept completely confidential and used only for flight safety purposes.

LOSAs should be undertaken under the guidance of a steering committee formed from all interested parties within the airline. The FAA suggests that an initial audit should not focus on any one particular aspect of operations but should be organisation-wide. Once completed, specific areas with potential safety-related shortcomings (e.g. particular routes or particular airports) can be identified and subject to later, more focused audits. For guidance, AC 120-90 indicates that a minimum of 50 or more observations per fleet should be undertaken. Below this figure there is a risk of the sample of observations made not being representative. This 'rule-of-thumb' also applies when, for example, auditing operations into a particular airport.

Observations only collect usable, meaningful data when they are collected using a structured framework where the relevant crew behaviours are captured. The proformas used to capture observations during flight should encompass an exhaustive range of well-defined categories and codes to ease the observer's data collection activities and there should also be space provided for a description of the context. The model of threat and error described in the earlier chapter should provide the theoretical underpinning for the data collection process. The composition of observers should also be agreed by pilots, management and the unions beforehand to ensure that they are acceptable to all concerned and regarded as being capable and impartial. This will further ensure 'buy-in'. All data collected should be kept confidential and be made anonymous. Analysis and interpretation of the LOSA data collected should be agreed by all interested parties before being finalised and disseminated back across the entire airline (and to specific departments for any potential targeted remedial actions required).

AC 120-90 notes that historically, in airlines, organisational safety changes have been driven largely by accident and incident investigation. LOSA provides the foundation for a flight deck-orientated safety change process on the basis of specific and quantified results of operational threats and errors prior to any untoward incidents occurring. It is a proactive, not a reactive process. The changes made on the basis of results from a LOSA should be followed up at a suitable interval to establish the efficacy of any changes implemented. Similar to the principles to be described in the following chapter on Safety Management, the FAA (2006) in AC 120-90 emphasises that the basic steps in a LOSA-driven safety change process are:

- Measurement (with LOSA) to obtain the targets.
- Making a detailed analysis of targeted issues.
- Listing potential changes for improvement.
- Performing a risk analysis and prioritising changes.
- Selecting and funding identified changes.
- Implementing these changes.
- Allowing time for changes to stabilise.
- Re-measurement to establish effectiveness of changes.

However, Thomas (2003) compared threats gathered from LOSA data to threats faced by aircrew in LOFT training scenarios. Almost 70 per cent of LOFT scenarios involved an aircraft malfunction but such problems occurred only in 14 per cent of instances in the LOSA data. The most frequent external safety threat in line operations was weather (almost 21 per cent) but this was incorporated in only 4 per cent of LOFT sessions. Other external threats to safety, such as operational pressures on crews, other air traffic and ground handling events occurred in no LOFT scenarios at all. In line with the later FAA guidance subsequently developed in AC 120-90, Thomas suggested that the LOSA data should be used to inform the scenario development in LOFT sessions to make it reflect the everyday safety problems faced by flight crew in operations. It was also observed that crew performance during LOFT sessions was considerably superior to that observed when flying on the line, a classic instance of training appearing effective at Kirkpatrick's (1976, 1998) learning level, but failing to transfer effectively to the workplace itself.

This approach of developing training needs directly from line operational requirements again reflects the training philosophy outlined by the FAA in the AQP process, discussed earlier (see AC 120-54; FAA, 1991a).

If instigated properly, LOSA can effectively serve to inform the CRM training process and help close the flight deck safety loop by directly evaluating the processes employed there. Many of the principles employed are identical in concept to those employed in airline-wide Safety Management processes, described anon. As stated earlier, LOSA and wider organisational Safety Management activities are entirely complementary and should not be considered as separate processes – they are merely the ends of a continuum.

FURTHER READING

Anyone with any interest in CRM must read:

Kanki, B.G., Helmreich, R.L. and Anca, J. (2010). *Crew Resource Management* (2nd edition). San Diego, CA: Academic Press.

After all, Bob Helmreich practically invented CRM and LOSA!

A historical perspective of the development of CRM and LOSA concepts can be found in the collection of essays collated by:

Salas, E., Wilson, K.A. and Edens, E. (2009). *Crew Resource Management: Critical Essays.* Aldershot: Ashgate.

The following, freely available publication from the UK CAA (http://www.caa.co.uk/docs/33/CAP737.PDF) is also an excellent source of practical advice and information:

Civil Aviation Authority (2006). *Crew Resource Management (CRM) Training: Guidance for Flight Crew, CRM Instructors (CRMIS) and CRM Instructor-Examiners (CRMIES)* (CAP 737). London: Civil Aviation Authority.

For a wider, non-aviation specific perspective on non-technical skills try:

Flin, R., O'Connor, P. and Crichton, M. (2008). *Safety at the Sharp End: A Guide to Non-Technical Skills*. Aldershot: Ashgate.

17
Airline Safety Management

This chapter looks at the airline-wide Safety *Management* function. Line Operations Safety Audits (LOSA) and Threat and Error Management (TEM) principles can be conceived of as flight-deck oriented Safety Management activities. These have been considered in the previous chapter, however many of the basic principles still apply across the airline as a whole.

Airline Safety Management encompasses and incorporates Flight Deck Safety Management. In common with the latter, emphasis is on risk management and avoiding (or managing) error across airline operations. Airline Safety Management processes are often driven by the investigation of accidents and incidents. The immediate results of such investigations are usually revisions in procedures and training, factors directly under the control of the airline.

Safety is a very difficult thing to define and hence, also measure. There are many conceptualisations of what safety is: for example having no accidents or serious incidents, the elimination of hazards or the elimination of error. The Safety Management system is an essential function in the *Management* of any airline and lives at the core of its operational philosophy. As a result, safety is viewed as a product resulting from the management of organisational processes aimed at identifying hazards, controlling risks and mitigating any potential adverse consequences. The International Civil Aviation Organization (2009), in its *Safety Management Manual* defines safety as:

> The state in which the possibility of harm to persons or of property damage is reduced to, and maintained at or below, an acceptable level through a continuing process of hazard identification and safety risk management.

The Safety Management function is also responsible for translating the regulatory requirements imposed by the wider societal *Medium* (such as those from ICAO or required in the national Air Operator's Certificates) into the operational procedures required for conducting the *Mission*.

The modern view of accident causation in high reliability organisations, such as airlines, is that there is rarely a single 'cause' to an accident. Accidents result from a combination of seemingly unrelated failures (or aspects of a system that are not optimal) that cannot be foreseen through prospective formal analysis. These are what Perrow (1984) terms 'non-linear' interactions. Accidents do not result solely from technical failures or human errors. The modern systemic view of accidents is that they are a result of organisational failures. However, this perspective is based on Reason's (1990) model of human error. In this model active failures (which are the errors proximal to the accident, associated with the performance of front-line operators in complex systems) and latent failures (distal errors and system misspecifications, which lie dormant within the system for a long time) serve to combine together with other factors to breach a system's defences. As Reason (1997) observed, complex systems are designed, operated, maintained and managed by human beings, so it is not surprising that human decisions and actions at an organisational level are implicated in all accidents. Active failures of operators have a direct impact on the safety of the systems. However, latent failures are often spawned in the upper levels of the organisation and are related to its management and regulatory structures.

Safety is paramount in the operation of airliners, although other organisational pressures aimed at enhancing economic performance significantly influence this goal. The balance between safety, performance and cost is a constant tension for airline managements. As aviation is a safety critical industry, though, it is extremely highly regulated but the vast majority of these national and international regulations simply serve to prescribe a *minimum* standard of safety. There is also a societal expectation to deliver a safe, economical and efficient product.

Figure 17.1, taken from Harris and Thomas (2005) but adapted from Diehl (1991b) provides an overview of the accident generation, investigation and prevention process. The shaded component delineates the aspects of this process within the remit of the airline.

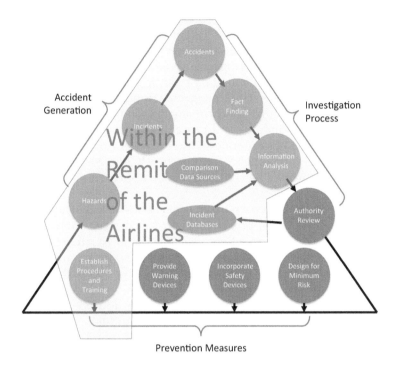

Figure 17.1   Accident generation, investigation and prevention elements (Harris and Thomas, 2005)

ICAO (2009) defines nine building blocks for Airline Safety Management:

- *Senior management's commitment to the management of safety* – Managing safety requires adequate allocation of resources.
- *Effective safety reporting* – To manage safety, data are required on hazards, incidents and accidents.
- *Continuous monitoring* – Collecting safety data is useless without continual analysis of trends.
- *Sharing information* – Once the data have been analysed it is essential to pass back relevant information to those who are in contact with the hazards identified.
- *Investigation of safety occurrences* – Identifying safety deficiencies as a result of establishing what happened and why it happened (rather than who was to blame).
- *Sharing safety lessons learned and best practices* – by promoting an active exchange of safety information across the industry so lessons can be learned from others.
- *Integration of safety training for operational personnel into all aspects of operation* – If 'safety is everybody's responsibility', to make this an effective mantra everyone needs help in achieving it.
- *Effective implementation* – of standard operating procedures, including the use of checklists and briefings on the flight deck and outside of it.
- *Continuous improvement of the overall level of safety* – Managing safety is an ongoing activity that requires continuous improvement.

Reason (1997) suggested that data collection in a Safety Management programme within a company may be: *Reactive* (collecting data from accidents and incidents that have already happened); *Proactive* (identifying hazards and safety risks before they result in an accident or incident, through using voluntary reporting systems, safety audits and safety surveys); or *Predictive* (where the Safety Management function actively seeks out emerging safety risks from a variety of sources and also engages in the real-time analysis of trends to ensure compliance with standards and procedures and check for 'drift'). Any Safety Management system needs to incorporate all three aspects, however the best systems put emphasis on the proactive and predictive approaches. ICAO (2009) developed these ideas a little further into a hierarchy of Safety Management approaches:

- *Insufficient* – A reactive approach based upon accident and incident reports.
- *Efficient* – Also a reactive approach but data are derived from reporting systems such as Mandatory Occurrence Reports (MORs) and Aviation Safety Reports (e.g. CHIRP – Confidential Human Factors Incident Report Programme; ASRS – Aviation Safety Reporting System; or in-house company schemes such as BASIS – British Airways Safety Information System).
- *Very efficient* – Where data are derived from ASRs in addition to safety surveys and safety audits (e.g. LOSA – Line Operations Safety Audits).
- *Highly efficient* – In which the safety data are derived from the continual monitoring of trends (such as flight data analysis programs) and the direct observation of operations.

The *Management* is responsible for providing the whole of the safety infrastructure for the operation of commercial aircraft, including all the 'M's lying within this sphere in the model and their interfaces with the Societal Medium. Indeed, the management structures and responsibilities in an airline must be approved by the airworthiness authorities before they will be allowed to operate commercially. There is, however, more to promoting safe operations than simply ensuring that all the regulations are adhered to.

REASON'S 'EMMENTAL CHEESE' MODEL OF ACCIDENT CAUSATION

About this time it is worth describing Reason's 'Emmental (Swiss) cheese' model of accident causation which also illustrates the relationship between system defences (barriers), chance and an accident. Reason's model is comprehensive as it also describes the relationship of organisational and personal psychological factors to the accident process. Many Safety Management models (including that promoted by ICAO) are based around the principles in Reason's approach. This model extends Reason's GEMS approach (see Chapter 6 on Error) back into the organisation to define the organisational and psychological precursors underlying inadequate performance. Surprisingly, Reason argues that there are relatively few psychological precursors, with even fewer organisational mis-specifications underlying them (see Figure 17.2).

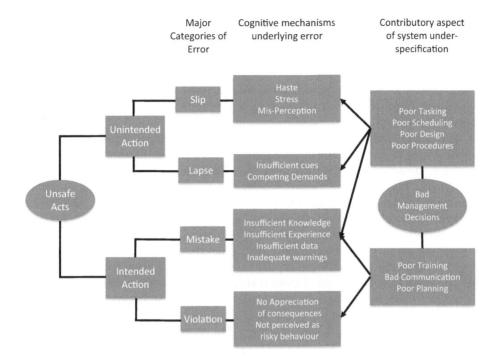

Figure 17.2   **An extension of Reason's model of human error (initially presented in Figure 6.1) back into the organisation to identify the generic failure types (adapted from Reason, 1990)**

Reason's model of the accident causation process starts at a point (organisationally) remote from the actual accident. He argues that the seeds for most accidents are sown at the higher levels of the organisation. The model starts with fallible decisions at the highest level. He suggests that even with the best intentions, higher levels of management will make mistakes in the priorities that they set which have a 'knock-on' effect down at the 'sharp end'. It is argued that top management decisions are more likely to be based around visible, rapid feedback (e.g. cost savings, efficiency increases) whereas safety gains always cost money, are unpredictable (prone to random fluctuations) and are often invisible.

Line managers are responsible for translating policy into practice (within constraints). Any deficiencies in these areas may have the direct effect of creating the psychological pre-cursors of unsafe acts. Unfortunately it is impossible to specify exactly what effect these organisational mis-specifcations will have (although formal error prediction techniques and some authors researching PSFs do try and specify these risk factors to some degree and forecast their effects on potential outcomes). Organisational culture can also have a detrimental effect. An overly 'can-do, always deliver on time no matter what' culture can increase time pressures on crews resulting in stress, corner cutting, low morale, etc, each one of which can make an unsafe act (error) more likely. These are the error enhancing conditions in the human operators (Figure 17.2).

Finally, even after an error has been made there is no reason why it should promulgate through the system if adequate defensive barriers are in place. In a well defended system barriers will be present to 'trap' errors before they propagate through the system to result in an accident. These are the elements on the bottom line in Figure 17.1 and are discussed in greater detail towards the end of this chapter. However, these barriers can only catch specific problems that they have been designed to intercept (through hazard and risk analysis). Reason argues that there is often more scope for Safety Management practices to be more effective at the higher levels of management than at the lower level of installing task specific barriers.

Reason describes these non-optimal elements in a system as 'pathogens'. They are always resident in a system but will only be 'triggered' when the conditions are right. In Reason's terminology, the actual error(s) themselves are designated as unsafe act tokens. There is a 'one-to-many' mapping of condition tokens (the things that increase the probability of an error being made) to unsafe act tokens. For example, high workload, perhaps caused by Air Traffic Control instructions or a minor aircraft malfunction (condition tokens) may contribute to any one of hundreds of errors of omission or commission occurring on the flight deck (unsafe act tokens). Because of the nature of this relationship, it is almost impossible to work backwards from unsafe act tokens to condition tokens.

This approach may be extended even further back in the accident causal sequence of events, to what are designated function types. It is at this stage where organisational factors (the types) are translated into the individual's experience (the tokens). There are fewer function types than there are condition tokens, but again, it is almost impossible to back track from a condition token to a function type due to the one-to-many correspondence between types and tokens. Examples of function types include poor tasking, deficient training, inadequate procedures, etc. To exemplify further, poor training may lead to any or all conditions, such as high workload, misdiagnosis or mis-perception of risks, etc. (Figure 17.3). One of the basic aims of Human Factors in a system safety programme is to identify and correct the condition tokens and function types (the contributory factors). It is not to attempt to investigate and remedy every individual error made (the unsafe act tokens)

Further examination of Figure 17.3 reveals why Reason's approach is widely known as the 'Emmental (Swiss) cheese' model. Non-optimisation of various aspects of the airline's organisational practices result in an increased likelihood of error. These are represented as 'holes' in Figure 17.3. The more function types that exist (and the worse that they are) the higher probability of a psychological pre-cursor (i.e. the bigger the hole in the system's defences). If adequate dependences are not put in place or are of poor quality, this will result in further (and larger) holes in the system defences. When and if all of these holes line up, there is the opportunity for an accident to occur.

Reason's Emmental Cheese model has been developed by other authors as the basis for an incident/accident investigation system (HFACS – Human Factors Analysis and Classification System; Shappell and Wiegmann, 2001; Wiegmann and Shappell, 2003). This is described in greater detail in Chapter 18 on Incident and Accident Investigation.

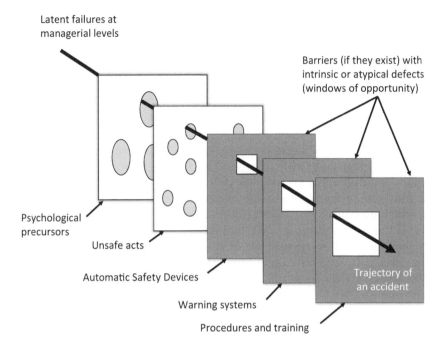

Latent failures at
managerial levels

Barriers (if they exist) with
intrinsic or atypical defects
(windows of opportunity)

Psychological
precursors

Unsafe acts

Automatic Safety Devices

Trajectory of
an accident

Warning systems

Procedures and training

**Figure 17.3   Reason's 'Emmental cheese' model of accident causation
(Reason, 1997)**

Accidents Do Just Happen

A traditional view of safety is that a system will be safe if it performs as per its design specifications, complies with all the relevant regulations and has no major deviations during routine operations (as only these deviations lead to unfortunate consequences). This is an outcome-oriented perspective (ICAO 2009). However, Dekker (2004b) has argued that organisations tend to 'drift' into failure. Even taking a wider view of organisational processes, technical malfunctions and human error, these accidents are difficult to predict. They are 'the effect of a systematic migration of organizational behavior toward accidents under the influence of pressure towards cost-effectiveness in an aggressive, competitive environment' (Rasmussen and Svedung, 2000, p. 14). Drift into failure is about people doing normal work in normal organisations and is not about failures and errors. The safety paradigm evolving, and the one actively promoted by ICAO, is based on the management of safety using a process control approach. The basic assumptions underpinning this approach to safety are that:

- The aviation system does not perform most of the time as per design specifications (everyday operational performance leads to practical drift in processes and procedures).
- Airline real-time performance is constantly monitored (performance-based) rather than relying on regulatory compliance to 'ensure' safety.
- Minor, seemingly inconsequential operational deviations are tracked and analysed (process oriented).

The remainder of this chapter examines further the manner and methods (from a Human Factors perspective) by which the Safety Management function deals with the identification, analysis and control of operational risk. However, before this a brief overview of organisational Safety Culture is required.

SAFETY CULTURE (AND SAFETY CLIMATE)

Activities that promote safety will only be engaged in by the people within an organisation if there is the collective mind and attitude to go further than just paying lip service to the regulatory requirements. Westrum and Adamski (1999) categorised organisations into three different types (Pathological, Bureaucratic or Generative) in terms of the manner in which they dealt with safety information. In Pathological organisations safety information is hidden, messengers with bad (safety) news are shot, management shirks safety responsibilities, bridging between organisational functions is discouraged, safety failures are covered up and new ideas to promote safety are crushed. In a Bureaucratic organisation safety information may be ignored but messengers with bad news are tolerated, responsibilities for safety are compartmentalized, bridging between functions is allowed but discouraged, however the organisation is just and merciful. Westrum and Adamski (1999) described Generative organisations as actively seeking safety information; in these organisations messengers are trained, responsibilities are shared, bridging between functions is rewarded, a failure in safety causes inquiry and new ideas to promote safety are welcomed. Hudson (2001) extended Westrum and Adamski's (1999) framework to include more organisational types and include reactive and proactive stages in the development of a Safety Culture.

Wiegmann, Zhang, von Thaden, Sharma and Mitchell (2002) identified four different 'eras' in the notion of accident causation (and the role of the human operator). The first three have already been alluded to in the preceding discussion. The Technical Period was the earliest where developments in new mechanical systems were rapid and most accidents were caused by mechanical malfunctions, particularly in the design, construction and reliability of equipment. This was followed by the Human Error Period. At this point systems were becoming more reliable and it was the faults of the human operator rather than equipment malfunctions that were seen as the root cause of accidents. Safety analysis became more focused on aspects of human error. Blame and responsibility were assigned to the person directly involved in the unsafe act. In the Socio-Technical Period the view of human error changed as it was now considered to be the result of the interaction of human and technical. This approach has now developed and evolved into the fourth era, the Organisational Culture Period. This approach recognises that operators are performing as part of a coordinated team of personnel, which is embedded within a particular organisational culture.

Table 17.1 Hudson's (2001) extension of Westrum and Adamski's (1999) categorisation of organisational Safety Culture

	Type of Organisation				
	Pathological	Reactive	Calculative	Proactive	Generative
Communication	• Nobody is informed • No feedback • Everybody is passive • No knowledge about safety • Don't actively seek safety issues • Only collect data that are legally required	• Management demands data on safety failures • Denial until forced to admit a problem • Lots of statistics collected (not analysed) • Safety only an issue after an accident	• Environment of command and control by management • Lots of statistical analysis but no follow up • Top-down flow of information with little feedback • Bottom-up reporting of accidents • Toolbox meetings promoted • Safety procedures exist • Action is delayed after knowledge	• Management seeks out and discusses safety issues • Management knows what to change and how to make changes • Feedback loop is both bottom-up and top-down) and is closed at supervisory level • Safety topics are part of all meetings • Management need detail to understand why accidents happen	• No threshold between management and workforce • Management participates/shares safety activities • Safety is top priority • All feedback loops are closed • Safety is integrated in other meetings; no special safety meetings are required • Workforce keeps itself up to date, and demands information to prevent problems
Organisational Attitudes	• No belief or trust • Environment of punishing, blaming and controlling the workforce	• Failures caused by individuals • No blame but responsibility • Workforce needs to be educated and follow the procedures • Management overreacts in eyes of workforce	• Workforce is more involved but has little input on procedures, designs and practices • Workforce does not understand the problem • Management is seen as obsessed with safety but not perceived as 'meaning it'	• Workforce involvement is promoted but ruled/organised by supervisory staff • Supervisory staff obsessed by safety statistics	• Management is recognised as a partner by workforce • Management respects the workforce • Management has to fix systematic failures – workforce has to identify them
Safety	• No status for safety issues • Safety issues are ignored • Minimal requirements for safety practices • No rewards for good performance • Safety is 'inherited' • Much reliance on experience	• Practices meets legal requirements • Management collects statistics but no follow up • Design/procedures are changed after accidents • Procedures are rewritten to prevent previous accidents • No regular updates or improvements	• Safety well accepted issue • Safety advisor collects data and undertakes own analyses • Rewards for positive performance • Quantitative methods applied • Procedures in place to solve unsolved problems • Standard procedures preferred from off the shelf • Large numbers of procedures but few checks on their use/knowledge	• Separate line safety advisors promoting improvement, but who try to reduce the inconvenience to operations • Career enhancement for good safety initiatives • Safety considerations incorporated in the early stages of design • Procedures rewritten by workforce and integrated with competencies • Some complaints about externally set targets	• Safety department is small, advising the management on strategy • No special rewards for good safety performance, just individual pride • Procedures are written by workforce • Policy and practice of continuous improvement • Small numbers of procedures which are integrated into training.
Organisational Behaviour	• Denial that anything is wrong • Avoids safety discussions • Management is hierarchical and stagnant to change • Focus on profits not on workforce	• Management holds workforce responsible for failures • Management often overreacts • Management states that it takes safety seriously but this is not believed by workforce	• Focus on detail and analysis of numbers • May believe that the company is doing well in spite of evidence to the contrary • Targets are not challenged • Inability to admit solutions may not work the first time	• Management knows the risks but is interested in safety • Takes culture into account • Safety given priority over production which may lead to incompatible organisational goals • Communication and assessment of accidents and incidents and their consequences	• Safety given equal priority to production • Enthusiastic communication between workforce and management (and vice versa) • Trust in the workforce leads to a great deal of freedom (trust)

The two terms 'Safety Culture' and 'Safety Climate' seem to be used almost interchangeably, however they are subtly different. Wiegmann, Zhang, von Thaden, Sharma and Mitchell (2002) identified seven basic commonalities to be found in all definitions of Safety Culture:

- Safety Culture is a concept defined at the group level or higher, which refers to the shared values among all the group or organisation members.
- It is concerned with formal safety issues in an organisation and closely related to, but not restricted to, management and supervisory systems.
- It emphasises the contribution from everyone at every level of an organisation.
- The Safety Culture of an organisation has an impact on its members' behaviour at work.
- It is reflected in the contingency between reward systems and safety performance.
- It is displayed in an organisation's willingness to learn from errors, incidents, and accidents.
- It is enduring, stable and resistant to change.

Safety Climate, however:

- Is a psychological phenomenon, which is usually defined as the perceptions of the state of safety at a particular time.
- It is closely concerned with intangible issues, such as situational and environmental factors.
- It is a temporal phenomenon, a 'snapshot' of Safety Culture and is relatively unstable and subject to change.

Glendon and Stanton (2000) suggest that the research methodology used in an investigation is a good indicator as to whether Culture or Climate is being measured. If a psychometric scale is the measurement instrument, then some aspect of Climate is being measured. A triangulated methodology might indicate that aspects of Culture were being tapped but this would also depend upon the depth and breadth of the measures used. Climate, they suggest refers to the perceived quality of an organisation's internal environment. It is a slightly more superficial concept than Culture, describing aspects of an organisation's current state: it measures and describes dimensions of organisational Culture within a limited range and, furthermore, this description of attitudes, beliefs and perceptions towards safety are limited as they can only report on these factors at the time that the measures are undertaken, hence the notion that they provide only a snapshot of an organisation's safety health.

Safety Culture is both more enduring than Climate and more elusive to pin down and define. Guldenmund (2000) suggests that Culture can only be investigated using an ethnographic approach. It is focused on the dynamic processes at work in an organisation and is qualitative and interpretative in nature. The UK Health and Safety Executive (2002) suggest that Safety Culture refers to the behavioural and situational facets of an organisation (i.e. 'what people do' and 'what the organisation has') and Safety Climate should be used to refer to the psychological characteristics of employees (i.e. 'how people feel'), in other words, their attitudes and perceptions with regard to safety.

Glick (1985) suggests that the root of this debate stems from the fact that these two concepts originate in different scientific disciplines. He argues that Climate-based research has developed primarily from a social psychological perspective whereas a Culture-based approach is rooted firmly in anthropology. Glick (1985) concludes that: '... the minor substantive differences between culture and climate may prove to be more apparent than real' (p. 612).

In the 2005 HSE review of Safety Culture and Climate literature, five basic dimensions of what constitutes good Safety Management were identified which should form the basis of the measurement (or assessment) of an organisation's safety health. These were:

- Leadership:
  - *Performance versus Safety Priority:* Senior management should give safety high status within the organisation. This can be can be demonstrated through providing sufficient:
    - Health and safety budget
    - * Opportunities for safety communication
    - * Health and safety training
    - * Support to personnel
    - * Health and safety specialists.
  - *High Visibility of Management's Commitment to Safety:* This can be demonstrated by the use of:
    - * Verbal communication
    - * Written safety communication (e.g. statements, newsletters).
  - Safety Management Systems
- Two-way communication:
  - *Top-down Communication:* From management to staff, for example by promulgating safety policy by newsletters describing safety news, safety issues, and major accident risks (see previous major point on leadership).
  - *Bottom-up Communication:* For example through the use of safety reporting systems for communicating problems and concerns. Feedback mechanisms should also be provided to respond with regard to remedial actions taken.
  - *Horizontal Communication:* A system should be available for the transfer of information between individuals and departments.
- Employee involvement:
  - *Ownership for Safety:* This can be encouraged through providing training and opportunities for employees to be responsible for specific aspects of safety.
  - *Safety Specialists:* These staff should play an advisory or consultancy role.
  - *Reporting:* It should be easy for all staff to report concerns about decisions that may affect them and feedback mechanisms should be in place to inform staff about any decisions that will affect them.
- Learning culture:
  - *Suggestion System:* A system should be in place that allows all employees to proactively contribute ideas for improvement.
  - *Analysis of Incidents and Accidents:* The results of these should be communicated with provision for feedback and sharing of information.
  - *Safety Climate Surveys:* All members of the workforce should be included in surveys to increase safety motivation and provide them with further opportunities to raise safety concerns. Feedback from the survey should be promulgated throughout the organisation.

- The existence of a just culture:
  - *Just Reporting Culture:* Organisations should move from a blame culture for attributing the causes of accidents and move toward a just culture or one of accountability. An agreed system should be in place that enables any degree of culpability to be assessed.
  - *Accident Investigators:* They should have a good understanding of the mechanisms of human error.
  - *Care and Concern for Staff:* Management should demonstrate care and concern towards employees.
  - *Reporting:* Employees should feel they can report issues or concerns without fear of reprisal as a result of coming forward.
  - *Confidentiality:* This should be maintained throughout the course of the investigation (and after, if required).

Many researchers (e.g. Merritt and Helmreich, 1995; Helmreich and Merritt, 1998; Glendon and Stanton, 2000) propose that Safety Culture is a sub-culture of organisational culture, which is itself a sub-culture of the industry culture, which in turn is a sub-culture of national culture (cf. Hofstede's (1991) conceptualisation of culture as an onion with many skins). Furthermore, an employee may simultaneously belong to many cultures or sub-cultures which at times will have conflicting values. Safety Culture cannot be separated from these other cultures hence any comprehensive model needs to extend beyond the boundaries of the organisation.

It has already been argued that all industries are open systems (i.e. they must interact with their environment). McDonald and Ryan (1992) and Simard and Marchand (1997) both argued that the development of safe working practices was dependent upon the control that management exercised over the work processes and other factors external to the organisation. Placing a boundary around an organisation limits the development of a comprehensive theory of Safety Culture, as influences beyond that boundary do not receive sufficient attention. As Lintern (2004) noted: 'Most approaches to Safety Management attempt to lock the system down so that it does not generate new properties .... However, open systems are infinitely generative'. In response, Morley and Harris (2006) developed an open system model of Safety Culture – the Ripple Model (see Figure 17.4). This model identified three threads running across people within (and without) an organisation, irrespective of their level and role. These were labelled 'Concerns', 'Influences' and 'Actions'. In this model it is argued that elements outside an organisation have a profound effect on the Safety Culture and the behaviour of people within that organisation.

- *Concerns* drive an individual to accept or reject safe working practices as a result of the prevailing culture. These concerns operate at an emotive (or affective) level and only drive safety behaviour to a relatively small degree as a result of the influences of other factors. For example, at line worker level, concerns centre on personal health and safety, job security, job satisfaction and well-being. Senior management's concerns are associated with meeting organisational and shareholder goals, maintaining stability and assuring the financial health of the organisation. From outwith the organisation, the regulator's principal safety concern is demonstrating control of the level of risk in its industry, its ability to respond to the economic and regulatory agenda set down by government, and its aptitude for balancing the conflicting demands of the public while allowing industry the leeway to operate in a cost effective manner.

- *Influences* refer to those factors that underpin the ability to actually act in the desired manner. For example, for line workers, influences include attitudes toward safety, the skills and knowledge to deal with safety issues, the motivation to employ safe working practices and the workforce's sense of ownership and empowerment (factors evident in many models of Safety Culture). For middle management, the influences behind promoting a sound Safety Culture include the supervisory style of management, its technical, knowledge and ability, leadership and communication skills, access to senior management, and the level of autonomy and empowerment afforded to it. At a much higher level, government's major influences stem from the societal level, for example the influences of national culture, the public perception of risk and economic imperatives.

- *Actions* are the behaviours themselves that have either a positive or negative impact upon safety. Descriptions of what constitutes good Safety Management fall into this category. In addition to actually working in a safe manner, at line worker level, safety actions include being vigilant for new operating hazards, participating in safety initiatives and communicating safety issues. Middle management should promote safety, being responsible for communicating information both up and down the organization, encouraging and monitoring working practices, and scheduling work appropriately. Senior management's actions include determining organisational values and goals to promote safety. The role of the regulator is to aid sharing safety information and integrating initiatives across the industry. The role of the government includes providing a clear mandate to the regulator and responding to society's safety concerns.

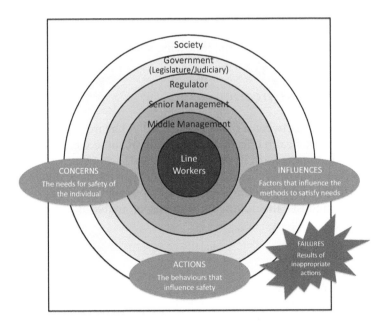

Figure 17.4   **Layers of influence and categories comprising the Ripple Model of Safety Culture (Morley and Harris, 2006)**

A high level of concern for safety (across both the organisation and society) coupled with appropriate influences promoting safety (particularly at higher levels of management and governmental levels) should lead to appropriate and effective actions. However, a high level of concern for safety associated with a low level of influence may result in ineffective or inappropriate safety actions being taken. Similarly, a low level of concern for safety associated with high levels of influence may result in potentially effective safety initiatives being ignored or only partially complied with.

The Ripple Model has been used to re-analyse the China Airlines Flight CI-611 accident in 2002 (Li and Harris, 2005c) and the Air Ontario accident at Dryden in 1989 (Harris, 2006b). In both cases it can be seen that the Safety Culture within the airlines involved was as much a product of external factors as their own internal problems and processes. You can't just draw notional boundaries around organisations operating in open systems and say 'the buck stops here'.

SAFETY MANAGEMENT SYSTEMS

The Safety Management system should be part of the culture of an organisation. It should be central to the way people do their jobs. A Safety Management system is simply a systematic, explicit and comprehensive process for managing safety risks. The report on Safety Management Systems (TP 13739) published by Transport Canada (2004) provides a succinct overview of the role, scope and processes within a Safety Management system. Any such management system should have three identifiable components to it:

- Hazards and risk identification.
- Development and Implementation of safety programmes.
- Monitoring (and revision of) the safety programme.

These are each briefly described in the following sub-sections.

Hazard and Risk Identification

There is a clear distinction between a risk and a hazard. A hazard is any source of potential damage, harm or adverse health effects on something or someone. Risk is not simply the probability of something untoward happening (a hazard turning into an accident); it is the product of an untoward event occurring *and* the severity of its likely outcome (probability x consequence). Thus, hazard identification needs to precede risk analysis.

Rausand (2004) suggests the following sources of material to aid in any hazard identification process:

- Similar existing systems.
- Previous hazard analyses.
- Hazard checklists and standards.
- Examination of the energy flow through the system.
- Identification of inherently hazardous materials.
- Interactions between system components.
- Brainstorming in teams.

- Analysis of the human–machine interface.
- Assessing any changes of use.
- Undertake a worst-case 'what-if' analysis.

Other sources of data that may aid in the identification of hazardous situations include:

- Accident reports/databases.
- Accident statistics.
- Near miss/dangerous occurrence reports.
- Reports from authorities or governmental bodies.
- Expert judgement.

Once the hazards have been identified to assess the overall level of risk the outcome of the frequency and likely severity of the event needs to be established. MIL-STD-882C (US Department of Defense, 1993) uses a matrix to combine frequency and severity into a single criticality index. This criticality assessment comprises of a number followed by a letter. The number represents the severity of an untoward event, where: (1) is death, system loss or irreversible environmental damage; (2) is severe injury, occupational illness, major system damage or reversible severe environmental damage; (3) is injury requiring medical attention, illness, system damage or environmental damage that may be mitigated; and (4) is possible minor injury, minor system damage or minimal environmental damage. The following letter represents the frequency of occurrence of the event, where: (A) denotes that some untoward event is expected to occur frequently; (B) such an event will occur several times in the life of an item; (C) an event is likely to occur sometime in the item's service life; (D) an event is unlikely but possible to occur sometime in the item's life; and, (E) the event is so unlikely that it can be assumed occurrence may not be experienced (Figure 17.5).

	Severity			
	(1) Catastrophic	(2) Critical	(3) Marginal	(4) Negligible
(A) Frequent	1A	2A	3A	4A
(B) Probable	1B	2B	3B	4B
(C) Occasional	1C	2C	3C	4C
(D) Remote	1D	2D	3D	4D
(E) Improbable	1E	2E	3E	4E

Risk Categories

High	Serious	Medium	Low

Figure 17.5   Risk assessment matrix (US Department of Defense, 1993)

Safety is not about the *elimination* of risk but the *control* of risk, reducing it to a level As Low As Reasonably Practicable (ALARP). The level of acceptable risk will also vary from system to system. For example, the level of acceptable risk in military operations in peacetime is considerably lower than that acceptable in time of war. Similarly, the level of risk acceptable in emergency paramedical operations is greater than that acceptable for civil airline operations. Any hazard that lives in the top left-hand corner of the matrix from MIL-STD-882C needs action: active risk-reducing measures are required. Anything in the bottom right-hand corner isn't worth worrying about. The difficult stuff is all the remainder in the middle, where 'appropriate' actions need to be taken to control the risk.

Implementation of Safety Programmes

The organisational structures and activities that make up a Safety Management system should completely permeate an organisation. The Safety Management activity should be integrated into 'the way things are done' throughout the establishment. This can only be achieved by the implementation and continuing support of a coherent safety policy which leads to well-designed procedures. The Safety Management systems document (TP 13739) from Transport Canada (2004) describes the four Ps of Safety Management:

- *Philosophy* – Safety Management starts with Management Philosophy:
    - Recognising that there will always be threats to safety.
    - Setting the organisation's standards.
    - Confirming that safety is everyone's responsibility.
- *Policy* – Specifying how safety will be achieved:
    - Clear statements of responsibility, authority and accountability.
    - Development of organisational processes and structures to incorporate safety goals into every aspect of the operation.
    - Development of the skills and knowledge necessary to do the job.
- *Procedures* – What management require employees to do to execute the policy:
    - Clear direction to all staff.
    - Means for planning, organising and controlling.
    - Means for monitoring and assessing safety status and processes.
- *Practices* – What really happens on the job?
    - Following well-designed, effective procedures.
    - Avoiding the shortcuts that can detract from safety.
    - Taking appropriate action when a safety concern is identified.

Wickens, Lee, Liu, and Gordon Becker (2004) explain the basic aspects for the implementation of a safety programme. These apply generally across an organisation. Safety Management issues relating specifically to the flight deck have been examined in preceding chapters (e.g. LOSA).

- *Management involvement* is the first (and perhaps most essential) element. Basically, this involves operationalising the Philosophy and Policy of the Safety Management programme into Procedures.
- *Safety rules* are a logical progression from the four Ps of Safety Management in that they aim to ensure that appropriate safety practices are followed. These are

not just SOPs on the flight deck but apply to safe working practices elsewhere in the organisation (e.g. on the ramp and in the hangar). These rules should be conspicuous and promulgated throughout. If necessary there should be policies and penalties prescribed for violation of safety rules. These should be outlined in the employee safety manual.

- *Employee training* should be provided on safe working practices and there should be regular safety and hazard awareness training for all employees. People cannot be expected to work safely and participate in the continuous improvement of safety unless they have been trained to do so.

- *Suggestion and safety promotion schemes* should reflect the increased awareness of potential hazards derived from training and experience, and provide employees with the opportunity to make a positive contribution toward improving safety. The suggestion scheme also provides the Safety Management programme with information concerning how updates or changes in operations affect safety practices. Suggestion schemes should be complemented with incentive schemes and other safety promotion schemes (posters, data trends, newsletters, etc.).

- Accident/incident reporting must be encouraged and when anything goes wrong it should be investigated by appropriately trained staff to identify root causes and suggest remedial actions. The results of these investigations and the actions to be taken subsequent to the event should be promulgated throughout the staff. Incidents should be logged and regularly analysed for trends – there is no point in logging data if you are not going to analyse it! Incidents occur more frequently than accidents and provide a good indicator of future problems. Some suggest there is anything between a 30:1 and 300:1 ratio of incidents to accidents. Incidents often share the same aetiology as accidents. The Human Factors aspects of the analysis of incidents and accidents is examined in greater detail in the following chapter.

Monitoring Safety Programmes

Safety Management programmes are of little or no use if they are not continually audited to assess their ongoing effectiveness. Indicators of safety performance fall into two basic categories: lagging indicators and positive performance measures.

Lagging indicators focus on common safety performance measures such as accidents and incidents, lost time injuries, number of production days lost, etc. in an effort to measure safety performance. However, this category of safety performance measure only really reflects a failure to control the situation in the past and gives no indication of positive risk management efforts which may take time to come to fruition, hence the term lagging indicators. The advantages and disadvantages of the use of lagging indicators are summarised in Table 17.2.

Table 17.2   Advantages and disadvantages of lagging indicators as markers of safety performance (adapted from Health and Safety Executive, 2001 and Consultnet, 2009)

Advantages	Disadvantages
• An accepted standard • Direct measure of what you are trying to achieve • Long history of use • Used by government agencies, industry associations • Easy to calculate • Indicate trends in performance (however, see 'Disadvantages') • Good for comparison (with self and others)	• Reactive • Often if a safety event results in an accident it is a matter of chance • May be biased (under-reporting; worker and/or management attitude to compensation system, e.g. safety awards and competitions) • Time off work does not reflect the severity of the event (accident potential) • Figures measured are typically low making it difficult to establish trends • Accident rates do not reflect accident potential • Statistics reflect outcomes not causes • Statistics reflect safety failure, not success

Positive performance measures adopt a more proactive approach to measuring safety based upon indicators rather than safety events (or rather lack safety events): these measures can reflect safety successes. However, it needs noting that there is no single reliable measure of safety performance in this category: what is required is a basket of measures to provide information on a range of safety activities. The general characteristics of good performance indicators are that they are:

• Controllable or able to be influenced.
• Relevant.
• Assessable or measurable.
• Understandable and clear.
• Accepted as true indicators of performance.
• Reliable, providing the same measures when assessed by different people.
• Sufficient to provide accurate information, but not too numerous.

A frequently used acronym for effective safety measures is that they should be SMART (Specific, Measurable, Attainable, Realistic/Relevant, and Timebound).

There are two basic types of positive performance measure – behavioural indicators and measures of Safety Management activity. General examples of behavioural indicators include such things as the percentage of employees wearing PPE on the ramp (hi-visibility jackets, safety boots) and the percentage of pre-start checks complete, etc. Management indicators include, for example, the percentage of managers attending safety meetings, the percentage of safety audits (LOSA) completed on time and the percentage risk assessments completed (Table 17.3). These measures should reflect the safety objectives and targets.

Table 17.3 **Some examples of Positive Safety Performance Indicators as markers of flight deck safety performance (Adapted from Consultnet, 2009)**

Indicator	Measure/monitor/ evaluate	Results
• Risk assessment	• Number of LOSAs completed • Control measures implemented	• Track reported LOSAs completed on a monthly basis by route/ fleet/department
• Work procedures	• Compliance with SOPs and checklists	• Track reported compliance with SOPs and checklists on a monthly basis by route/fleet/ department
• Behaviour based observations	• CRM processes	• Track CRM evaluations on a monthly basis by fleet/ department
• Timeliness of incident reporting	• Number of incidents reported • Number of incidents reported within 24 hours	• Track reported numbers on a monthly basis by fleet/ department
• Incident investigation effectiveness	• Investigations complete on time • Percentage of corrective actions implemented	• Track reported numbers on a monthly basis by route/fleet/ department

REPORTING SYSTEMS

Employee reporting schemes are a vital part of any Safety Management system. Not only do they provide a check on the safety health of an operation they are also a vital component in the hazard identification process. These schemes are normally organised confidentially (but not necessarily anonymously) and may be operated either as an in-house (company) scheme or on a national basis (e.g. the UK CHIRP system or the US ASRS).

There are advantages and disadvantages of both types of system. In-house confidential reporting systems can be tailored specifically to the needs of the airline; they are a direct check on the operation of the Safety Management system, and by monitoring their content closely it is possible to react quickly to newly emerging hazards. The disadvantage of this type of scheme is that it is more difficult to maintain employee confidentiality when reporting (or promulgating safety information through the publication of incident reports) especially if the operation is small. A great deal of integrity is required to ensure personnel have confidence in the confidentiality of the scheme, its impartiality and its utility. Only in this way will it continue to receive reports and hence maintain its usefulness. In many cases, to help maintain employee confidentiality, company safety reporting schemes are operated by a third party. National schemes do not have these issues. They also attract a great many more reports allowing the larger scale analysis of trends which may be associated with, for example, a particular aircraft type or approach procedure. However, they cannot deal with issues within a company. In the last decade or so, several national schemes have made available anonymised versions of their reports on line with a search engine (e.g. http://asrs.arc.nasa.gov/search/database.html). In this way users can perform their own specific searches on these national databases which can supplement in-house reporting schemes to provide a wider perspective.

O'Leary (1995) suggests that if the Safety Culture in an airline is not exemplary (e.g. an organisation where messengers are shot) confidentiality is vital if reports are to be encouraged. Even in this case, though, crew may still be reluctant to report. Confidential reports also hold greater candour in their content compared to more formal occurrence reports (O'Leary and Pidgeon, 1995). When safety reporting systems were first introduced they were promoted as part of a 'no blame' culture, aimed both at encouraging incident reporting and enhancing safety. This has been particularly successful in national schemes. The US NASA ASRS now typically receives over 30,000 reports each year. ASRS is impartial and independent from the regulator, enforcement agencies and the reporters' own airlines. Furthermore, to promote reporting, in certain circumstances (e.g. altitude 'busts') the FAA guarantees immunity from sanction if an ASRS report is filed.

Furthermore, active involvement in the Safety Management process can be motivating for personnel, helping to create a Safety Culture. Crew are often the best-placed personnel to identify many operational hazards (this is the fundamental basis of the LOSA and TEM processes). However, this motivation can only be sustained if the Safety Management department acknowledges and acts on contributions received. Safety reporting schemes give personnel part ownership of the airline's safety record.

However the 'no-blame' concept has drawbacks. Reports from individuals who either wilfully or repeatedly engage in dangerous behaviours ('violations' in Reason's error taxonomy) are treated just the same as those from people who make genuine errors, and under the terms of the reporting scheme, they are held not to be to blame. Blame-free schemes do not distinguish between culpable and non-culpable unsafe acts. In the case of deliberate or persistent transgressors, many people require a level of accountability for their actions when a mishap occurs. A failure to punish such offences can also suppress the willingness of others to report incidents. This has led to 'no-blame' cultures being replaced by 'Just' cultures.

Reason (1997) describes a Just culture as:

> … an atmosphere of trust in which people are encouraged (even rewarded) for providing essential safety-related information, but in which they are also clear about where the line must be drawn between acceptable and unacceptable behaviour. (p. 195)

The Global Aviation Information Network (2004) developed various guidelines for the development of a 'Just' reporting culture. These included:

- Ease of making a safety report.
- Professional handling of investigations.
- Rapid feedback to the reporting community.
- Separation of the department collecting and analysing safety reports from that with the authority to institute disciplinary proceedings and impose sanctions.
- Independence of the managers from the reporting system.
- Clear procedures for determining culpability and follow-up action.

It is, however, difficult to draw the line between culpable and non-culpable safety events. Marx (2001) identified four categories of behaviour that result in unsafe acts:

- *Human error* – where an undesirable outcome was inadvertently caused.
- *Negligent conduct* – where a person's conduct falls below the normal standard required but where they were capable of performing to the standard required. It applies when a person fails to use the reasonable level of skill expected of someone engaged in that particular activity.
- *Reckless conduct (gross negligence)* is more culpable than negligence. To be reckless, the risk has to be obvious to a reasonable person. Recklessness is a conscious disregard of an obvious risk.
- *Intentional violations* – when a person knew or foresaw the result of their action but still elected to proceed.

Reason (1997) proposed various tests to establish negligence, one of which was labelled the 'substitution test'. In these circumstances the essential question to answer was would another person (well motivated, equally competent and comparably qualified) have made the same error under similar circumstances? If 'yes', then the error was probably blameless; if 'no', it should be established if there were system-induced reasons for the error, such as insufficient training, selection or experience (Reason's function types). If there were no such reasons, negligent behaviour was possible.

REMEDIAL ACTIONS

A Safety Management system is of no use if it does not also have the capability to recommend or implement safety actions other than producing newsletters providing feedback on safety events and offering words of caution.

The bottom line of Figure 17.1 (and potentially the system defences shown in Figure 17.3, if present) outline the safety activities described in the US Department of Defense system safety standard MIL-STD-882C (US Department of Defense, 1993). However not all aspects fall within the influence of the airlines (designated by the shaded area) and it would be unfair to suggest that all these components are solely (or even primarily) the responsibility of the operators. Only the last element, development of procedures and training, lies comfortably within the remit of the airlines. The other elements are more concerned with aspects of the design of the aircraft and so lie in the province of the manufacturer. Nor should it be inferred that it is solely airlines that generate hazards, incidents and accidents, although the operator does have it within their power to identify and/or manage many hazards (see the sections on LOSA and TEM). Commercial aviation is an open system and hazards cross organisational boundaries.

The US Department of Defense specifies four generic categories of barrier. The hierarchy of safety barriers from MIL-STD-882C is:

- *Design for minimum risk* – Eliminate the hazard from the system if possible. Design the system so the accident cannot happen. In Human Factors terms, make it 'idiot proof'.
- *Incorporate safety devices* – Design into the system automatic devices which, when a specified hazard occurs, prevent the system from entering a dangerous state.
- *Provide warning devices* – These should activate early, leaving the pilot time to stop a critical system state developing.

- *Develop procedures and training* – Provide adequate training in procedures to operate the aircraft in a safe manner.

It can be seen that as you descend the hierarchy, the likelihood of a barrier being effective reduces from a being a certainty to merely being a probability (especially in the last case). Automatic safety devices are designed, built and maintained by human beings, so they will probably work; warning devices are also designed, built and maintained by human beings but they also require that the pilots see or hear them, identify them correctly and react appropriately. Procedures and training require that the pilot recognises that it is all going horribly wrong, identifies the problem correctly and reacts appropriately. Developing procedures and training is a quick and cheap fix, but the one that is ultimately least likely to be effective.

A safety critical system (to protect from a high risk event) should be defended at many levels (e.g. automatic cut-off; if the automatic cut-off fails a warning device should trigger; if both of these fail continual operator monitoring should pick up the problem). Not all barriers are possible in all components of the system. For example, to prevent an aircraft from stalling barriers can be inserted at the second, third and fourth levels (automatic stick-push system, stall warning horn and stall recognition and recovery training). However in a maintenance task barriers may only be possible at either the top and/or bottom level. A top-level barrier would be making the components in such a way so they could not be mis-assembled. The lowest level barriers are dependent upon following the procedures outlined in the manuals.

The size of the holes in Reason's Emmental Cheese model (Figure 17.3) relates to the probability of a fix from MIL-STD-882C being effective. Automatic safety devices have small holes (windows of opportunity) associated with them. Procedures have very big 'windows of opportunity' in them. Nevertheless, if you defend a system in depth all the 'holes' (irrespective of their size) need to line up for the accident to actually happen, making it less likely that it will occur. However, hardware barriers can only catch the specific problems that they have been purposely designed to intercept.

The Safety Management system is the nexus of all the disparate Human Factors activities across the airline. Basically, in this context the job of the Human Factors practitioner is to identify, eliminate or minimise hazardous situations. However, if this is not possible efforts have to be put in to make personnel aware of the hazards, understand the nature of the hazard, and give them the information to be able to make decisions about how to avoid the hazard and have the ability to do so (see Figure 17.6, taken from Ramsey, 1985). The Safety Management function should have links to the selection, training, occupational medicine and engineering functions. However, don't forget that pure luck (either good or bad) plays a significant role in accident causation – you can't attribute everything to the management!

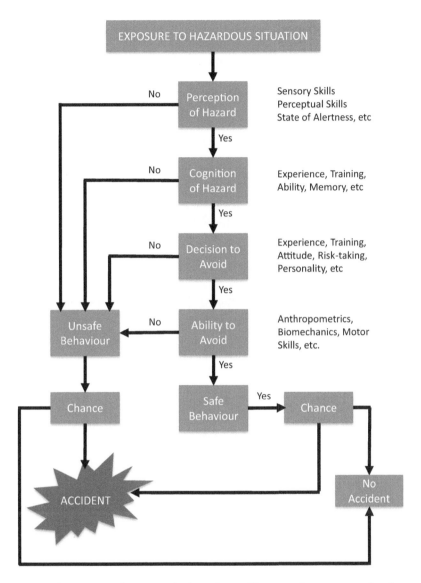

Figure 17.6   Operator characteristics that affect various steps in the accident sequence (Ramsey, 1985)

FURTHER READING

The ICAO *Safety Management Manual* is an excellent document to start with, containing practical advice, a regulatory perspective and lots of background theory:

International Civil Aviation Organization (2009). *Safety Management Manual* (2nd edition) (ICAO Doc 9859). Montreal: ICAO.

This is freely available on line at:

http://www.icao.int/anb/SafetyManagement/DOC_9859_FULL_EN_V2.pdf

For those who prefer books, James Reason's work is essential reading (after all, half this chapter was inspired by him):

Reason, J.T. (1997). *Managing the Risks of Organizational Accidents*. Aldershot, UK: Ashgate.

Or there is also:

Stolzer, A.J., Halford, C.D. and Goglia, J.J. (2009). *Safety Management Systems in Aviation*. Aldershot, UK: Ashgate.

From a non-aviation perspective, try:

Glendon, A.I., Clarke, S.G. and McKenna, E.F. (2006). *Human Safety and Risk Management* (2nd Edition). Boca Raton, FL. CRC Press.

18
Incident and Accident Investigation

When the worst comes to the worst, incidents and accidents need to be thoroughly investigated. Accidents vary rarely just 'happen' and they are hardly ever just the consequence of a single isolated failure, be it in either the human or the technical component of the system. Accidents are usually the final stage in a chain of events.

Incidents and accidents can be viewed as a product of a series of errors partly resulting from a failure in Flight Deck or Airline Safety Management, or induced by design. The most immediate remedial actions resulting from such investigations are usually revisions in procedures and training, although changes in airline operating practices to reduce stress and/or workload may also be implemented.

From a safety perspective the amount of damage and the number and severity of injuries is of little interest. What is of central interest is *why* the accident happened (i.e. the sequence of events leading to the accident). Aircraft incidents can often provide a better, wider insight into the problems underlying aspects of the system of aircraft operations that require remedy. In what follows the emphasis will be on the processes to investigate the Human Factors aspects of an aircraft accident, however this will be only one aspect of the investigation process. Always remember, not only should the incident/accident investigation establish what happened and why, the ultimate objective is to identify aspects of the system for improvement to make it less likely that a similar type of accident will happen again. An accident is just the final part of (usually) a long list of other, related events. The 'best' remedies (most efficient or most wide ranging) may not be in close proximity to the sequence of events leading up to the actual accident.

The aetiology of accidents and incidents may be very similar – they are often just categorised as one or the other as a result of the consequences rather than their causes. The role of chance should never be underestimated: this can provide the dividing line. A helicopter gearbox that fails at 6 feet AGL just after take-off will cause the aircraft to come to the ground with a bit of a 'thump', a lot of injured pride and little else. A gearbox that fails at 6,000 feet over the North Sea will have considerably more disappointing consequences. Although the root cause of the gearbox failure may be the same in each case (which may itself be a consequence of human error during maintenance) only the chance factor separates the accident from the incident. Always remember that luck comes in two 'flavours' – good and bad. From an examination of incident reports, it is not an unknown for pilots to land at the wrong airport in the opposite direction to the active runway and not to realise what they have done until they try to taxi to the gate. Again, it is simply good luck that separates an incident from an accident. Accidents and incidents should be investigated on the basis of their potential, *not* on their outcome.

There are two basic issues in the investigation of incidents and accidents: collecting the data and analysing the data. It is essential to have some sort of theoretical accident causation model within which to analyse data, as this will help to provide an explanation for what happened and why. It will also help to guide the collection of data. However, a lot more has been written about the analysis of incident and accident data rather than its collection, probably because this is a lot easier! But the findings from investigations are only as good as the quality of data obtained.

Incident and accident investigation can be regarded as part of the Safety *Management* system (which it is), however, the need to investigate incidents and accidents can also be regarded as a failure in Safety Management. Indeed the investigation of accidents is mandated by ICAO so can be regarded as a societal requirement (a product of the social *Medium*) however a great deal of the time and money spent on accident investigation may be better spent on the investigation of incidents, and that money in turn may be better spent on proactive safety measures. Just a thought …

COLLECTING HUMAN FACTORS INCIDENT AND ACCIDENT DATA

When investigating an incident or accident there are two basic categories of underlying factors to consider:

- *Causative factors* – failures of critical pieces of equipment or human errors.
- *Contributory factors* – aspects of the task or conditions in which the task is undertaken that promote human failures.

Causative factors definitely are related to the sequence of events in an accident. If a wing falls off an aircraft it will definitely fall out of the sky; if a pilot fails to lower the landing gear it will definitely crash on landing. Contributory factors are more difficult. These are only probabilistically related to the accident outcome. Tiredness, fatigue, high blood-alcohol levels will not always result in an accident but they will make an error more likely. Poor ergonomic working conditions or poor display design will not always result an error, and that error will not always result in an accident, but these contributory factors will make an error more likely. Unfortunately, for the Human Factors accident investigator, most of the factors that you are looking for are contributory rather than causative, which tends to make this aspect of the investigation more difficult and the findings can be regarded as being less robust as a result.

Primary sources of data are directly related to the circumstances surrounding the accident but often these are not sources of especially 'good' data. However, there are many other sources of data to explore. The following is by no means an exhaustive list of primary data sources:

- The aircraft wreckage.
- Flight Data Recorder (the 'black box' – which is actually orange).
- Cockpit Voice Recorder (CVR – which is also orange …).
- Personal records (pilot's log books/operational and training records, etc.).
- Aircraft servicing and maintenance records.
- Air Traffic Control records and tapes.
- Aircraft flight manuals and Standard Operating Procedures.
- Aircraft from the same airline with identical equipment fit.
- Flight briefing materials (nature of the flight, flight plan, meteorological information, etc.).
- Medical records (including post-mortems).
- Interview data.

From a Human Factors perspective it is often the case that the more reliable the data are, the less helpful they will be to you! For example, the data from the aircraft wreckage will give you good information about speed and attitude at the time of impact (*what* happened) but it will give you little insight into *why* it happened. Flight manuals and SOPs will tell you what the pilots *should* have been doing but they will provide little information about what they *actually* were doing. With regard to any interviews, remember that interviewees are neither flight data recorders nor cockpit voice recorders. Their memories are unreliable and they are prone to biases (remember all that stuff about scripts and schemata in LTM from Chapter 2 dealing with Human Information Processing)?

After the field investigation is complete the investigator will need to obtain supporting empirical evidence to facilitate the interpretation of the primary data and to support any assertions concerning the Human Factors causes and contributions to the sequence of events in the accident (see Figure 17.1 in the previous chapter on Airline Safety Management). Typical secondary sources of data include:

- Texts in psychology, Human Factors, aviation medicine and organisational behaviour.
- Databases of previous accidents and incidents. The aircraft manufacturer may also hold a database containing useful data.
- Professional advice from specialists in areas which may potentially be related to the factors involved in the accident.

Remember that as an investigator you need to be a generalist. Do not be afraid to seek out specialist advice (in fact this should be encouraged). Always maintain a sense of perspective, though. This is supporting information that may give you an insight into *why* a pilot or maintenance engineer made a mistake. It cannot be regarded as a definitive statement of the truth. Also, whenever you seek expert advice, always remember that the information you receive is usually opinion. Flight crew and doctors have an uncanny knack of making their opinions sound like definitive statements of fact!

Once an investigation starts the investigator needs to be aware that they will have their own biases: there is a tendency to look a little harder for those (potential) aspects of an accident that are in line with a certain area of expertise. The key role of the Human Factors investigator is to gather all the Human Factors evidence that she/he feels is pertinent. To make sure that all the possible Human Factors issues are at least considered (irrespective of whether they actually play a part or not) various checklists have been developed. These are general frameworks to guide the investigator: they are not exhaustive lists of issues. Furthermore, most checklists have been developed with the pilots as their prime target. However, these people may not be the only human aspect of the system under scrutiny, for example maintenance or ATC may be a major issue in the investigation of any accident. The aircrew may just be the victims of the accident, not the cause. Many of the factors in these checklists, though, can be applied to other aspects of the investigation with a bit of judicious modification.

The Human Factors checklist for accident investigation developed by Feggetter (1982) is a comprehensive checklist very much aimed at investigating the psychological factors underlying flight crew performance in a military context. It only deals with events and factors proximal to the accident (an issue that will be discussed further when it comes to frameworks for analysing and interpreting the data obtained).

- Cognitive System
  - Human Information Processing system
    * Senses   *Check within sensory limits*
    * Perception   *Check order of reception of information*
    * Attention   *Check for distractions*
    * Memory   *Check frequency of events*
    * Decision
    * Action
    * Monitoring
    * Feedback
  - *Visual illusions*   Likely conditions
                               Refraction
                               Fog
                               Ground texture
                               Autokinetic

– *False hypothesis*	Likely conditions
	Expectancy high
	Attention elsewhere under stress
	After period of high concentration
– *Habits*	Previous experience
	Positive and negative transfer
– *Motivation*	Arousal levels (high/low)
	Real objective of interviewee
– *Training*	Check experience and skill
– *Personality*	Check interaction with:
	Colleagues and family
	Attitudes to job
	Self awareness/ambition

- Social System
  - *Social pressure* — Pressure from crew, trans-cockpit gradient, passengers or company
  - *Role* — Role conflict (e.g. 'can do' attitude versus safety)
  - *Life stresses* — Death of relative/friend, divorce, financial worries, etc.

- Situational System
  - Physical Stress
    - Physical condition
    - State of nutrition — *Food and water*
    - Drugs — *History of medication, prescribed drugs (and non-prescribed drugs)*
    - Smoking
    - Alcohol — *Time, type and amount*
    - Fatigue — *Work/rest cycle; duty times and activities*
    - Sleep — *Quality and quantity*
  - Environmental Stress
    - Altitude
    - Speed and motion
    - Ambient light
    - Glare
    - Disorientation
    - Temperature
    - Lighting levels
    - Noise
    - Vibration
  - Ergonomic aspects
    - Design of controls — *Check stereotype responses*
    - Design of Displays — *Check compatibility*
    - Anthropometrics
    - Presentation of material — *Check for ambiguities in:* Labelling legibility / Position of information / Text layout

–   Policy and procedures
    *   Normal   *Check for existence and understanding*
    *   Non-normal

The Human Performance Investigation Factors framework from the National Transportation Safety Board (Figure 18.1) is a little less comprehensive than that proposed by Feggetter (1982), however it does include in it a category to look explicitly at the task and task/time components.

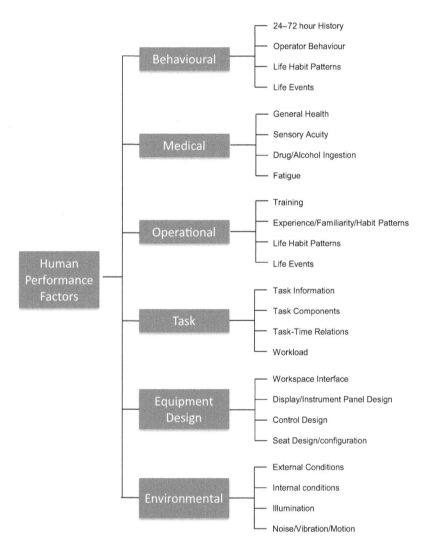

Figure 18.1   Taxonomy of Human Performance Investigation Factors (National Transportation Safety Board, 1983)

The quality of the analysis of incident or accident data is determined by the quality of the data collected. It is better to collect too much data rather than too little. Also, it is vital to remember that the analysis framework chosen will also determine the type of data collected. Finally, it is also worth re-visiting very briefly two of the key principles discussed in the earlier chapters on decision making and human error. Reason (1990) suggested that decision making was affected by cognitive biases, including the availability heuristic (where undue weight is given to facts that come readily to mind) and confirmatory bias (where the decision maker decides upon one solution early in the decision-making process and then gathers further data/information to support this view, discounting contradictory interpretations and failing to regard negative information). The accident investigator can be very prone to both of these biases. As a result, always work through all the components in any data collection checklist used – do not stop when you have found the 'culprit'. Furthermore, when aggregating incident and accident data to look for safety trends (an essential part of the Safety Management function) it is equally important to know what factors *are not* related to certain types of failure. These data are often missing in incident/accident reports and in incident databases.

ANALYSING INCIDENT AND ACCIDENT DATA

Resilient systems, such as aeronautics, are characterised by the need for several breaches of system defences to occur before an accident may result. One view is that this should require many errors/failure before an accident can happen – not just a single point of failure. Another view becoming more common (as espoused in the previous chapter) is that in many cases complex systems simply drift into failure over a period of time. To an extent, these two views are reflected in the two major approaches to analysing incident/accident data. There is an approach that undertakes the analysis process from what may be broadly classified as a 'cause and influence' perspective; the other approach takes a broader, systemic view. It would be wrong to think of these approaches as conflicting, though; both have their strengths and weaknesses. However, when used in combination an even wider perspective may be obtained.

The data collected proximal to an accident tend to be of 'higher quality' (in terms of apparent causation) than those data collected relating to events weeks and months before the event. This proximal data collected can often be represented on a time line describing events and their associated causative or contributory factors (e.g. Multilinear Events Sequencing – Benner, 1975b; or Sequentially Timed Events Plotting – Hendrick and Benner, 1987). However, such event-chain based analyses can supplement an individual accident analysis but cannot be used to aggregate data over a number of analyses to identify wider-ranging, systemic issues. Furthermore, some factors cannot be put neatly on a time line (e.g. a 'poor Safety Culture').

In other cases such a fine-grained analysis is either not possible or is inappropriate, in which case a generic approach is more suitable as a result of the paucity or the quality/nature of data available (e.g. in the analysis of general aviation accidents or accidents occurring to military aircraft where sophisticated flight data recorders and cockpit voice recorders are not carried). These structured systemic approaches, such as HFACS; Shappell and Wiegmann, 2001; Wiegmann and Shappell, 2003) can also aggregate data over many incidents and accidents to begin to isolate trends. They adopt a more organisationally-oriented perspective on incident and accident causation.

Cause and Influence Approaches to Analysis

All of these approaches commence with arranging events upon a timeline leading up to the accident. There may be a single sequence of events or several sequences of related events, all of which converge before or at the time of the final failure. Each event may be exacerbated (or ameliorated) by other conditions (contributory factors). A simple, outline representation of this approach is given in Figure 18.2.

In Multilinear Events Sequencing (MES – Benner, 1975b) all actors involved in the sequence of events leading to the accident have their own time line. In this approach an actor is defined as anything that can bring about an event; they can be human or machine (automation). In Figure 18.2, the line depicting the secondary sequence of events could be attributed to the First Officer and the line depicting the primary sequence of events may be attributable to the Captain. However, in MES the automatic system detecting that the landing gear was not deployed would be given its own event line (this would now be designated as an actor using MES).

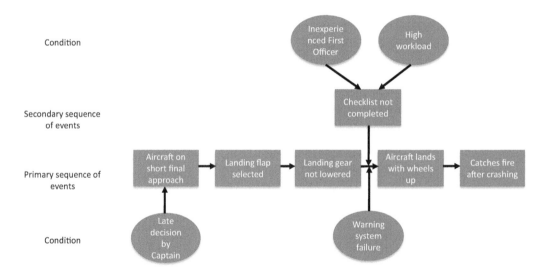

Figure 18.2   Generic cause and influence time line analytical approach to accident analysis (including contributory conditions)

Sequentially Timed Events Plotting (STEP – Hendrick and Benner, 1987) is essentially a refinement of Benner's MES technique. Each actor's actions are depicted on the timeline traced from the start of the incident/accident sequence to its conclusion. These events are positioned relative to one another and causal links are represented by connections between the events on the timelines. When gathering data for a STEP analysis the following information for each event need to be collected:

- *Actor* – who (or what) was (or was not) performing the action?
- *Action* – what did the actor do (or perhaps what did they not do) to what?
- *Time event began* – self explanatory.
- *Event duration* – self explanatory.
- *Event location* – where did this happen (place and equipment)?
- *Description* – narrative description of what occurred.
- *Data source/evidence* – what evidence is there to support these observations?

It has been argued that when using the STEP technique the condition factors do not need to be added to each event as the same factors will apply throughout the event (Hendrick and Benner, 1987). However, this takes a very narrow view of what a condition actually is (M it would exclude many of Reason's psychological pre-cursors of unsafe acts – Reason, 1990) as these apply to an individual, and may apply differentially to different actors in the sequence. It also assumes that conditions remain the same throughout the incident/accident sequence. However, this is also an unsustainable view, as the sequence of events may extend over a period of several days (or even weeks). In other words, in addition to the above factors, collect the conditions affecting performance as well. Note that how the majority of the items in the checklists from Feggetter and the NTSB are actually condition factors in MES and STEP parlance.

One of the hardest decisions to be made when undertaking this type of analysis is to determine what to include and what not to include, and how far back to extend the process. To this end, consider the National Transportation Safety Board (NTSB, 1983) definition of 'Probable Cause':

> ... condition(s) and/or event(s), or the collective sequence of condition(s) or event(s) that most probably caused the accident to occur. Had the condition(s) or event(s) been prevented or had one or more of the condition(s) or event(s) been omitted from the sequence the accident would not have occurred.

To this end, for each event identified a substitution test should be undertaken. Would the incident/accident still have happened if this event was removed from the sequence? If the answer is 'no', then the event is a key occurrence which warrants further analysis (i.e. it is directly implicated). If the answer is 'yes', then the event is probably not directly related to the incident/accident.

One essential output of accident investigation should be in identifying where safety barriers were either non-existent, inadequate or breached, with a view to recommending changes in the system to make similar accidents less likely. Once the sequence of events and their associated condition factors have been mapped, a barrier analysis may then be performed.

Paradies, Unger, Haas and Terranova (1993) suggested the following five questions for completing a barrier analysis:

- What physical, natural, human action and/or administrative controls were in place as barriers to prevent this accident?
- Where in the sequence of events would these barriers have prevented the accident?
- Which barriers failed?
- Which barriers succeeded?

- Were there any other physical, natural, human action and/or administrative controls that might have prevented this accident if they had been in place?

The generic format of barriers has been described in the immediately previous chapter on Airline Safety Management (from MIL-STD-882C): design for minimum risk, incorporate automatic safety devices, provide warning devices and, finally, develop procedures and training.

The analytical approaches based upon time line analyses are better utilised for events proximal to the incident or accident itself, however they are poor at identifying managerial/ organisational aspects underlying aviation mishaps. MORT (Management Oversight Risk Tree – Johnson, 1980) is different from the cause and influence approaches used in MES and STEP as it uses an event-tree based approach. It was originally developed for application in the nuclear energy industry but has also subsequently been used in the analysis of helicopter incidents and accidents (see Ulleberg, Sten, Rosness, Ingstadt, Hudson, Harris and Elwell, 1990). When using the approach as an incident/accident investigation tool the top level node is represented by the final event which results in 'Damage, destruction, other costs, lost production or reduced credibility of the enterprise in the eyes of society' (Johnson, 1980). The structured analysis requires the analyst to trace the evolution of events using a pre-determined tree-based tool to isolate the causes of the accident resulting from management oversights and omissions, or from assumed risks, or as a product of a combination of both. The MORT template consists of eight interconnecting trees, containing 98 generic problems underpinned by 200 basic causes. All aspects that are assessed to be 'Less Than Adequate' in the analysis are subject to further de-composition and investigation. The majority of elements in the event tree are related to failures in management functions (e.g. training; supervision; maintenance planning or risk assessment). The key fact here, though, is that unlike MES and STEP, MORT explicitly acknowledges the role of management in incident/accident causation and incorporates it into the analysis. However, management and supervisory issues are still very much included on a 'cause and effect' basis. Often it is difficult to assess organisational issues and their role in accident causation in such a manner. Indeed, many researchers have argued that it is also inappropriate to attempt to isolate and analyse systemic contributory factors in this way.

Systemic Approaches to Analysis

At the time of writing HFACS (Shappell and Wiegmann, 2001; Wiegmann and Shappell, 2003) is perhaps the most widely used Human Factors incident and accident analysis framework. HFACS is a generic human error-coding framework, based upon Jim Reason's models of human error and of the organisational root causes of accidents (both described earlier in this book). It was originally developed for US naval aviation as a tool for the analysis of the Human Factors aspects of accidents. However, the approach has now also been used for the analysis of large scale data sets of incidents and accidents, for example the analysis of accidents in US commercial aviation (Wiegmann and Shappell, 2001a, 2001b) and Taiwanese civil aviation (Li, Harris and Yu, 2008). It has also provided a structured framework for cross-sector and cross-cultural comparisons of accident causation (e.g. Shappell and Weigmann, 2004; Li, Harris and Chen, 2007) and has been used in the process for the prospective assessment of the effectiveness of aviation safety products developed as part of the NASA Aviation Safety Program (e.g. Andres, Luxhøj and Coit, 2005; Luxhøj and Hadjimichael, 2006).

In Reason's model active failures (which are the errors proximal to the accident, associated with the performance of the pilots) and latent failures (distal errors and system misspecifications, which lie dormant within the system for a long time) serve to combine together with other factors to breach a system's defences. As Reason (1997) observed, complex systems are designed, operated, maintained and managed by human beings, so it is not surprising that human decisions and actions at many organisational levels are implicated in all accidents. Active failures on the flight deck have a direct impact on the safety of the systems. However, latent failures are often spawned in the upper levels of the airline and are related to its management and regulatory structures.

HFACS examines human error at four levels. Each higher level is assumed to affect the next downward level in the framework. The HFACS framework is described diagrammatically in Figure 18.3.

- *Level-1: 'Unsafe acts of operators'* (active failures proximal to the accident) – This level is where the majority of causes in the investigation of accidents is located. Such causes can be classified into the two basic categories of errors and violations.
- *Level-2: 'Preconditions for unsafe acts'* (latent/active failures) – This level addresses the latent failures within the causal sequence of events as well as more obvious active failures. It also describes the context of substandard conditions of operators and the substandard practices they adopt.
- *Level-3: 'Unsafe supervision'* (latent failures) – This level traces the causal chain of events producing the unsafe acts up to the front-line supervisors.
- *Level-4: 'Organisational influences'* (latent failures and system misspecifications, distal to the accident) – This level encompasses the most elusive of latent failures, fallible decisions of upper levels of management which directly affect supervisory practices and which indirectly affect the actions of front-line operators.

At the first level of 'unsafe acts of operators', errors represent the mental/physical activities of an individual that fail to achieve the intended outcomes. Violations refer to the wilful disregard for the rules and regulations that provide safety of flight (Reason, 1990). However, errors and violations do not provide the level of granularity required of most accident investigations. Wiegmann and Shappell (2003) expanded errors further into the four sub-categories of 'skilled-based errors', 'decision errors', 'perceptual errors', and 'routine and exceptional violations' (Figure 18.3). Routine violations tend to be habitual by nature and are often tolerated by the governing authority. On the other hand, exceptional violations appear as isolated departures from authority, and are not necessarily indicative of an individual's typical behaviour pattern, nor condoned by management (Reason, 1990).

Simply focusing on the 'unsafe acts of operator', linked to the majority of accidents, is like focusing on a fever without understanding the underlying illness that is causing it. Wiegmann and Shappell (2003) classified 'preconditions for unsafe acts' into seven sub-categories of 'adverse mental states', 'adverse physiological states', 'physical/mental limitations', 'crew resource management' (CRM), 'personal readiness', 'physical environment' and 'technological environment'. Most of these sub-categories, with the exception of those referring to environmental factors, can be regarded as what Reason (1990) described as psychological precursors for unsafe acts. In subsequent versions of

HFACS (e.g. US Department of Defense, n.d.; Shappell, Detwiler, Holcomb, Hackworth, Boquet and Wiegmann, 2007) the category of 'crew resource management' was renamed to 'communication, coordination, and planning' and the category of 'personal readiness' was re-labelled 'fitness for duty' to make them more descriptive of their content. The US Department of Defense version also adds two further Level-2 preconditions, 'psycho-behavioural factors' and 'cognitive' factors (see Figure 18.3).

The role of supervisors is to provide their personnel with the facilities and capability to succeed and to ensure the job is done safely and efficiently. Level-3 in HFACS is primarily concerned with the supervisory influence both on the condition of pilots and the operational environment. HFACS contains four categories at this level of 'inadequate supervision', 'planned inappropriate operation', 'failure to correct a known problem' and 'supervisory violation' (Figure 18.3).

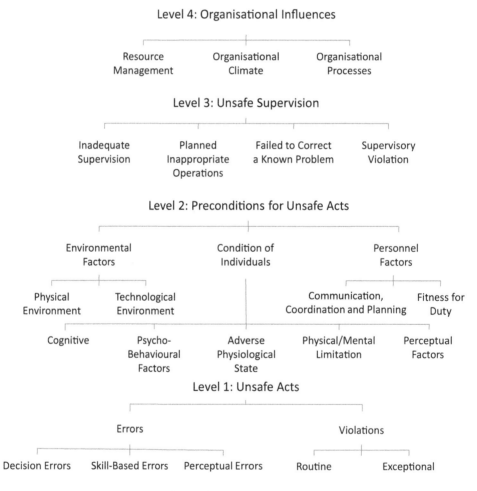

Figure 18.3   Department of Defense's modified form of Shappell and Wiegmann's HFACS framework. Each upper level is proposed to affect items at the lower levels

The corporate decisions about resource management are based on two conflicting objectives, the goal of safety and the goal of on-time and cost-effective operations. It needs to be noted, though, that the decisions of upper-level management can affect supervisory practices, as well as the conditions and actions of operators. However, these organisational errors often go unnoticed due to the lack of framework to investigate them. These elusive latent failures were identified by Wiegmann and Shappell (2003) as failures in 'resource management', 'organisational climate' and 'organisational process'.

There have been criticisms of HFACS. Beaubien and Baker (2002) noted that it was often difficult to collect information about the latent conditions from incident or accident reports and Dekker (2001) suggested that the HFACS framework has only a slight link between human error and the working environment. It was also suggested that there is confusion between categorisation and analysis. He added that the framework merely repositioned human errors by shifting them from the forefront to higher up in the organisation instead of finding solutions for them. Although HFACS is based directly on the organisational theory of failure promoted by Reason (1990, 1997) at the time it was derived there was little or no quantitative data to support the theoretical model upon which it was based nor were there any mechanisms proposed by which the categories/levels in the incident/ accident analysis system influenced each other. Recent work has, however, established relatively strong statistical relationships describing empirically the relationships between various components at the four different organisational levels in the analysis and classification system, giving support to the underpinning theory behind the framework (see Tvaranyas, Thompson and Constable, 2006; Li and Harris 2006a; Li and Harris 2006b and Li, Harris and Yu, 2008). These studies have provided an understanding, based upon empirical evidence, of how actions and decisions at higher managerial levels within an airline may promulgate to result in operational errors and accidents. These associations between levels and components in the HFACS model, though, should not be interpreted as 'paths of causality', as in an event chain model of accident causation. They are better interpreted as 'paths of influence'.

Harris and Li (2011) have further modified and extended the HFACS system. It has already been stated on several occasions that commercial aviation should be regarded as an Open System, but the original conceptualisation of HFACS cannot accommodate this. Accidents in civil aviation are often characterised by errors promulgating across organisational boundaries: it is relatively rare that accidents involve just a single organisational entity. However, there is a fundamental assumption inherent in the architecture of HFACS that the root causes of any accident or incident are all internal to the airline. Wiegmann and Shappell (2003) suggest that external influences to the organisation all act via Level-4 (Organisational Influences). While this may be true in the operation of military aircraft (which fly in relatively closed systems) this is certainly not the case in the civil world. But, it should be noted that HFACS was originally developed as a military (US naval aviation) Human Factors accident investigation tool.

Reason's approach to the causation of human error upon which HFACS is based also needs to be placed within the historical context in which it was developed. HFACS was developed within the 'closed' environment of naval aviation but Reason's organisationally-based model of error was also expounded during a time when organisations themselves were much more 'closed' than they are today. Reason initially undertook his work during the 1980s, however the nature of business has changed dramatically during the last two decades (or so). At the beginning of the twenty-first century there is a great deal more outsourcing, off-shoring and sub-contracting of functions previously undertaken within an airline. Reason's model implicitly assumes a semi-'closed' system. As a result HFACS could not easily incorporate errors which migrated across organisational boundaries. STAMP (Systems-Theoretical Accident Model and Processes; Leveson, 2002, 2004) on the other hand, has been developed using a systems engineering approach and hence incorporates the involvement of multiple systems through the notion of the control and communication of constraints. In STAMP, accidents are considered to result from inadequate control or enforcement of safety-related constraints (occurring during the design, development or operation of the system) and not from individual or component failures. Leveson proposed that systems are interrelated components kept in a state of dynamic equilibrium by feedback loops of information and control. Safety is a product of control structures embedded in an adaptive socio-technical system. Accidents are viewed as control failures.

Harris and Li (2011) developed HFACS-S (STAMP) – an extension to the HFACS methodology – that can accommodate errors promulgating across organisational boundaries, thereby acknowledging the Open System nature of modern airline operations. What was developed is essentially a hybrid model for accident and incident analysis extending HFACS by using elements and concepts borrowed from STAMP. STAMP is comprised of three basic concepts: hierarchical levels of control, process control models and constraints. All systems are made up of hierarchically-arranged control structures which require effective communication channels. In a complex system, such as an airline, there may be several, hierarchically-arranged, control functions (cf. Reason's model of organisationally-based error inherent in the HFACS framework). Safety-related constraints specify the relationships that constitute safe system states. The control processes enforce these constraints on system behaviour.

In an Open System, control processes and constraints also need to be imposed between components in different organisations, for example, between Air Traffic Control and the crew on an aircraft's flight deck. Leveson (2002, 2004) proposed three basic categories of control flaws (with sub-categories) which may result in accidents (see Table 18.1). Leveson has applied this approach to the analysis of several accidents where multiple agencies were involved (e.g. the shooting down of the Black Hawk helicopters over northern Iraq – Leveson, Allen and Storey, 2002; the Boeing 757 Cali accident in Columbia – Leveson, 2004; and the loss of the Ariane 5 Launch vehicle – Leveson, 2002).

HFACS-S commences with a basic HFACS analysis, however each organisation (or sub-unit within an organisation) implicated in the accident causal sequence is subject to an individual HFACS analysis. In HFACS-S all errors are transmitted from one organisation to another via an active failure of an operator (Level-1: 'Unsafe Acts'). Nevertheless they may be received into another organisation at any level in the HFACS hierarchy (not just at the top level). Furthermore, as the promulgation of errors across organisations requires communication (this is an Open Systems approach) the nature of this communication shortfall is characterised using Levenson's categories of control flaw. Furthermore, for the HFACS analyses conducted within each organisation, the linkages between categories are also described using the STAMP control flaws taxonomy. Harris and Li (in press) have used this approach to re-analyse the events from the Uberlingen mid-air collision on 1 July 2002 (Bundesstelle für Flugundfalluntersuchung – BFU, 2004).

Table 18.1   Leveson's (2002) three basic categories of control flaws (with sub-categories)

Inadequate control actions (enforcement of constraints)
Unidentified hazards
Inappropriate, ineffective or missing control actions for identified hazards
Design of control process do not enforce constraints
Process models that are inconsistent, incomplete or incorrect
Inadequate coordination among controllers and decision makers
Inadequate execution of control action
Communication flaw
Inadequate actuator (operator) operation
Time lag
Inadequate or missing feedback
Not provided in system design
Communication flaw
Time lag
Inadequate sensor operation (incorrect or no information provided)

FINAL THOUGHTS

Several authors have suggested a pyramidal relationship between hazards, incidents and accidents of between 100:10:1 to 300:30:1. All these claims are somewhat trite and are almost completely unverifiable. However, there are many more incidents than there are accidents and better quality data can be obtained from their investigation to identify their root causes. Careful, structured analysis of incidents can reveal underlying trends pointing to organisational or equipment deficiencies before an accident occurs. These shortcomings may never be found from an accident investigation, which is the analysis of a one-off, 'freak' event. However, incident and accident investigation needs to be subject to the same 'rules' as confidential reporting schemes if complete cooperation is to be elicited from all those involved. The purpose of any incident/accident is to provide information for flight safety purposes to prevent reoccurrence of a similar type of accident. It is not to establish liability or culpability. Furthermore, there are no 'hard and fast' guidelines for incident/accident investigation and analysis. What has been described is for guidance and should be used with discretion. Rudyard Kipling would have made a half decent accident investigator:

> I keep six honest serving men
> They taught me all I knew
> Their names are *What* and *Why* and *When*
> And *How* and *Where* and *Who*

(Rudyard Kipling: *The Elephant's Child*)

FURTHER READING

There are many books available on Human Factors in accident investigation and the human error root causes of accidents. For two contrasting approaches to the topic try:

Strauch, B. (2004). *Investigating Human Error: Incidents, Accidents, and Complex Systems*. Aldershot: Ashgate.

Or, of course:

Dekker, S.W.A. (2006). *The Field Guide to Understanding Human Error*. Aldershot: Ashgate.

There are also a couple of excellent web-books available free of charge(!) containing a wealth of information about accident and incident analysis.

Johnson, C.W. (2003). *Failure in Safety-Critical Systems: A Handbook of Accident and Incident Reporting*. Glasgow: University of Glasgow Press. Available at: http://www.dcs.gla.ac.uk/~johnson/book/.

and

Leveson, N.G. (2009). *Engineering a Safer World: System Safety for the 21st Century (or Systems Thinking Applied to Safety)*. Boston, MA: Massachusetts Institute of Technology. Available at: http://sunnyday.mit.edu/book2.pdf.

19
Concluding Thoughts: Human Factors in Aviation as a Route to Increased Operational Efficiency

Human Factors in aviation has its historical roots firmly in the pursuit of safety, initially in the design of the flight crew interfaces to avoid inadvertent operation. This was then followed by advances in training and flight deck management, followed by an evolution in safety management processes. Within the military there has been more of an emphasis on the use of Human Factors to improve performance, particularly in the areas of training and interface design. It is probably only in the application domain of pilot selection that in the world of commercial aviation the emphasis has been firmly placed upon enhancing performance while reducing costs (specifically, as a result of those finally selected failing to complete training).

Within commercial aviation Human Factors has almost become regarded as a 'hygiene factor'. It has become a 'cost' associated with the necessity to maintain and improve safety but has added little to the bottom line on the balance books. It has tended to be a net 'debit' to the organisation rather than justifying itself by demonstrating that it can be a 'credit'. In the opening chapter it was described how, from a flight deck design perspective, producing a 'pilot unfriendly' interface will result in a product which is difficult to use and promotes error. However, from a manufacturer's perspective, providing a 'user friendly' flight deck does not mean that it is possible to charge more for the aircraft! Minor issues in the design of the flight deck become issues in training to be dealt with by the airline. Similarly, improvements in safety as a result of training all cost money. The natural inclination is always to call for more training and more training facilities (in the name of safety). Human Factors practitioners rarely ask for less training!

However, the Human Factors discipline has now matured to such an extent that it can also be used to enhance the operational efficiency of airlines. In other words, it can be used in such a manner as to become a financial credit to the organisation rather than an irrecoverable cost. Yet to truly enhance operating efficiency the human component in any system cannot be examined in isolation from all the other components, Human Factors or otherwise. A wider, socio-technical perspective needs to be adopted before any true efficiency gains incorporating good Human Factors can be realised. The application of Human Factors to isolated problems cannot pay its way, but by taking an integrated, through-life approach to tracking Human Factors costs (and savings) the benefits that accrue can be demonstrated. First, though, a change of attitude is required – one in which cost savings and increased performance are regarded as acceptable objectives for Human Factors.

It is still doubtful that Human Factors can 'add value' but it can reduce costs by promoting efficiency. The term 'efficiency' in this context is somewhat difficult to define. Traditionally, discussion of this type should always start with a definition, for example:

Efficiency (n). The quality or degree of being efficient

Efficient (a). ...productive of desired effects; especially: productive without waste.

(Webster's *Third New International Dictionary*)

In this context 'efficiency' and 'waste' can either directly or indirectly, be related to time or money. For example, efficiency can be enhanced either by providing training that results in the same level of competence in the trainees but by using cheaper training methods or through completing the training more quickly. Alternatively, efficiency can also be enhanced by spending the same amount of time and/or money on the training but producing a higher standard of graduate at the end of it.

Personnel costs are one of the most significant costs in the operation of an airline. British Airways accounts for the year 2009/10 (http://www.britishairways.com/cms/global/microsites/ba_reports0910/overview/cfo3.html) show that staff costs are the second highest cost after fuel and represent nearly 25 per cent of operating costs. Even for a low cost operator (such as RyanAir) which outsources a great deal of its activities, staff costs still represent 11 per cent of its operating cost (RyanAir, 2009: http://www.ryanair.com/doc/investor/2009/q4_2009_doc.pdf). Another way of looking at it is that airline efficiency can be enhanced by doing the same amount with fewer people, or doing more with the same amount of people. Whenever the efficiency equation is presented in this way it can be seen that Human Factors is central to its enhancement.

Good Human Factors will not significantly increase efficiency by itself. On its own, only small incremental gains may be made. However, to make efficiency gains the discipline must coalesce once again in order that the maximum benefit from an integrated, through-life approach can be realised. While increasing levels of specialisation (as described in each of the preceding chapters) have served to develop the science base it has also simultaneously mitigated against its coherent application in commercial aviation.

An integrated, system-wide approach considers the users, their tools, task requirements, organisation and the environmental context. The Human Factors National Advisory Committee for Defence and Aerospace (2003) describes some examples of the benefits of taking a wider socio-technical/HFI approach to equipment design. For instance, the developer of an aircraft engine who adopted an HFI approach reduced the number of tools required for the line maintenance of a new turbine from over 100 to just 10; fewer specialist skills were needed further allowing a consolidation in the number of maintenance trades required which also resulted in an overall reduction in training time.

In 1997 the cost of accidents and incidents on the airport ramp was estimated to be $2 billion, much of which was uninsured losses. However, these direct costs represented only a small proportion of the overall cost (e.g. damage to aircraft; repair and replacement of damaged parts). Indirect costs are far more substantial, for example compensation, re-scheduling of services, service fees, replacement aircraft, loss of perishable cargo, etc. Airports Council International (1993) reported that 84 per cent of ramp accidents occurred when ground equipment struck an aircraft (known as 'ramp rash'). An observational study at a major UK airport (Harris and Thomas, 2001) showed that deviations from recommended procedures when servicing aircraft were commonplace, including in 68 per cent of all turnarounds, ground equipment being positioned around the aircraft without the help of a banksman and in 19 per cent of turnarounds at least one vehicle being reversed up to the side of the aircraft without external guidance. However, these observations merely describe *what* happened, not *why*. When the reasons underlying such behaviours were proffered a more complex picture emerged. Aircraft servicing and provisioning is usually provided by several sub-contractors. To keep prices down and margins high, sub-contractors are under considerable time and financial pressures. Competition is encouraged by the airlines. To remain competitive, sub-contractors operate with the minimum number of personnel. As a result, a common explanation for procedural violations was that 'no one was available to see me back' or 'there was no time to get someone to help me reverse'. Reason (1997) suggested that such violations are quietly overlooked by management (until something goes wrong). The root cause for such behaviours, though, resides within the contracts branches of the airlines. Competition between suppliers keeps prices low and punitive clauses are written into contracts to punish late or non-delivery. There is an argument to be made that if margins were not eroded to the bare minimum, then incidences of ramp rash would begin to decrease. There would be someone available to guide back the driver of a reversing truck. The provision of supplies may cost slightly more but the airline may save large amounts from the reduced requirement for aircraft repair, delays, insurance premiums, etc. Arguing that a sub-contractor is responsible for these aspects of ramp safety is only a partial solution and does not address the root cause of the problem. It merely transfers the blame. Taking a wider view of Human Factors may be beneficial in terms of efficiency and safety. The action of people writing and negotiating contracts *does* influence human behaviour. Paying more may ultimately promote efficiency and cost less.

What these examples illustrate is that organisations can realise significant economic benefits by actively considering the wider organisational impact of procurement and task redesign. But, to do so requires taking both a through-life and an organisation-wide perspective.

The most obvious area in which Human Factors can make an immediate cost impact is in the selection of personnel. It has already been noted in Chapter 8 that even ten years ago (at the time of writing) the cost to an airline of a cadet who failed training was €50,000 against a figure of €3,900 (each) for the cost of the selection process for each candidate (Goeters and Maschke, 2002). The net utility of a rigorous selection procedure is dependent upon the cost of the selection process, the number of applicants required versus the number of applicants in the applicant pool, the cost of training (and the cost of failing training) and other indirect costs associated with subsequent performance and employee turnover in the company (see Holling, 1998). However, the selection process is only the beginning of producing a safe and competent pilot.

Training a cadet pilot to occupy the left-hand seat is expensive. The biggest cost savings (or efficiency gains, depending upon how you look at it) are in the training process itself. Using lower levels of fidelity of simulation (matched to the cognitive requirements of the training) may produce significant cost savings while not damaging the quality of the graduates produced. A more radical approach, with potentially much greater cost efficiencies, questions the whole basis of each stage of the training. For example, the value of the initial stages of training in a piston-engined light aircraft has already been questioned when the objective is to produce crew for high technology, high performance commercial aircraft (Harris, 2009).

An even more radical approach to achieving cost efficiencies may be taken if the process starts with the design and development of the aircraft itself. As a first step consider simply reducing the number of VNAV modes in a modern highly automated aircraft. From a navigation viewpoint the majority of VNAV modes do not relate directly to the three-dimensional navigation solution as they are aircraft referenced and not Earth referenced (only approach mode is referenced to a fixed point on the planet). In other words, they do not directly support the pilots' primary task. There are several aircraft-referenced modes (vertical speed, flight path angle, climb-to-speed) but not all of these are required for operations. Simplification of the number of vertical modes will not only reduce the aircraft development (automation) costs, it will also considerably reduce training costs and will simultaneously reduce the potential for error: one common perspective on pilot error is that every mode that is available on the flight deck is potentially the wrong mode to be in.

Dekker (2004a) has disabused the notion that high levels of automation reduce labour costs. Taking an even broader view, the certification and maintenance of highly automated aircraft is also undoubtedly far more expensive than that for a simpler machine. Across the whole lifecycle of the aircraft, it is doubtful if these extra costs associated with highly automated aircraft outweigh the operational efficiency gains they offer. However, even the approach of simplifying certain flight deck functions will only produce relatively modest cost savings. Greater cost savings can be achieved with a more radical, human-centred approach.

The trend in flight deck design over the past half century has been one of progressive 'de-crewing'. The common flight deck complement is now that of two pilots; 50 years ago, it was not uncommon for there to be five crew on the cockpit of a civil airliner (two pilots, Flight Engineer, Navigator and Radio Operator). Just two pilots, with much increased levels of assistance from the aircraft, now accomplish the same job once undertaken by five. Many of the functions once performed by flight crew are now wholly (or partially) performed by the aircraft itself. The emphasis in the role of the pilot has changed from one of being a 'flyer' to one of being a systems (or flight deck) manager. The aircraft and its systems are now more usually under supervisory control, rather than manual control. Emphasis is more based around crew and automation management rather than flight path control. Research into the safety implications and procedures required for the operation of existing long-haul aircraft during the cruise phase using just a single member of flight deck crew is already underway. This will reduce the operating costs associated with the need to carry a third pilot during ultra-long haul operations specifically for the approach and landing phase.

For smaller regional aircraft (up to 50 seats) it has been estimated that over a 200-nautical-mile leg, approximately 15 per cent of the direct operating costs can be accounted for by crew costs, although some estimates place these as high as 35 per cent of direct operating costs (Alcock, 2004). This is increased even further on smaller aircraft. Annual accounts from a typical low-cost operator (EasyJet, 2009) suggest that even for a larger aircraft, overall crew costs represent nearly 12 per cent of total operating costs (http://2009annualreport.easyjet.com/business-review/financial-review.html). As a result, considerable savings are possible with a reduction in the number of flight deck crew, particularly with small passenger and cargo aircraft. With the judicious use of the range of technologies developed during the past decade there are no major reasons why a single pilot operated commercial aircraft is not feasible. Indeed, the individual technologies required have now all reached a suitable level of maturity. The military have been operating complex, high performance single-crew aircraft for many years and Uninhabited Air Vehicles (UAVs) are now a regular part of operations. It is time for these technologies to be spun-out further into the commercial domain where they may be applied to financial advantage. The greatest obstacle to the operation of civil, single pilot aircraft is not the technology *per se*. It is combining the technology, designing the user interfaces and developing a new concept of operations to make such an aeroplane safe and useable in a wide range of normal and non-normal operating situations when flown by a typical commercial pilot. The Human Factors requirements are the prime driver *not* the hardware and software technologies. Architectures for the design and operation of single-crew aircraft have been proposed where the control (and crewing) of the aircraft is distributed in real time across both the flight deck and ground stations (Harris, 2007). The second crew member is not replaced by on-board Artificial Intelligence or Intelligent Knowledge-Based Systems (these would be difficult to develop and very challenging to demonstrate the required level of reliability from a certification perspective) but is merely displaced. Furthermore, the ground-based component's primary functions are modified to become a support for the pilot (e.g. navigation, system management or fault diagnosis) not a duplicate of the skills and functions of the pilot on board. A great deal of the time, two crew members are not required to fly an aircraft and when two are required there is no reason why the people performing the functions need to be physically co-located. A single crew-member on the ground may simultaneously support several aircraft.

This approach begins to blur further the distinction between the ground and the airborne functions in the operation of commercial aircraft. This process commenced with the initial introduction of Air Traffic Control and has continued further with the introduction of direct routing as an air traffic concept. This concept significantly affects the pilot's role and responsibilities. Responsibility for separation and most aspects of routing will be delegated to the flight deck. Aircraft will fly direct routes and manoeuvre freely at their optimum speed and altitude. In this way, they will spend less time in the air and use less fuel, significantly increasing efficiency. The impetus to move to such a system is also driven by the fact that the current system is inefficient in its use of the airspace available and unless changes are made, it will be impossible to cope with the increasing growth in air traffic in many parts of the world.

However, changes to the physical airspace demand wide-ranging changes throughout all other components of the system. To optimise efficiency aircraft need to be re-equipped with new navigation and surveillance equipment, the crew need to be trained to use this equipment and associated procedures, and the company management has to be responsible for integrating all these aspects and conforming to the international regulatory agreements for the approval of equipment and its operation in Free Flight airspace. Simply considering this emphasises the need for an integrated, system-wide approach to the application of Human Factors across all aspects of an airline's operation. Considering each of these aspects alone will not produce the required level of system integration and operation necessary. And it will be more expensive.

Taking this concept one step further, there seems to be a logical progression to devolving many flight planning activities to Air Traffic providers (basically airlines buy a quality assured, guaranteed de-conflicted route from them) which is uploaded directly to the aircraft, removing the requirement for this function in the airline. Remember, the aircraft is now responsible for separation when in the air (not ATC). These roles would seem to have reversed, once again, between the ground and the air. Perhaps even greater efficiencies can be enacted if the separation between these two components is further reduced. However, that is maybe one step too far for this book.

Possibly the greatest obstacle to increasing efficiency across the airline system as a whole is neither technological, associated with Human Factors or associated with the management of airlines but is a product of the nature of the regulatory system which now serves to inhibit innovation and may also be stifling potential advances in safety. The regulatory framework within which commercial aviation operates has its roots in military acceptance and airworthiness criteria developed from just after World War II. These criteria were essentially aimed at ensuring the structural integrity of aircraft and specifying a degree of system redundancy to enhance engineering integrity. As the civil (commercial) aviation system began to develop rapidly during the late 1940s and 1950s the rule system became ever more diverse to accommodate all the different facets of airline operations. However, as a result, the regulatory system also became increasingly fragmented. For example, from a Human Factors perspective Part 25 of Title 14 (Aeronautics and Space) of the Federal Aviation Regulations deals with aspects of the pilot interface (e.g. controls, displays and automation, etc.). These aspects are covered across a number of separate regulations but with the advent of CS 25.1302 (in the EASA version of the airworthiness regulations) there is the beginning of a coherent approach to ensuring the consistent application of Human Factors principles across the flight deck. The basics of pilot licensing and training (covering flying and navigating an aircraft) are covered Part 61 of the Federal Aviation Regulations. These basics are then supplemented by further licence endorsements to fly an aircraft at night, fly an aircraft with more than one engine, fly in controlled airspace, etc. To fly for hire or reward requires a commercial pilot's licence (certificate in the US) or to fly fare-paying passengers requires an Airline Transport Pilot Licence (certificate). All of these are generic requirements. However, once the aircraft weighs over 5,700 kg a specific type rating is required. This essentially ensures that there is a good 'fit' between the *Machine* and the hu*Man* (in terms of specific training and demonstration of knowledge and ability). The technology approved to deliver such training is covered in Part 60 of the Regulations (Flight Simulators and Training Devices).

However, if you wish to operate an airline to make money from an aircraft the requirements to do this are covered in Parts 91 of the Federal Aviation Regulations (the basic rules for the operation of aircraft) and Part 119. This latter part specifies the airline-wide management functions and processes required for an Air Operator's (or Air Carrier's) Certificate – the *Management* component. The manner in which the *Mission* of carrying passengers from one place to another is specified in Part 121 of the Regulations (Operating requirements: Domestic, flag, and supplemental operations); Part 125 (Certification and Operations: airplanes having a seating capacity of 20 or more passengers or a maximum payload capacity of 6,000 pounds or more; and rules governing persons on board such aircraft) or Part 135 (Operating requirements: Commuter and on demand operations and rules). The wider environmental context of operations (not considered in Figure 19.1 – part of the *Physical Medium*) includes yet more parts of the Regulations, such as Part 71 (Designation of Class A, B, C, D, and E Airspace Areas; Air Traffic Service Routes; and Reporting Points); Part 77 (Objects Affecting Navigable Airspace); Part 139 (Certification of Airports) and Part 153 (Airport Operations).

This brief description merely scratches the surface of the regulatory (*Social Medium*) fragmentation, with reference to the 5Ms model (which is where we came in). The rules have evolved separately over the years (see Harris, 2011). Summarising the previous paragraphs: Part 25 is about the Aircraft; Parts 60 and 61 about the people; Parts 91 and 119 about the Management; Parts 121, 125 and 135 are about Operations; and Parts 77, 139 and 153 are about the wider operating environment external to the aircraft (see Figure 19.1). The one thing that the rules are not specifically concerned with is the *system* of transporting people and cargo safely from A to B. However, this is their intent.

Take the Human Factors-oriented examples of workload and error. Workload (as a concept) appears in 37 different FAA Advisory Circulars across 19 parts of the Regulations. The *system* objective should be to manage the pilot's workload but this is handled in different ways across the various parts of the rules. Pilot workload is a product of the number and difficulty of the tasks to be performed in the available time; the ease of use of the equipment on the flight deck and interactions with complicating factors and stressors in the environment. Error appears in 45 different Advisory Circulars across 24 parts. Again, the system objective is to manage error but it is widely regarded in the scientific literature that error is a product of equipment design, procedures, training and the environment which are all covered by different parts of the Regulations. If these issues cannot be separated how can they be regulated separately?

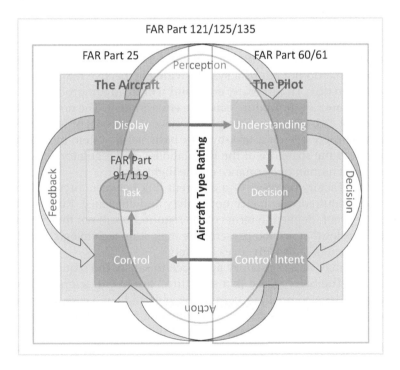

Figure 19.1   Parts of the Federal Aviation Regulations superimposed over a representation of the classical 'Perception–Decision–Action–Feedback' control loop (see also Figure iii.i) to illustrate the fragmentation of the regulatory structures from a Human Factors perspective

Activity on the flight deck proceeds on a task-by-task basis (as does Human Factors design itself). Pilots interact with several systems simultaneously when performing a task, thus inconsistencies in interfaces are much more obvious to them and procedures tend to become very noticeable. Many Human Factors problems and issues lie not within an individual regulation but between the regulations. The new European Human Factors flight deck certification rule CS 25.1302 tries to take a task-based approach to certification of the flight deck but is limited by the scope of Part 25 itself. Factors outside Part 25 cannot be considered when assessing compliance, nor is it permitted that the regulation itself can address issues outside those associated with the design and structure of the aircraft. The probability of design-induced error on the flight deck will be significantly reduced but the level of risk in flight operations may only be marginally reduced as a result of not taking a systemic perspective Revisions to training practices (e.g. the US AQP programme) provide targeted training for the specific requirements of the airline but are again limited in their scope as a result of the regulatory structures.

When the Regulations are described in this manner not only does it become apparent that their structure impedes system-wide improvements in efficiency, it may also be the case that it is the regulatory system itself that is preventing further improvements in safety (hence the observation in Chapter 1 that during the last decade the serious accident rate has plateaued at approximately one per million departures; Boeing Commercial Airplanes, 2009). Adding another *local* regulation to fix one specific aspect of a much wider *system* is unlikely to have any major effect. A new systemic approach is required. Fragmented rules that do not afford a system-wide perspective may not increase safety to the degree anticipated. To re-iterate, error and workload are products of interactions between the pilots, aircraft, procedures and the environment. The notion of human error having a single root cause is an oversimplified view of the roots of failure. Furthermore, flying an aircraft progresses on a task-by-task basis *not* a system-by-system basis (the approach implicit in the structure of the Regulations). The regulatory structures are not commensurate with human performance on the flight deck.

From a Human Factors perspective a coherent link between aircraft design, training and operations is required to enhance both safety and efficiency. Rules and regulations need to be future proof: defined in terms of the required result not the method to achieve it. Airworthiness rules that are too prescriptive may stifle technological and operational innovation and potentially also advances in safety.

The Safety Case may be a regulatory approach to satisfy the requirements of modern airline operations. The Safety Case is a structured argument, supported by a body of evidence that provides a compelling, comprehensible and valid case that a system is safe for a given application in an operating environment. Safety Cases are *not* prescriptive (the aim is simply to demonstrate systems meet the safety goal); they do not separate the human from the system; they are evidence-based (data driven) and are subject to continual revision. From a Human Factors perspective, specific topics under consideration in Safety Case presentations normally cover: the competencies required to perform the work; training and training needs analysis; development and maintenance of procedures; communication processes; manning levels; automation and allocation of function; supervision of staff; shift patterns; hardware and software layout; environmental PSFs; and human error potential and Safety Culture.

Safety Cases are also not static – they change in response to changes in the nature of operations. Such an approach is becoming used much more frequently in defence aerospace, for example Eurofighter Aircraft Avionics, the Hawk Aircraft Safety Justification and in

Military Air Traffic Management Systems. A similar approach is being use for the safety evaluation of civil Unmanned Air Systems (CAA, 2010). Furthermore, the basis for Safety Cases is already being used by all airlines as part of their Safety Management processes.

If safety regulation is to progress in a manner compatible with the management of Human Factors issues it has to progress on a systemic basis, not a system-by-system basis. A Safety Case-based approach provides this opportunity. This is not to say that it should replace the current set of Regulations as this would be completely impractical. However, it can be used as an adjunct and/or alternative where a suitable waiver to existing regulations is granted.

Aviation Human Factors as a discipline has come of age. Despite what has been written in the preceding paragraphs, it must avoid its natural inclination to define itself solely within the domain of safety. The discipline must also coalesce once again in order that the maximum benefit from an integrated, through-life approach can be realised. To a large degree, while increasing levels of specialisation have served to develop the science it has also simultaneously mitigated against its coherent application in commercial aviation. The opportunity now exists to capitalise on the developments made by this relatively new discipline, which was originally born in the aviation domain just half a century ago to improve both operational efficiency and safety. These objectives are not mutually exclusive and may well prove to be completely compatible.

References

Ackoff, R.L. (1989). From Data to Wisdom. *Journal of Applied Systems Analysis*, 16, 3–9.

Air Accident Investigation Branch (2002). *Airbus A321-211 G-JSJX, AAIB Bulletin no. 3/2002*. Farnborough, Hants: UK Air Accident Investigation Branch.

Air Navigation Order (2000). *Statutory Instrument 2000, No. 1562*. London: The Stationery Office Limited.

Aircraft Accident Investigation Commission, Ministry of Transport, Japan (1994). *Aircraft Accident Investigation Report 96-5: China Airlines, Airbus Industrie A300B4-622R, B1816, Nagoya Airport, April 26, 1994*. (Translated by H. Sogame and P. Ladkin.) (Available from: URL: http://sunnyday. mit.edu/accidents/nag-3.html Accessed 12 December 2007).

Airports Council International (1993). *Ramp Safety Handbook*. Geneva, Switzerland: Author.

Alcock, C. (2004). New turboprop push tied to rising fuel costs. *Aviation International News*, 29 September 2004. Accessed at: http://www.ainonline.com/Publications/era/era_04/era_ newturbop18.html (30 January 2007).

Allan, J.R. (1988). Thermal Protection. In: J. Ernsting and P. King (eds) *Aviation Medicine* (2nd edition) (pp. 247–60). London: Butterworths.

Allen, J.A. (1986). Maintenance Training Simulator Fidelity and Individual Differences in Transfer of Training. *Human Factors*, 28, 497–509.

Allerton, D. (2009). *Principles of Flight Simulation*. Chichester: John Wiley.

American Society of Heating, Refrigeration and Air Conditioning Engineers (1997). *Handbook of Fundamentals*. Atlanta: Author.

Anastasi, A. and Urbina, S. (1997). *Psychological Testing* (7th edition). Upper Saddle River, NJ: Prentice Hall.

Anderson, J.R. (1982). Acquisition of Cognitive Skill. *Psychological Review*, 89, 369–406.

Anderson, J.R. (1983). A Spreading Activation Theory of Memory. *Journal of Verbal Learning and Verbal Behavior*, 22, 261–95.

Andres, D.M., Luxhøj, J.T. and Coit, D.W. (2005). Modelling of Human-System Risk and Safety: Aviation Case Studies as Exemplars. *Human Factors and Aerospace Safety*, 5, 137–68.

Annett, J. and Stanton, N.A. (2000). *Task Analysis*. London: Taylor and Francis.

Anthony, E. (1988). Psychiatry. In: J. Ernsting and P. King (eds) *Aviation Medicine* (2nd edition) (pp. 619–43). London: Butterworths.

Applegate, J.D. and Graeber, R.C. (2001). Integrated Safety Systems Design and Human Factors Considerations for Jet Transport Aeroplanes. *Human Factors and Aerospace Safety*, 1, 201–21.

Arnold, J. (2005). *Work Psychology* (5th edition). Harlow: Prentice Hall.

Atkinson, R. and Shiffrin, R. (1968). Human Memory: A Proposed System and Its Control Processes. In: K. Spence and J. Spence (eds) *The Psychology of Learning and Motivation: Advances in Research and Theory* (Volume 2). New York: Academic Press.

Aviation Safety Council (2002). *In-flight Breakup Over Taiwan Strait Northeast of Makung, Penghu Island China Airlines Flight CI-611, Report no. ASCAOR-05-02-001*. Taipei, ROC: Aviation Safety Council.

Backs, R.W. and Seljos, K.A. (1994). Metabolic and Cardiorespiratory Measures of Mental Effort: The Effects of Level of Difficulty in a Working Memory Task. *International Journal of Psychophysiology*, 16, 57–68.

Baddeley, A.D. (1986). *Working Memory*. Oxford: Oxford University Press.

Baddeley, A.D. (1996). Exploring the Central Executive. *Quarterly Journal of Experimental Psychology*, 49A, 5–28.

Baddeley, A.D. (2000). The Episodic Buffer: A New Component of Working Memory? *Trends in Cognitive Sciences*, 4, 417–23.

Baddeley, A.D. and Hitch, G.J. (1974). Working Memory. In: G.H. Bower (ed.) *The Psychology of Learning and Motivation* (Volume 8) (pp. 47–89). New York: Academic Press.

Bailey, N.R. and Scerbo, M.W. (2007). Automation-Induced Complacency for Monitoring Highly Reliable Systems: The Role of Task Complexity, System Experience, and Operator Trust. *Theoretical Issues in Ergonomics Science*, 8, 321–48.

Bainbridge, L. (1987). Ironies of Automation. In: J. Rasmussen, K. Duncan, and J. Leplat (eds) *New Technology and Human Error* (pp. 271–83). Chichester: Wiley.

Banbury, S. and Trebblay, S. (2004). *A Cognitive Approach to Situation Awareness: Theory and Application*. Aldershot: Ashgate.

Barnett, B.J. and Wickens, C.D. (1988). Display Proximity in Multicue Information Integration: The Benefits of Boxes. *Human Factors*, 30, 15–24.

Baron, J. and Hershey, J.C. (1988). Outcome Bias in Decision Evaluation. *Journal of Personality and Society Psychology*, 54, 569–79.

Baron, S. (1988). Pilot Control. In: E.L. Wiener and D.C. Nagel (eds). *Human Factors in Aviation* (pp. 347–85). San Diego, CA: Academic Press.

Barrick, M.R. and Mount, M.K. (1991). The Big Five Personality Dimensions and Job Performance: A Meta-Analysis. *Personnel Psychology*, 44, 1–26.

Bartram, D. and Baxter, P. (1996). Validation of the Cathay Pacific Airways Pilot Selection Program. *International Journal of Aviation Psychology*, 6, 149–69.

Bates, J.E.W. (2002). An Examination of Hangover Effects on Pilot Performance. *Dissertation Abstracts International: Section B: The Sciences and Engineering*, 62(9-B), 4257.

Battelle Memorial Institute. (1998). *A Review of Issues Concerning Duty Period Limitations, Flight Time Limitations, and Rest Requirements. Federal Aviation Administration Report (AAR-100)*. Washington, DC: US Department of Transportation.

Beach, L.R. and Lipshitz, R. (1993). Why Classical Decision Theory is an Inappropriate Standard for Evaluating and Aiding Most Human Decision Making. In: G. Klein, J. Orasanu, R. Calderwood and C.E. Zsambok (eds) *Decision Making in Action: Models and Methods*. Norwood, NJ: Ablex Publishing Corporation.

Beatty, J. (1982). Task-Evoked Pupillary Responses, Processing Load, and the Structure of Processing Resources. *Psychological Bulletin*, 91, 276–92.

Beaubien, J.M. and Baker, D.P. (2002). A Review of Selected Aviation Human Factors Taxonomies, Accident/Incident Reporting Systems and Data Collection Tools. *International Journal of Applied Aviation Studies*, 2, 11–36.

Bell, H.H. and Lyon, D.R. (2000). Using Observer Ratings to Assess Situation Awareness. In: M.R. Endsley and D.J. Garland (eds) *Situation Awareness Analysis and Measurement* (pp. 129–46). Mahwah, NJ: Lawrence Erlbaum.

Bell, H.H. and Waag, W.L. (2000). Using Observer Ratings to Assess Situation Awareness in Tactical Air Environments. In: M.R. Endsley and D.J. Garland (eds) *Situation Awareness Analysis and Measurement* (pp. 93–9). Mahwah, NJ: Lawrence Erlbaum.

Benassi, V.A., Sweeney, P.D. and Dufour, C.L. (1988). Is There a Relationship Between Locus of Control Orientation and Depression? *Journal of Abnormal Psychology*, 97, 357–67.

Benner, L. (1975a). DECIDE in Hazardous Materials Emergencies. *Fire Journal* (July, 1975).

Benner, L. (1975b). Accident Investigations: Multilinear Events Sequencing Methods. *Journal of Safety Research*, 7, 67–73.

Bennett, S.A. (2003). Flight Crew Stress and Fatigue in Low-Cost Commercial Air Operations – An Appraisal. *International Journal of Risk Assessment and Management*, 4, 207–31.

Bensel, C.K. and Santee, W.R. (1997). Climate and Clothing. In: G. Salvendy (ed.) *Handbook of Human Factors and Ergonomics* (2nd edition) (pp. 909–34). New York: John Wiley.

Benson, A.J. (1988). Motion Sickness. In: J. Ernsting and P. King (eds) *Aviation Medicine* (2nd edition) (pp. 318–38). London: Butterworths.

Beringer, D.B. and Ball, J. (2001). General Aviation Pilot Visual Performance Using Conformal and Non-Conformal Head-Up and Head-Down Highway-In-The-Sky Displays. In: R. Jensen (ed.) *Proceedings of the International Symposium on Aviation Psychology*. Columbus, OH: Ohio State University.

Besco, R. (1994, January). Pilot Personality Testing and the Emperor's New Clothes. *Ergonomics in Design*, 24–9.

Biederman, I. (1981). On The Semantics of a Glance at a Scene. In: M. Kubovy and J. Pomerantz (eds) *Perceptual Organization*. Hillsdale, NJ: Erlbaum.

Billings, C.E. (1997). *Flight Deck Automation*. Mahwah, NJ: Lawrence Erlbaum Associates.

Billings, C.E., Lauber, J.K., Funkhouser, H., Lyman, G. and Huff, E.M. (1976). *NASA Aviation Safety Reporting System, Technical Report No. TM-X-3445*. Moffett Field, CA: NASA Ames Research Center.

Blake, R.R. and Mouton, J.S. (1978). *The Managerial Grid*. Houston TX: Gukf Publishing Company.

Boeing Commercial Airplanes (2009). *Statistical Summary of Commercial Jet Airplane Accidents Worldwide Operations 1959–2008*. Seattle, WA: Author.

Bonner, M., Taylor, R., Fletcher, K. and Miller, C. (2000). Adaptive Automation and Decision Aiding in the Military Fast Jet Domain. In: D.B. Kaber and M.R. Endsley (eds) *Human Performance, Situation Awareness and Automation: User-Centred Design for the New Millennium* (pp. 154–9). Madison, WI: Omnipress.

Bourgeois-Bourgrine, S., Cabon, P., Mollard, R., Coblentz, A. and Speyer, J.J. (2003). Fatigue of Short-Haul Flights Aircrews in Civil Aviation: Effects of Work Schedules. *Human Factors and Aerospace Safety*, 3, 117–87.

Boy, G.A. and Ferro, D. (2004). Using Cognitive Function Analysis to Prevent Controlled Flight into Terrain. In: D. Harris (ed.) *Human Factors for Civil Flight Deck Design* (pp. 55–68). Aldershot, UK: Ashgate.

Braby, C.D., Harris, D. and Muir, H.C. (1993). A Psychophysiological Approach to the Assessment of Work Underload. *Ergonomics*, 36, 1035–42.

Braithwaite, M.G. and Braithwaite, B.D. (1990). Simulator Sickness in an Army Simulator. *Occupational Medicine*, 40, 105–10.

Branson, R.K., Rayner, G.T., Cox, L., Furman, J.P., King, F.J. and Hannum, W. (1975). *Interservice Procedures for Instructional Systems Development: Executive Summary and Model*. Tallahasee FL: Center for Educational Technology, Florida State University.

British Airways (2010). 2009/10 Annual Report and Accounts. Available from: http://www. britishairways.com/cms/global/microsites/ba_reports0910/overview/cfo3.html (accessed 17 August 2010).

Broadbent, D. (1958). *Perception and Communication*. London: Pergamon Press.

Broadbent, D.E. (1976). Noise and the Details of Experiments: A Reply to Poulton. *Applied Ergonomics*, 7, 231–5.

Brooks, P. and Arthur, J. (1997). Computer-Based Motorcycle Training: The Concept of Motivational Fidelity. In: D. Harris (ed.) *Engineering Psychology and Cognitive Ergonomics* (Volume 1): *Transportation Systems* (pp. 403–10). Aldershot: Ashgate.

Brown, A. and Bartram, D. (2009). *Supplement to the OPQ32 Technical Manual*. Thames Ditton: SHL Group Limited.

Brown, R.V., Kahr, A.S. and Peterson, C. (1974). *Decision Analysis for the Manager*. New York, NY: Holt, Rinehart and Winston.

Bryan, L.A., Stonecipher, J.W. and Aron, K. (1954). 180-Degree Turn Experiment. *University of Illinois Bulletin*, 54, 1–52.

Buch, G. and Diehl, A. (1984). An Investigation of the Effectiveness of Pilot Judgment Training. *Human Factors*, 26, 557–64.

Bundesstelle Für Flugundfalluntersuchung (2004). *Investigation Report AX001-1-2/02*. Braunschweig: Author.

Bureau d'Enquêtes et d'Analyses pour la Sécurité de l'Aviation Civile (1992). *Rapport de la commission d'enquête sur l'accident survenu le 20 janvier 1992 près du Mont Sainte-Odile (Bas Rhin) à l'Airbus A 320 immatriculé F-GGED exploité par la compagnie Air Inter (F-ED920120)*. Le Bourget: Author.

Bureau of National Affairs (1987). *Stress in the Workplace: Costs, Liability and Prevention*. Arlington, VA: Author.

Burke, E., Hobson, C. and Linsky, C. (1997). Large Sample Validations of Three General Predictors of Pilot Training Success. *International Journal of Aviation Psychology*, 7, 225–34.

Cabon, P., Mollard, R., Coblentz, A., Fouillot, J.P. and Speyer, J.J. (1995). Recommandations pour le maintien du niveau d'éveil et la gestion du sommeil des pilotes d'avions long-courriers. *Médecine Aéronautique et Spatiale*, 34, 19–28.

Caird-Daley, A., Harris, D., Bessell, K. and Lowe, M. (2007). *Training Decision Making Using Serious Games. Human Factors Integration Defence Technology Centre Report (HFIDTC/WP 4.6.1)*. Yeovil: Aerosystems International/HFI-DTC.

Caldwell, J.A. and Caldwell, J.L. (2003). *Fatigue in Aviation*. Aldershot: Ashgate.

Campion, M.A., Campion, J.E. and Hudson, J.P. (1994). Structured Interviewing: A Note on Incremental Validity and Alternative Question Types. *Journal of Applied Psychology*, 79, 998–1002.

Canada Market Research (1990). *Substance Use and Transportation Safety: Aviation Mode*. Toronto: Canada Market Research.

Carlson, R.A., Sullivan, M.A. and Schneider, W. (1989). Component Fluency in a Problem-Solving Context. *Human Factors*, 31, 489–502.

Carretta, T.R. (1992). Understanding the Relations Between Selection Factors and Pilot Training Performance: Does the Criterion Make a Difference? *International Journal of Aviation Psychology*, 2, 95–105.

Carretta, T.R. and Ree, M.J. (1993). Basic Attributes Test (BAT): Psychomoteric Equating of a Computer-Based Test. *International Journal of Aviation Psychology*, 3, 189–201.

Carretta, T.R. and Ree, M.J. (2003). Pilot Selection Methods. In: P.S. Tsang and M.A. Vidulich (eds) *Principles and Practice of Aviation Psychology* (pp. 357–96). Mahwah, NJ: Lawrence Erlbaum Associates.

Casali, J.G. and Wierwille, W.W. (1984). On the Comparison of Pilot Perceptual Workload: A Comparison of Assessment Techniques Addressing Sensitivity and Intrusion Issues. *Ergonomics*, 27, 1033–50.

Casner, S.M. (2003). *Learning about Cockpit Automation: From Piston Trainer to Jet Transport. NASA Report NASA/TM-2003-212260*. Moffett Field CA: NASA Ames Research Center.

Cattell, R.B. (1946). *The Description and Measurement of Personality*. New York: World Books.

Cattell, R.B., Cattell, A.K., and Cattell, H.E.P. (1993). *16PF Fifth Edition Questionnaire*. Champaign, IL: Institute for Personality and Ability Testing.

Chapanis, A. (1999). *The Chapanis Chronicles: 50 years of Human Factors Research, Education, and Design*. Santa Barbara, CA: Aegean Publishing Company.

Chappelow, J.W. and Churchill, M. (1988). Selection and Training. In: J. Ernsting and P. King (eds) *Aviation Medicine* (2nd edition) (pp. 423–34). London: Butterworths.

Chen, S.S., Jacobsen, A.R., Hofer, E., Turner, B.L. and Wiedemann, J. (2000). *Vertical Profile Display Development (Paper 2000-01-5613)*. SAE World Aviation Congress, San Diego, CA, 2000.

Chidester, T.R. (1990). Trends and Individual Differences in Response to Short-Haul Flight Operations. *Aviation Space and Environmental Medicine*, 61, 132–8.

Chidester, T.R., Helmreich, R.L., Gregorich, S.E. and Geis, C.E. (1991). Pilot Personality and Crew Coordination: Implications for Training and Selection. *International Journal of Aviation Psychology*, 1, 25–44.

Childs, J.M. and Spears, W.D. (1986). Flight-skill Decay and Recurrent Training. *Perceptual and Motor Skills*, 62, 235–42.

Cholewiak, R.W. and McGrath, C.M. (2006). *Vibrotactile Targeting in Multimodal Systems: Accuracy and Interaction*. Paper presented at the Virtual Reality 2006 Conference (in conjunction with the IEEE Conference). Arlington VA (22–23 March 2006).

Christoffersen, K. and Woods, D.D. (2000). *How to Make Automated Systems Team Players*. Columbus, OH: Institute for Ergonomics: The Ohio State University.

Civil Aviation Authority (1978). *Joint Airworthiness Requirements (JAR 25 – Large Aeroplanes)*. London: Civil Aviation Authority.

Civil Aviation Authority (1998). *Global Fatal Accident Review 1980-96 (CAP 681)*. London: Civil Aviation Authority.

Civil Aviation Authority (2003). *Joint Aviation Requirements Flight Crew Licensing Notes for the Qualification and Approval of Flight Navigation Procedures Trainers (FNPTs) and Basic Instrument Training Devices (BITDs). Standards Document 18 (Version 2)*. London: Civil Aviation Authority.

Civil Aviation Authority (2004). *The Avoidance of Fatigue in Aircrews* (CAP 371). Civil Aviation Authority: London.

Civil Aviation Authority (2006). *Crew Resource Management (CRM) Training: Guidance for Flight Crew, CRM Instructors (CRMIS) and CRM Instructor-Examiners (CRMIES)* (CAP 737). London: Civil Aviation Authority.

Civil Aviation Authority (2008). *Global Fatal Accident Review 1997–2006 (CAP 776)*. London: Civil Aviation Authority.

Civil Aviation Authority (2010). *Unmanned Aircraft System Operations in UK Airspace – Guidance* (CAP 722). London: Civil Aviation Authority.

Clothier, C. (1991). Behavioral Interactions across Various Aircraft Types: Results of Systematic Observations of Line Operations and Simulations. In: R.S. Jensen (ed.) *Proceedings of the Sixth International Symposium on Aviation Safety* (pp. 332–7). Columbus, OH: Ohio State University.

Code of Federal Regulations (2003). *Title 14: Aeronautics and Space. Part 25: Airworthiness standards: Transport Category Airplanes*. Washington, DC: National Archives and Records Administration. (Available at: http://www.gpo.gov/nara/cfr.)

Connolly, T.J., Blackwell, B.B. and Lester, L.F. (1989). A Simulator-Based Approach to Training in Aeronautical Decision Making. *Aviation Space and Environmental Medicine*, 60, 50–2.

Consultnet (2009). *Safety Performance Measurement*. (Retrieved from: http://www.consultnet.ie/Safety%20Performance%20Measurement.htm 27 April 2011.)

Cook, M. (2009). *Personnel Selection: Adding Value Through People*. Chichester: Wiley-Blackwell.

Cook, M., Noyes, J. and Masakowski, Y. (2007). *Decision Making in Complex Environments*. Aldershot: Ashgate.

Coombs, L.F.E. (1990). *The Aircraft Cockpit*. Wellingborough, UK: Patrick Stephens Ltd.

Cooper, C. (1999). *Intelligence and Abilities*. London: Routledge.

Cooper, C. (2010). *Individual Differences and Personality*. London: Hodder Education.

Cooper, C.L. and Sloane, S.J. (1987). Coping with Pilot Stress: Resting at Home Compared with Resting Away From Home. *Aviation, Space, and Environmental Medicine*, 58, 1175–82.

Cooper, G.E. and Harper, R.P. (1969). *The Use of Pilot Rating in the Evaluation of Aircraft Handling Qualities (NASA TN-D-5153)*. Washington, DC: National Aeronautics and Space Administration.

Courteney, H.Y. (1999). Assessing Error Tolerance in Flight Management Systems. In: D. Harris (ed.) *Engineering Psychology and Cognitive Ergonomics* (Volume 3) (pp.43–50). Aldershot: Ashgate.

Cox, T. and Mackay, C.J. (1976). Transactional model of stress. In: T. Cox (ed.) *Stress*. Basingstoke: Macmillan.

Craik, K.J.W. (1940). *The Fatigue Apparatus (Cambridge cockpit) (Report 119)*. London: British Air Ministry, Flying Personnel Research Committee.

Craik, K.J.W. and Vince, M.A. (1945). A Note on the Design and Manipulation of Instrument-Knobs. Applied Psychology Laboratory, Cambridge University Report. Cambridge University.

Crandall, B., Klein, G. and Hoffman, R.R. (2006). *Working Minds: A Practitioner's Guide to Cognitive Task Analysis*. Cambridge, MA: MIT Press.

Curry, R.E. (1985). *The Introduction of New Cockpit Technology: A Human Factors Study. NASA Technical Memorandum 86659, 1-68*. Moffett Field, CA: NASA Ames Research Center.

Cushman, W.H. and Rosenberg, D. (1991). *Human Factors in Product Design*. Amsterdam: Elsevier Science Publishers B.V.

Cuthbert, J.W. (1997). Alcoholism in the Aviation Industry. *Proceedings of the Royal Society of Medicine*, 70, 116–18.

D'Oliveira, T.C. (2004) Dynamic Spatial Ability: An Exploratory Analysis and a Confirmatory Study. *International Journal of Aviation Psychology*, 14, 19–38.

Dahlström N., Dekker, S.W.A. and Nählinder, S. (2006). Introduction of Technically Advanced Aircraft in Ab-Initio Flight Training. *International Journal of Applied Aviation Studies*, 6, 131–44.

Daiper, D. and Stanton, N.A. (2004). *Handbook of Task Analysis in Human Computer Interaction*. Chichester: Ellis Horwood.

Damkot, D.K. and Osga, G.A. (1978). Survey of Pilots' Attitudes and Opinions about Drinking and Flying. *Aviation, Space and Environmental Medicine*, 49, 390–4.

Davenport, M.D. and Harris, D. (1992). The Effect of Low Blood Alcohol Levels on Pilot Performance in a Series of Simulated Approach and Landing Trials. *International Journal of Aviation Psychology*, 2, 271–80.

Davies, A.K., Tomoszeck, A., Hicks, M.R. and White, J. (1995). AWAS (Aircrew Workload Assessment System): Issues of Theory Implementation and Validation. In: R. Fuller, N. Johnston and N. McDonald (eds) *Human Factors in Aviation Operations*. Aldershot: Avebury.

Dekker, S.W.A. (2001). The Re-Invention of Human Error. *Human Factors and Aerospace Safety*, 1, 247–66.

Dekker, S.W.A. (2004a). On the Other Side of a Promise: What Should we Automate Today? In: D. Harris (ed.) *Human Factors for Civil Flight Deck Design* (pp. 183–98). Aldershot, UK: Ashgate

Dekker, S.W.A. (2004b). Why We Need New Accident Models. *Human Factors and Aerospace Safety*, 3, 1–18.

Dekker, S.W.A. (2005). *Ten Questions about Human Error: A New View of Human Factors and System Safety*. Mahwah, NJ: Lawrence Erlbaum.

Dekker, S.W.A. (2006). *The Field Guide to Understanding Human Error*. Aldershot: Ashgate.

Dekker, S.W.A. and Hollnagel, E. (eds) (1999). *Coping with Computers in the Cockpit*. Aldershot: Ashgate.

Dekker, S.W.A. and Hollnagel, E. (2004). Human Factors and Folk Models. *Cognition, Technology and Work*, 6, 79–86.

Dekker, S.W.A. and Woods, D.D. (2002). MABA-MABA or Abracadabra: Progress On Human-Automation Cooperation. *Cognition, Technology and Work*, 4, 240–4.

Dennis, K.A. and Harris, D. (1998). Computer-Based Simulation as an Adjunct to Ab Initio Flight Training. *International Journal of Aviation Psychology*, 8, 261–76.

Department of Trade and Industry, Accident Investigation Branch (1973). *Civil Aircraft Accident Report 4/73: Trident I G-ARPI: Report of the Public Inquiry into the Causes and Circumstances of the Accident near Staines on 18 June 1972*. London: HMSO.

Desmond, P.A., Hancock, P.A. and Monette, J.L. (1998). Fatigue and Automation-Induced Impairments in Simulated Driving Performance. *Transportation Research Record*, 1628, 8–14.

Deutsch, J. and Deutsch, D. (1963). Attention: Some Theoretical Considerations. *Psychological Review*, 70, 80–90.

Diehl, A.E. (1991a). The Effectiveness of Training Programs for Preventing Aircrew Error. In: R.S. Jensen (ed.) *Sixth International Symposium on Aviation Psychology* (Volume 2) (pp. 640–55). Columbus, OH: Ohio State University.

Diehl, A.E. (1991b). Human Performance and Systems Safety Considerations in Aviation Mishaps. *International Journal of Aviation Psychology*, 1, 97–106.

Digman, J.M. (1990). Personality Structure: Emergence of the Five-Factor Model. *Annual Review of Psychology*, 41, 417–40.

Dismukes, R.K., Berman, B.A. and Loukopoulos, L.D. (2007). *The Limits of Expertise: Rethinking Pilot Error and the Causes of Airline Accidents*. Aldershot: Ashgate.

Dismukes, R.K., McDonnell, L.K. and Jobe, K.K. (2000). Facilitating LOFT Debriefings: Instructor Techniques and Crew Participation. *International Journal of Aviation Psychology*, 10, 35–57.

Drew, G.C. (1940). An Experimental Study of Mental Fatigue. *British Flying. Personnel Research Committee Memorandum No. 227*. London: British Air Ministry, Flying Personnel Research Committee.

Dudfield, H.J. (1991). Colour Head-Up Displays: Help or Hindrance. In: *Proceedings of the 35th Human Factors and Ergonomics Society Annual Conference* (pp. 146–50). Santa Monica, CA: Human Factors and Ergonomics Society.

Duggan, S.J. and Harris, D. (2001). Modelling Naturalistic Decision Making Using an Artificial Neural Network: Pilots' Responses to a Disruptive Passenger Incident. *Human Factors and Aerospace Safety*, 1, 145–67.

Durso, F.T., Hackworth, C.A., Truitt, T.R., Crutchfield, J., Nikolic, D. and Manning, C.A. (1998). Situation Awareness as a Predictor of Performance for En Route Air Traffic Controllers. *Air Traffic Control Quarterly*, 6, 1–20.

Easterbrook, J.A. (1959). The Effect of Emotion on Cue Utilization and the Organization of Behavior. *Psychological Review*, 66, 183–201.

EasyJet plc (2009). Annual Report and Accounts, 2009. (Available from: http://2009annualreport.easyjet.com/business-review/financial-review.html Accessed 17 August 2010.)

Ebbatson, M. (2006). Practice Makes Imperfect: Common Factors in Recent Manual Approach Incidents. *Human Factors and Aerospace Safety*, 6, 275–8.

Ebbatson, M., Harris, D., Huddlestone, J. and Sears, R. (2008). Combining Control Input with Flight Path Data to Evaluate Pilot Performance in Transport Aircraft. *Aviation Space and Environmental Medicine*, 79, 1061–4.

Ebbatson, M., Harris, D., Huddlestone, J. and Sears, R. (2010). The Relationship between Manual Handling Performance and Recent Flying Experience in Air Transport Pilots. *Ergonomics*, 53, 268–77.

Edkins, G.D. (2002). A Review of the Benefits of Aviation Human Factors Training. *Human Factors and Aerospace Safety*, 2, 201–16.

Edwards, E. (1972). Man and Machine: Systems for Safety. In: *Proceedings of British Airline Pilots Association Technical Symposium* (pp. 21–36). London: British Airline Pilots Association.

Edwards, E. (1975). *Stress and the Airline Pilot*. Paper presented to BALPA Technical Symposium: Aviation Medicine and the Airline Pilot. Department of Human Sciences, Loughborough University, October 1975.

Edwards, W. (1954). The Theory of Decision Making. *Psychological Bulletin*, 51, 380–417.

Edworthy, J. and Patterson, R.D. (1985). Ergonomic Factors in Auditory Warnings. In: I.D. Brown, R. Goldsmith, K. Coombes and M.A. Sinclair (eds) *Ergonomics International 85* (pp. 232–4). London: Taylor and Francis.

Ellis, G.A. and Roscoe, A.H. (1982). *The Airline Pilot's View of Flightdeck Workload: A Preliminary Study Using a Questionnaire. Technical Memorandum FS(B) 465*. Bedford: Royal Aircraft Establishment.

Embrey, D.E. (1986). *SHERPA: A Systematic Human Error Reduction and Prediction Approach*. Paper presented at the International Meeting on Advances in Nuclear Power Systems, Knoxville, Tennessee, 1986.

Endsley, M.R. (1988). Design and Evaluation for Situation Awareness Enhancement. *Proceedings of the Human Factors Society 32nd Annual Meeting* (pp. 97–101). Santa Monica, CA: Human Factors Society.

Endsley, M.R. (1995). Toward a Theory of Situation Awareness in Dynamic Systems. *Human Factors*, 37, 32–64.

Endsley, M.R. (1997). The Role of Situation Awareness in Naturalistic Decision Making. In: C.E. Zsambok and G. Klein (eds) *Naturalistic Decision Making* (pp. 269–84). Mahwah, NJ: Lawrence Erlbaum.

Endsley, M.R. (2000). Direct Measurement of Situation Awareness: Validity and Use of SAGAT. In: M.R. Endsley and D.J. Garland (eds) *Situation Awareness Analysis and Measurement* (pp. 147–73). Mahwah, NJ: Lawrence Erlbaum.

Endsley, M.R. and Bolstad, C.A. (1994). Individual Differences in Pilot Situation Awareness. *International Journal of Aviation Psychology*, 4, 241–64.

Endsley, M.R. and Garland, D.J. (eds) (2000). *Situation Awareness: Analysis and Measurement*. Mahwah, NJ: Lawrence Erlbaum Associates.

Endsley, M.R. and Jones, W.M. (2001). A Model of Inter and Intra-Team Situation Awareness: Implications for Design, Training and Measurement. In: M. McNeese, E. Salas and M.R. Endsley (eds) *New Trends in Cooperative Activities: Understanding System Dynamics in Complex Environments* (pp. 46–67). Santa Monica, CA: Human Factors and Ergonomics Society.

Endsley, M.R. and Kaber, D.B. (1999). Level of Automation Effects on Performance, Situation Awareness and Workload in a Dynamic Control Task. *Ergonomics*, 42, 462–92.

Endsley, M.R., Selcon, S.J., Hardiman, T.D. and Croft, D.G. (1998). A Comparative Evaluation of SAGAT and SART for Evaluations of Situation Awareness. In: *Proceedings of the Human Factors and Ergonomics Society 32nd Annual Meeting* (Volume 1) (pp. 82–6). Santa Monica, CA: Human Factors and Ergonomics Society.

Ernsting, J., Sharp, G.R. and Harding, R.M. (1988). Hypoxia and Hyperventilation. In: J. Ernsting and P. King (eds) *Aviation Medicine* (2nd edition) (pp. 45–59). London: Butterworths.

European Aviation Safety Agency (2008). *Certification Specifications for Large Aeroplanes (CS- 25): Amendment 5*. Cologne: EASA. (Available from: http://www.easa.europa.eu/ws_prod/g/rg_certspecs.php#CS-25.)

European Aviation Safety Agency (2009). *EU-OPS 1: Commercial Air Transportation (Aeroplanes)*. Cologne: EASA.

Eysenck, H.J. (1967). *The Biological Basis of Personality*. Springfield, IL: Charles C. Thomas Publishing.

Eysenck, H.J. (1991). Dimensions of Personality: 16: 5 or 3? Criteria for a Taxonomic Paradigm. *Personality and Individual Differences*, 12, 773–90.

Eysenck, H.J. and Kamin, L. (1981). *Intelligence: The Battle for the Mind*. London: Pan.

Eysenck, M.W. and Keane, M.T. (2005). *Cognitive Psychology: A Student's Handbook* (5th edition). Hove: Psychology Press.

Fadden, S., Ververs, P.M. and Wickens, C.D. (1998). Costs and Benefits of Head-Up Display Use: A Meta-Analytic Approach. *Proceedings of the 42nd Annual Meeting of the Human Factors and Ergonomics Society* (pp. 16–20). Santa Monica, CA: Human Factors and Ergonomics Society.

Farmer, E., van Rooij, J., Riemersma, J., Jorna, P. and Moraal, J. (1999). *Handbook of Simulator Based Training*. Aldershot: Ashgate.

Federal Aviation Administration (1987). *Advisory Circular AC 25-11 Transport Category Airplane Electronic Display Systems*. Washington, DC: US Department of Transportation.

Federal Aviation Administration (1990). *Line Operational Simulation: Line Oriented Flight Training, Special Purpose Operational Training, Line Operational Evaluation (Advisory Circular AC 120-35B)*. Washington, DC: US Department of Transportation.

Federal Aviation Administration (1991a). *Advanced Qualification Program (Advisory Circular AC 120-54)*. Washington, DC: US Department of Transportation.

Federal Aviation Administration (1991b). *Airplane Simulator Qualification (Advisory Circular AC 120-40B)*. Washington, DC: US Department of Transportation.

Federal Aviation Administration (1993). *Advisory Circular AC 25-1523-1 Minimum Flight Crew*. Washington, DC: US Department of Transportation.

Federal Aviation Administration (1996a). *Human Factors Design Guide (Version 1.0)*. William J. Hughes Technical Center, Arlington VA: Author.

Federal Aviation Administration (1996b). *Report on the Interfaces between Flightcrews and Modern Flight Deck Systems*. Washington, DC: US Department of Transportation.

Federal Aviation Administration (2005). *Advisory Circular AC 23-1523 Minimum Flight Crew*. Washington, DC: US Department of Transportation.

Federal Aviation Administration (2006). *Line Operations Safety Audits (AC 120-90)*. Washington, DC: US Department of Transportation.

Feggetter, A.J. (1982). A Method for Investigating Human Factors Aspects of Aircraft Accidents and Incidents. *Ergonomics*, 11, 1065–75.

Fiedler, F.E. (1967). *A Theory of Leadership Effectiveness*. New York, NY: McGraw-Hill.

Field, E. (2004). Handling Qualities and Their Implications for Flight Deck Design. In: D. Harris (ed.) *Human Factors for Civil Flight Deck Design* (pp. 157–81). Aldershot, UK: Ashgate.

Field, E. and Harris, D. (1998). The Implications of the Deletion of the Cross-Cockpit Control Linkage in Fly-By-Wire Aircraft: A Communication Analysis. *Ergonomics*, 41, 1462–77.

Fischer, E., Haines, R.F. and Price, T.A. (1980). *Cognitive Issues in Head-Up Displays (NASA Technical Paper 1711)*. Moffett Field, CA: NASA Ames Research Center.

Fitts, P.M. and Jones, R.E. (1947). *Analysis of 270 'Pilot Error' Experiences in Reading and Interpreting Aircraft Instruments (Report TSEAA-694-12A)*. Wright–Patterson Air Force Base, OH: Aeromedical Laboratory.

Fitts, P.M. and Jones, R.E. (1961). Analysis of Factors Contributing to 460 'Pilot-Error' Experiences in Operating Aircraft Controls. In: W.H. Sinaiko (ed.) *Selected Papers on Human Factors in the Design and Use of Control Systems*. New York, NY: Dover.

Fitts, P.M. and Posner, M.I. (1967). *Human Performance*. Belmont, CA: Brooks-Cole.

Fitzgibbons, A., Davis, D. and Schutte, P.C. (2004). *Pilot Personality Profile Using the NEO-PI-R. NASA Technical Report L-18379: NASA/TM-2004-21323*. Langley Research Center, VA: NASA.

Flanagan, J.C. (1954). The Critical Incident Technique. *Psychological Bulletin*, 51, 327–58.

Flin, R., Martin, L., Goeters, K-M., Hörmann, H-J., Amalberti, R., Valot, C. and Nijhuis, H. (2003). Development of the NOTECHS (Non-Technical Skills) System for Assessing Pilots' CRM Skills. *Human Factors and Aerospace Safety*, 3, 97-119.

Flin, R., O'Connor, P. and Crichton, M. (2008). *Safety at the Sharp End: A Guide to Non-Technical Skills*. Aldershot: Ashgate.

Flynn, C.F., Sturges, M.S., Swarsen, R.J. and Kohn, G.M. (1993). Alcoholism and Treatment in Airline Aviators: One Company's Results. *Aviation, Space and Environmental Medicine*, 59, 314–18.

Foushee, H.C. and Helmreich, R.L. (1988). Group Interaction and Flight Crew Performance. In: E.L. Wiener and D.C. Nagel (eds) *Human Factors in Aviation* (pp. 189–227). San Diego, CA: Academic Press.

Foushee, H.C. and Manos, K.L. (1981). *Within Cockpit Communication Patterns and Flight Crew Performance (NASA Technical Paper 1875)*. Moffett Field, CA: NASA Ames Research Center.

Foushee, H.C., Lauber, J.K., Baetge, M.M. and Acomb, L.B. (1986). *Crew Factors in Flight Operations III: The Operational Significance to Short-Haul Air Transport Operation. (NASA Technical Memorandum 88322)*. Moffett Field, CA: NASA Ames Research Center.

Friedman, M. and Rosenman, R.H. (1974). *Type A Behavior*. New York, NY: Alfred A. Knopf.

Furedy, J.J. (1987). Beyond Heart Rate in the Cardiac Psychophysiological Assessment of Mental Effort: The T-Wave Amplitude Component of the Electrocardiogram. *Human Factors*, 29, 183–94.

Galer, I.A.R. (1987). *Applied Ergonomics Handbook* (2nd edition). London: Butterworths.

Gander, P.H., Gregory, K.B., Graeber, R.C., Connell, L.J., Miller, D.L. and Rosekind, M.R. (1998). Flight Crew Fatigue II: Short Haul, Fixed Wing Air Transport Operations. *Aviation, Space and Environmental Medicine*, 69 (9 supplement), B8–B15.

Gaur, D. (2005). Human Factors Analysis and Classification System Applied to Civil Aircraft Accidents in India. *Aviation, Space and Environmental Medicine*, 76, 501–5.

Gaver, W.W. (1989). The SonicFinder: An Interface that Uses Auditory Icons. *Human Computer Interaction*, 4, 67–94.

Gawron, V.J. (2008). *Human Performance, Workload, and Situational Awareness Measures Handbook*. London: CRC Press.

Ginnett, R.G. (1993). Crews as Groups: Their Formation and their Leadership. In: E.L. Wiener, B.G. Kanki and R.L. Helmreich (eds) *Cockpit Resource Management* (pp. 71–98). San Diego, CA: Academic Press.

Glendon, A.I. and Stanton, N.A. (2000). Perspectives on Safety Culture. *Safety Science*, 34, 193–214.

Glendon, A.I., Clarke, S.G. and McKenna, E.F. (2006). *Human Safety and Risk Management* (2nd edition). Boca Raton, FL. CRC Press.

Glick, W.H. (1985). Conceptualizing and Measuring Organizational and Psychological Climate: Pitfalls in Multilevel Research. *Academy of Management Review*, 10, 601–16.

Global Aviation Information Network (2004). *A Roadmap to a Just Culture: Enhancing the Safety Environment. Report of GAIN Working Group E, Flight Ops/ATC Ops Safety Information Sharing*. (Available from: http://www.coloradofirecamp.com/just-culture/index.htm Accessed 22 December 2009.)

Goeters, K-M. (1995). Psychological Evaluation of Pilots: The Present Regulations and Arguments for their Application. In: N. Johnston, R. Fuller and N. McDonald (eds) *Aviation Psychology: Selection and Training* (pp. 149–56). Aldershot: Avebury.

Goeters, K-M. (2002). Evaluation of the Effects of CRM Training by the Assessment of Non-Technical Skills under LOFT. *Human Factors and Aerospace Safety*, 2, 71–86.

Goeters, K-M. and Maschke, P. (2002). *Cost-Benefit Analysis: Is the Psychological Selection of Pilots Worth the Money?* Paper presented at European Association for Aviation Psychology Conference 2002. Warsaw 16–20 September 2002.

Goettl, B.P. and Shute, V.J. (1996). Analysis of Part-Task Training Using the Backward-Transfer Technique. *Journal of Experimental Psychology: Applied*, 2, 227–49.

Göteman, O. (1999). Automation Policy or Philosophy? – Management of Automation in the Operational Reality. In: S.W.A. Dekker and E. Hollnagel (eds) *Coping With Computers in the Cockpit* (pp. 215–24). Aldershot: Ashgate.

Gottfredson, L. (1997). Mainstream Science on Intelligence. *Intelligence*, 24, 13–23.

Gough, H.G. and Bradley, P. (1996/2002). *CPITM Manual* (3rd edition). Mountain View, CA: CPP, Inc.

Graeber, R.C. (1986). Sleep and Wakefulness in International Aircrews. *Aviation, Space and Environmental Medicine*, 57 (12, supplement), B1–B64.

Graeber, R.C. (1988). Air Crew Fatigue and Circadian Rhythmicity. In: E.L. Wiener and D.C. Nagel (eds) *Human Factors in Aviation* (pp. 305–44). San Francisco: Academic Press.

Green, R.G. and Farmer, E.W. (1988). Ergonomics. In: J. Ernsting and P. King (eds) *Aviation Medicine* (2nd edition) (pp. 445–57). London: Butterworths.

Gregorich, S.E., Helmreich, R.L. and Wilhelm, J.A. (1990). The Structure of Cockpit Management Attitudes. *Journal of Applied Psychology*, 75, 682–90.

Grether, W.F. (1949). The Design of Long-Scale Indicators for Speed and Accuracy of Quantitative Reading. *Journal of Applied Psychology*, 33, 363–72.

Grether, W.F. (1973). Human Performance at Elevated Environmental Temperatures. *Aerospace Medicine*, 44, 747–55.

Griffin, G.R. and Koonce, J.M. (1996). Review of Psychomotor Skills in Pilot Selection Research of the US Military Services. *International Journal of Aviation Psychology*, 6, 125–48.

Gross, D.C. (1999). Report from the Fidelity Implementation Study Group (Paper 99S-SIW-167). In: *Proceedings of Spring 1999 Simulation Interoperability Workshop*. Orlando, FL: Simulation Interoperability Standards Organization.

Guldenmund, F.W. (2000). The Nature of Safety Culture: A Review of Theory and Research. *Safety Science*, 34, 215–57.

Gundel, A., Marsalek, K. and ten Thoren, C. (2005). Support of Mission and Work Scheduling by a Biomedical Fatigue Model. In: *Strategies to Maintain Combat Readiness during Extended Deployments – A Human Systems Approach* (pp. 28-1–28-12). Meeting Proceedings RTO-MP-HFM-124, Paper 28. Neuilly-sur-Seine, France: NATO/OTAN.

Hall, J.R. (1989). *The Need for Platform Motion in Modern Piloted Flight Training Simulators (Technical Memorandum FM 35)*. Bedford: Royal Aerospace Establishment.

Hamman, W.R., Seamster, T.L., Smith, K.M. and Lofaro, R.J. (1993). The Future of LOFT Scenario Design and Validation. In: R.S. Jensen (ed.) *Proceedings of the Seventh International Symposium on Aviation Psychology* (pp. 589–94). Columbus, OH: Ohio State University Press.

Hancock, P. (1981). The Limitation of Human Performance in Extreme Heat Conditions. In: *Proceedings of the Human Factors Society* (pp. 74–8). Santa Monica, CA: Human Factors Society.

Hancock, P.A. and Chignell, M.H. (1987). Adaptive Control in Human–Machine Systems. In: P.A. Hancock (ed.) *Human Factors Psychology* (pp. 305–45). Amsterdam: North Holland, Elsevier.

Hancock, P.A. and Meshkati, N. (1988). *Human Mental Workload*. Amsterdam: Elsevier Science.

Hancock, P.A. and Parasuraman, R. (1992). Human Factors and Safety Issues in Intelligent Vehicle Highway Systems (IVHS). *Journal of Safety Research*, 23, 181–98.

Hansen, J.S. and Oster, C.V. (1997). *Taking Flight: Education and Training in Aviation Careers*. Washington, DC: National Academy Press.

Hardy, R. (1990). *Callback: National Aeronautics and Space Administration's Aviation Safety Reporting System*. Shrewsbury: Airlife.

Harper, C.R. (1983). Airline Pilot Alcoholism: One Airline's Experience. *Aviation, Space and Environmental Medicine*, 54, 590–91.

Harris, D. (2002). Drinking and Flying: Causes, Effects and the Development of Effective Countermeasures. *Human Factors and Aerospace Safety*, 2, 297–317.

Harris, D. (2003). The Human Factors of Fully Automatic Flight. *Measurement and Control*, 36, 184–7.

Harris, D. (2004a). Head-Down Flight Deck Display Design. In: D. Harris (ed.) *Human Factors for Civil Flight Deck Design* (pp. 69–102). Aldershot, UK: Ashgate.

Harris, D. (2004b). *Human Factors for Civil Flight Deck Design*. Aldershot; Ashgate.

Harris, D. (2005). Drinking and Flying: Causes, Effects and the Development of Effective Countermeasures. In: D. Harris, and H.C. Muir (eds). *Contemporary Issues in Human Factors and Aviation Safety* (pp. 199–219). Aldershot: Ashgate.

Harris, D. (2006a). The Influence of Human Factors on Operational Efficiency. *Aircraft Engineering and Aerospace Technology*, 78, 20–25.

Harris, D. (2006b). Keynote Address: An Open Systems Approach to Safety Culture: Actions, Influences and Concerns. In: *Proceedings of Australian Aviation Psychology Association (AAvPA) International Conference – Evolving System Safety 2006*. Sydney, Australia 9–12 November.

Harris, D. (2007). A Human-Centred Design Agenda for the Development of a Single Crew Operated Commercial Aircraft. *Aircraft Engineering and Aerospace Technology*, 79, 518–26.

Harris, D. (2008). Human Factors Integration in Defence. *Cognition, Technology and Work*, 10, 169–72.

Harris, D. (2009). A Design and Training Agenda for the Next Generation of Commercial Aircraft Flight Decks. In: D. Harris (ed.) *Engineering Psychology and Cognitive Ergonomics – EPCE 2009 (LNAI 5639)*. Berlin: Springer-Verlag.

Harris, D. (2010). Human Factors for Flight Deck Certification: Issues in Compliance with the New European Aviation Safety Agency Certification Specification 25.1302. *Journal of Aeronautics, Astronautics and Aviation* (Series A), 42(1), 11–20.

Harris, D. (2011). Rule Fragmentation in the Airworthiness Regulations: A Human Factors Perspective. In: D. Harris (ed.) *Engineering Psychology and Cognitive Ergonomics – EPCE 2011*. Berlin: Springer-Verlag.

Harris, D. and Harris, F.J. (2004). Predicting the Successful Transfer of Technology between Application Areas; a Critical Evaluation of the Human Component in the System. *Technology in Society*, 26, 551–65.

Harris, D. and Li, W-C. (2008). Cockpit Design and Cross-Cultural Issues Underlying Failures in Crew Resource Management. *Aviation Space and Environmental Medicine*, 79, 537–8.

Harris, D. and Li, W-C. (2011). An Extension of the Human Factors Analysis and Classification System (HFACS) for use in Open Systems. *Theoretical Issues in Ergonomics Science*, 12, 108–28.

Harris, D. and Maxwell, E. (2001). Some Considerations for the Development of Effective Countermeasures to Aircrew Use of Alcohol While Flying. *International Journal of Aviation Psychology*, 11, 237–52.

Harris, D. and Smith, F.J. (1997). What Can Be Done Versus What Should Be Done: A Critical Evaluation of the Transfer of Human Engineering Solutions between Application Domains. In: D. Harris (ed.) *Engineering Psychology and Cognitive Ergonomics* (Volume 1) (pp. 339–46). Aldershot: Ashgate.

Harris, D. and Stanton, N.A. (2010). Aviation as a System of Systems. *Ergonomics*, 53, 145–8.

Harris, D. and Thomas, L. (2005). The Contribution of Industrial and Organizational Psychology to Safety in Commercial Aircraft. In: G. Hodgkinson and K. Ford (eds) *International Review of Industrial and Organizational Psychology, 2005* (pp. 177–219). London: John Wiley.

Harris, D. and Thomas, T.J. (2001). Ramp Rash: Paying More May Cost You Less. *Human Factors and Aerospace Safety*, 1, 295–8.

Harris, D., Stanton, A., Marshall, A., Young, M.S., Demagalski, J.M. and Salmon, P. (2005). Using SHERPA to Predict Design-Induced Error on the Flight Deck. *Aerospace Science and Technology*, 9, 525–32.

Hart, S.G. and Staveland, L.E. (1988). Development of NASA-TLX (Task Load Index): Results of Empirical and Theoretical Research. In: P.A. Hancock and N. Meshkati (eds) *Human Mental Workload* (pp. 139–83). Amsterdam: North-Holland.

Haugli, L., Skogstad, A. and Hellesoy, O.H. (1994). Health, Sleep and Mood Perceptions Reported by Airline Crews Flying Short and Long Haul Flights. *Aviation, Space and Environmental Medicine*, 65, 27–34.

Health and Safety Executive (2001). *A Guide to Measuring Health and Safety Performance*. Sudbury, Suffolk: HSE Books.

Health and Safety Executive (2002). *Safety Climate Measurement User Guide and Toolkit*. Report prepared by Loughborough University for the Offshore Safety Division of the HSE. (Available from: http://www.lboro.ac.uk/departments/bs/safety/document.pdf Retrieved 18 December 2009.)

Health and Safety Executive (2005). *A Review of Safety Culture and Safety Climate Literature for the Development of the Safety Culture Inspection Toolkit (Research Report 367)*. Sudbury, Suffolk: HSE Books.

Health and Safety Executive (2008). *HSE Leaflet INDG424: Working Together to Reduce Stress at Work: A Guide for Employees*. Sudbury, Suffolk: HSE Books.

Hedge, J.W., Bruskiewicz, K.T., Borman, W.C., Hanson, M.A., Logan, K.K. and Siem, F.M. (2000). Selecting Pilots with Crew Resource Management Skills. *International Journal of Aviation Psychology*, 10, 377–92.

Hellenic Accident Investigation and Aviation Safety Board (2006). *Accident Investigation Report 11/2006: Accident of the a/c 5B-DBY of Helios Airways, Flight HCY522 on August 14, 2005, in the area of Grammatiko, Attikis, 33 km Northwest of Athens International Airport*. Athens: Hellenic Accident Investigation and Aviation Safety Board.

Helmreich, R.L. (1984). Cockpit Management Attitudes. *Human Factors*, 26, 583–9.

Helmreich, R.L. (1994). The Anatomy of a System Accident: The Crash of Avianca Flight 052. *International Journal of Aviation Psychology*, 4, 265–84.

Helmreich, R.L. (2000). On Error Management: Lessons from Aviation. *British Medical Journal*, 320, 781–5.

Helmreich, R.L. and Foushee, H.C. (1993). Why Crew Resource Management? Empirical and Theoretical Bases of Human Factors Training in Aviation. In: E.L. Wiener, B.G. Kanki and R.L. Helmreich (eds) *Cockpit Resource Management* (pp. 3–45). San Diego, CA: Academic Press.

Helmreich, R.L. and Merritt, A.C. (1998). *Culture at Work in Aviation and Medicine*. Aldershot: Ashgate.

Helmreich, R.L. and Schaefer, H.G. (1994) Team Performance in the Operating Room. In: M. Bogner (ed.) *Human Error in Medicine*. Hillsdale, NJ: Lawrence Erlbaum and Associates.

Helmreich, R.L. and Wilhelm, J.A. (1991). Outcomes of Crew Resource Management Training. *International Journal of Aviation Psychology*, 1, 287–300.

Helmreich, R.L., Klinect, J.R., and Wilhelm, J.A. (1999). Models of Threat, Error, and CRM in Flight Operations. In: R.S. Jensen (ed.) *Proceedings of the Tenth International Symposium on Aviation Psychology* (pp. 677–82). Columbus, OH: The Ohio State University.

Helmreich, R.L., Merritt, A.C., and Wilhelm, J.A. (1999). The Evolution of Crew Resource Management Training in Commercial Aviation. *International Journal of Aviation Psychology*, 9, 19–32.

Helmreich, R.L., Wilhelm, J., Kello, J., Taggart, E. and Butler, R. (1990). *Reinforcing and Evaluating Crew Resource Management: Evaluator/LOS Instructor Manual*. Austin, TX: NASA/University of Texas/FAA Aerospace Group.

Helmreich, R.L., Wilhelm, J.A., Klinect, J.R. and Merritt, A.C. (2001). Culture, Error, and Crew Resource Management. In: E. Salas, C.A. Bowers and E. Edens (eds) *Improving Teamwork in Organizations* (pp. 305–31). Hillsdale, NJ: Erlbaum.

Hendrick, K. and Benner, L. (1987). *Investigating Accidents with S-T-E-P*. New York, NY: Marcel Dekker.

Herzberg, F. (1964). The Motivation-Hygiene Concept and Problems of Manpower. *Personnel Administration*, 27, 3–7.

Ho, C. and Spence, C. (2008). *The Multisensory Driver*. Aldershot: Ashgate.

Ho, C., Reed, N. and Spence, C. (2006). Assessing the Effectiveness of 'Intuitive' Vibrotactile Warning Signals in Preventing Front-To-Rear-End Collisions in a Driving Simulator. *Accident Analysis and Prevention*, 38, 989–97.

Ho, C., Tan, H.Z., and Spence, C. (2005). Using Spatial Vibrotactile Cues to Direct Visual Attention in Driving Scenes. *Transportation Research Part F: Traffic Psychology and Behaviour*, 8, 397–412.

Hockey, G.R.J. (1970). Effect of Loud Noise on Attentional Selectivity. *Quarterly Journal of Experimental Psychology*, 22, 28–36.

Hofstede, G. (1984). *Culture's Consequences: International Differences in Work-Related Values*. Beverly Hills, CA: Sage Publications.

Hofstede, G. (1991). *Cultures and Organizations: Software of the Mind*. London: McGraw-Hill Book Company.

Hofstede, G. (2001). *Culture's Consequences: Comparing Values, Behaviors, Institutions, and Organizations across Nations*. Beverly Hills, CA: Sage Publications.

Holling, H. (1998). Utility Analysis of Personnel Selection: An Overview and Empirical Study Based on Objective Performance Measures. *Methods of Psychological Research Online*, 3, 1–24. (Available at: http://www.pabst-publishers.de/mpr/).

Hollnagel, E. (1993). *Human Reliability Analysis – Context and Control*. London: Academic Press.

Hollnagel, E. (1998). *Cognitive Reliability and Error Analysis Method: CREAM*. Oxford: Elsevier Science.

Holmes, T.H. and Rahe, R.H. (1967). The Social Readjustment Rating Scale. *Journal of Psychosomatic Research*, 11, 213–18.

Hörmann, H-J. (1995). FOR-DEC: A Perspective Model for Aeronautical Decision Making. In: R. Fuller, R. Johnston and N. McDonald (eds) *Human Factors in Aviation Operations* (pp. 17–23). Aldershot: Ashgate.

Hörmann, H-J. and Maschke, P. (1996). On the Relation Between Personality and Job Performance of Airline Pilots. *International Journal of Aviation Psychology*, 6, 171–8.

Hornick, R. (1973). Vibration. In: *Bioastronautics Data Book* (2nd edition). NASA Technical Report SP 3006. Washington DC: NASA.

Hornick, R.J. and Lefritz, N.M. (1966). A Study and Review of Human Response to Prolonged Random Vibration. *Human Factors*, 8, 481–92.

Hosman, R.J.A.W. and van der Vaart, J.C. (1981). Effects of Vestibular and Visual Motion Perception on Task Performance. *Acta Psychologica*, 48, 271–87.

Hubbard, D.C. (1987). Inadequacy of Root Mean Square Error as a Performance Measure. In: R.S. Jensen (ed.) *Proceedings of the Fourth International Symposium on Aviation Psychology* (pp. 698–704). Columbus, OH: Ohio State University.

Huddleston, J.H.F. (1964). *Human Performance and Behaviour in Vertical Sinusoidal Vibration*. Institute of Aviation Medicine Report 303. Farnborough: RAF Institute of Aviation Medicine.

Huddlestone, J.A. and Harris, D. (2003). *Air Combat Student Performance Modelling Using Grounded Theory Techniques*. Paper presented at Interservice/Industry Training, Simulation and Education Conference (I/ITSEC). Orlando, FL. 1–4 December 2003.

Hudson, P.T.W. (2001). Safety Culture: Theory and Practice. In: *The Human Factor in System Reliability*. NATO Series RTO-MP-032. Brussels: North Atlantic Treaty Organisation.

Huffcutt, A.I. and Arthur, W. (1994). Hunter and Hunter (1984) Revisited: Interview Validity for Entry-Level Jobs. *Journal of Applied Psychology*, 79, 184–90.

Huitt, W. (2003). *The Information Processing Approach to Cognition*. Educational Psychology Interactive. Valdosta, GA: Valdosta State University.

Human Factors Integration Defence Technology Centre (2006a). *Cost Arguments and Evidence for Human Factors Integration*. London: UK Ministry of Defence.

Human Factors Integration Defence Technology Centre (2006b). *Cost-Benefit Analysis for Human Factors Integration: A Practical Guide*. London: UK Ministry of Defence.

Human Factors National Advisory Committee for Defence and Aerospace (2003). *Gaining Competitive Advantage Through Human Factors: A Guide For The Civil Aerospace Industry*. London: Department of Trade and Industry.

Hunter, D.R. (2003). Measuring General Aviation Pilot Judgment Using a Situational Judgment Technique. *International Journal of Aviation Psychology*, 13, 373–86.

Hunter, D.R. and Burke, E.F. (1994). Predicting Aircraft Pilot-Training Success: A Meta-Analysis of Published Research. *International Journal of Aviation Psychology*, 4, 297–313.

Hunter, D.R. and Burke, E.F. (1995). *Handbook of Pilot Selection*. Aldershot: Avebury Aviation.

Hutchins, E. (1995a). *Cognition in the Wild*. Cambridge, MA: MIT Press.

Hutchins, E. (1995b). How a Cockpit Remembers its Speeds. *Cognitive Science*, 19, 265–88.

Ikomi, P.A., Boehm-Davis, D., Holt, R.W. and Incalcaterra, K.A. (1999). Jump Seat Observations of Advanced Crew Resource Management (ACRM) Effectiveness. In: R.S. Jensen (ed.) *Proceedings of the Tenth International Symposium on Aviation Psychology* (pp. 292–7). Columbus, OH: Ohio State University.

International Civil Aviation Organization (2009). *Safety Management Manual* (2nd edition) (ICAO Doc 9859). Montreal: ICAO.

International Labor Office (2007). *Encyclopaedia of Occupational Safety and Health*. Geneva: International Labor Organization. (Available from: http://www.ilocis.org/en/contilo.html. Accessed 30 March 2010.)

Irwin, C.M. (1991). The Impact of Initial and Recurrent Cockpit Resource Management Training on Attitudes. In: R.S. Jensen (ed.) *Proceedings of the Sixth International Symposium on Aviation Psychology* (pp. 344–9). Columbus, OH: Ohio State University.

Jaeger, R.J., Agarwal, G.C. and Gottlieb, G.L. (1980). Predictor Operators in Pursuit and Compensatory Tracking. *Human Factors*, 22, 497–506.

Jahns, D.W. (1973). Operator Workload: What Is It and How Should It be Measured? In: K.D. Gross and J.J. McGrath (eds) *Crew Systems Design*. Santa Barbara, CA: Anacapa Sciences Inc.

James, M., Birch, C., McClumpha, A. and Belyavin, A. (1993). *The Perception of Workload on the Automated Flight Deck*. Farnborough: Defence Research Agency.

Janis, I.L. and Mann, L. (1977). *Decision Making*. New York, NY: The Free Press.

Jarrett, D.N. (2005). *Cockpit Engineering*. Aldershot: Ashgate.

Jarvis, S. and Ebbatson, M. (2007). *Human Factors and Safety Review of RNP Procedures and Pilot Training for the Boeing 737-300 ZQN Operations*. Final Report. Air New Zealand, Auckland, New Zealand.

Jenkins, C.D., Zyzanski, S.J. and Rosenman, R.H. (1979). *Jenkins Activity Survey*. New York, NY: The Psychological Corporation.

Jenkins, D.P., Stanton, N.A., Salmon, P.M. and Walker, G.H. (2008). *Decision Making Training for Synthetic Environments: Using the Decision Ladder to Extract Specifications for Synthetic Environments Design and Evaluation (HFIDTC/2/WP4.6.2/2)*. Yeovil: Aerosystems International/HFI-DTC.

Jensen, R.S. (1995). *Pilot Judgement and Crew Resource Management*. Aldershot: Ashgate.

Jensen, R.S. (1997). The Boundaries of Aviation Psychology, Human Factors, Aeronautical Decision Making, Situation Awareness, and Crew Resource Management. *International Journal of Aviation Psychology*, 7, 259–68.

Jensen, R.S. and Benel, R. (1977). *Judgment Evaluation and Instruction in Civil Pilot Training*. Washington, DC: Federal Aviation Administration.

Jensen, R.S., Guilke, J. and Tigner, R. (1997). Understanding Expert Aviator Judgment. In: R. Flin, E. Salas, M. Strub and L. Martin (eds) *Decision Making Under Stress: Emerging Themes and Applications* (pp. 233–42). Aldershot: Ashgate.

Jing, H-S. and Yang, L-C. (2006). Authoritarianism: A Construct of Chinese Culture Described Using a Neural Network. *Human Factors and Aerospace Safety*, 5, 309–22.

Jing, H-S., Lu, C-J. and Peng, S-J. (2001). Culture, Authoritarianism and Commercial Aircraft Accidents. *Human Factors and Aerospace Safety*, 1, 341–59.

Jing, H-S., Lu, P-J., Yong, K. and Wang, H-C (2002). The Dragon in the Cockpit: the Faces of Chinese Authoritarianism. *Human Factors and Aerospace Safety*, 2, 257–75.

Johannsen, G. (1979). Workload and Workload Measurement. In: N. Moray (ed.) *Mental Workload: Its Theory and Measurement* (pp. 3–11). New York, NY: Plenum.

Johnson, C.W. (2003). *Failure in Safety-Critical Systems: A Handbook of Accident and Incident Reporting*. Glasgow: University of Glasgow Press. Available at: http://www.dcs.gla.ac.uk/~johnson/book/.

Johnson, W.G. (1980). *MORT, Safety Assurances Systems*. New York, NY: Marcel Dekker.

Johnson, W.W., Battiste, V. and Holland, S. (1999). A Cockpit Display Designed to Enable Limited Flight Deck Separation Responsibility. *Proceedings of the 1999 SAE/AIAA World Aviation Congress*. Anaheim, CA: Society of Automotive Engineers/American Institute for Aeronautics and Astronautics.

Johnson, W.W., Battiste, V., Delzell, S., Holland, S., Belcher, S. and Jordan, K. (1997). Development and Demonstration of a Prototype Free Flight Cockpit Display of Traffic Information. *Proceedings of the 1997 SAE/AIAA World Aviation Congress*. Anaheim, CA: Society of Automotive Engineers/American Institute for Aeronautics and Astronautics.

Johnston, N. (1993). CRM: Cross-Cultural Perspectives. In: E.L. Wiener, B.G. Kanki and R.L. Helmreich (eds) *Cockpit Resource Management* (pp.367–98). San Diego, CA: Academic Press.

Johnston, N. (1996). Psychological Testing and Pilot Licensing. *International Journal of Aviation Psychology*, 6, 179–97.

Johnston, N., McDonald, N. and Fuller, R. (1997). *Aviation Psychology in Practice*. Aldershot: Avebury Aviation.

Joint Aviation Authorities (1998). *Crew Resource Management – Flight Crew. Temporary Guidance Leaflet 5 (JAR-OPS)*. Administrative and Guidance Material (Section 4 – Operations). Hoofdorp, NL: JAA.

Joint Aviation Authorities (1999). Joint *Airworthiness Requirement – Flight Crew Licensing - Aeroplanes (JAR-FCL 1)*. Hoofdorp, NL: Joint Aviation Authorities.

Joint Aviation Authorities (2000). *Joint Airworthiness Requirement – Operations (JAR-OPS)*. Hoofdorp, NL: Joint Aviation Authorities.

Joint Aviation Authorities (2008). *JAR-FSTD 2A – Aeroplane Flight Training Devices*. Hoofdorp, NL: Joint Aviation Authorities.

Jones, D.G. (2000). Subjective Measures of Situation Awareness. In: M.R. Endsley and D.J. Garland (eds) *Situation Awareness Analysis and Measurement*. Mahwah, NJ. Lawrence Erlbaum Associates.

Jorna, P.G.A.M. (1992). Spectral Analysis of Heart Rate and Psychological State: A Review of Its Validity as a Workload Index. *Biological Psychology*, 34, 237–57.

Jorna, P.G.A.M. (1993). Heart Rate and Workload Variations in Actual and Simulated Flight. *Ergonomics*, 36, 1043–54.

Kaempf, G.L. and Orasanu, J. (1997). Current and Future Applications of Naturalistic Decision Making. In: C.E. Zsambok and G. Klein (eds) *Naturalistic Decision Making* (pp. 81–90). Mahwah, NJ: Lawrence Erlbaum.

Kahneman, D. (1973). *Attention and Effort*. Englewood Cliffs, NJ: Prentice-Hall.

Kahneman, D., Slovic, P. and Tversky, A. (1982). *Judgement under Uncertainty: Heuristics and Biases*. Cambridge: Cambridge University Press.

Kaiser, M.K. and Schroeder, J.A. (2003). Flights of Fancy: The Art and Science of Flight Simulation. In: P.S. Tsang and M.A. Vidulich (eds) *Principles and Practice of Aviation Psychology* (pp. 435–71). Mahwah, NJ: Lawrence Erlbaum Associates.

Kanki, B.G. and Palmer, M.T. (1993). In: E.L. Wiener, B.G. Kanki and R.L. Helmreich (eds) *Cockpit Resource Management* (pp. 99–136). San Diego, CA: Academic Press.

Kanki, B.G., Folk, V.G. and Irwin, C.M. (1991). Communication Variations and Aircrew Performance. *International Journal of Aviation Psychology*, 1, 149–62.

Kanki, B.G., Helmreich, R.L. and Anca, J. (2010). *Crew Resource Management* (2nd edition). San Diego, CA: Academic Press.

Karasek, R. (1979). Job Demands, Job Decision Latitude, and Mental Strain: Implication for Job Redesign. *Administrative Science Quarterly*, 24, 285–308.

Karlins, M., Koh, F. and McCully, L. (1989). The Spousal Factor in Pilot Stress. *Aviation, Space and Environmental Medicine*, 60, 1112–15.

Katz, D. and Kahn, R.L. (1978). *The Social Psychology of Organizations* (2nd edition). New York, NY: Wiley.

Kazdin, A.E. (ed.) (2000). *Encyclopedia of Psychology*. New York, NY: Oxford University Press, USA.

Keeney, R.L. and Raiffa, H. (1976). *Decisions with Multiple Objectives: Preferences and Value Trade-Offs*. New York, NY: Wiley.

Keinan, G. (1987). Decision-making under Stress: Scanning Alternatives Under Controllable and Uncontrollable Threats. *Journal of Personality and Social Psychology*, 52, 639–44.

Kelly, B.D. (2004). Flight Deck Design and Integration for Commercial Transports. In: D. Harris (ed.) *Human Factors for Civil Flight Deck Design* (pp. 3–31). Aldershot: Ashgate.

Kelly, B.D., Graeber, R.C. and Fadden, D.M. (1992). Applying Crew-Centred Concepts to Flight Deck Technology: The Boeing 777. *Proceedings of the Flight Safety Foundation 45th International Air Safety Seminar*. Long Beach, CA: Flight Safety Foundation.

Kennedy, R.S. and Fowlkes, J.E. (1992). Simulator Sickness is Polygenic and Polysymptomatic: Implications for Research. *International Journal of Aviation Psychology*, 2, 23–38.

Kiger, J.L. (1984). The Depth/Breadth Trade-Off in the Design of Menu-Driven User Interfaces. *International Journal of Man–Machine Studies*, 20, 201–13.

Kirkpatrick, D.L. (1976). Evaluation of Training. In: R.L. Craig and L.R. Bitel (eds) *Training and Development Handbook* (pp. 18.1–18.27). New York, NY: McGraw Hill.

Kirkpatrick, D.L. (1998). *Evaluation of Training Programmes*. San Francisco, CA: Berrett-Koehler.

Kirwan, B., Evans, A., Donohoe, L., Kilner, A., Lamoureux, T., Atkinson, T. and MacKendrick, H. (1997). *Human Factors in the System Design Lifecycle*. Paris: FAA/Eurocontrol ATM RandD Seminar.

Klein, G.A. (1989) Recognition-Primed Decisions. In: W.B. Rouse (ed.) *Advances in Man–Machine Systems Research* (Volume 5) (pp. 47–92). Greenwich, CT: JAI Press.

Klein, G.A. (1993a). Sources of Error in Naturalistic Decision Making. Cognitive and Contextual Factors in Aviation Accidents: Decision Errors. In: E. Salas. and G. Klein (eds) *Linking Expertise and Naturalistic Decision Making*. Mahwah, NJ: Lawrence Erlbaum Associates, Inc.

Klein, G.A. (1993b). A Recognition-Primed Decision (RPD) Model of Rapid Decision Making. In: G. Klein, J. Orasanu, R. Calderwood and C. Zsambok (eds) *Decision Making in Action: Models and Methods* (pp. 138–47). Norwood, NJ: Ablex Publishing Corporation.

Klein, G.A. (1997). The Current Status of the Naturalistic Decision Making Framework. In: R. Flin, E. Salas, M. Strub and L. Martin (eds) *Decision Making Under Stress: Emerging Themes and Applications* (pp. 11–28). Aldershot: Ashgate.

Klein, G.A. (2000a). How Can we Train Pilots to Make Better Decisions? In: H. O'Neil and D. Andrews (eds) *Aircrew Training and Assessment* (pp. 165–95). Mahwah, NJ: Lawrence Erlbaum Associates.

Klein, G.A. (2000b). Analysis of Situation Awareness from Critical Incident Reports. In: M.R. Endsley and D.J. Garland (eds) *Situation Awareness Analysis and Measurement* (pp. 51–71). Mahwah, NJ: Lawrence Erlbaum.

Klein, G.A. and Klinger, D. (1991a) Naturalistic Decision Making. *Human Systems IAC Gateway*, Volume XI, 16–19.

Klein, G.A. and Klinger, D. (1991b). Naturalistic Decision-Making. *CSERIAC Gateway*, 2, 1–4.

Klein, K.E. and Wegmann, H.M. (1980). *Significance of Circadian Rhythms in Aerospace Operations. NATO AGARDograph Number 247*. Neuilly-sur-Seine: NATO AGARD.

Klinect, J.R., Wilhelm, J.A. and Helmreich, R.L. (1999). Threat and Error Management: Data from Line Operational Safety Audits. In: R.S. Jensen (ed.) *Proceedings of the Tenth International Symposium on Aviation Safety* (pp. 683–8). Columbus, OH: Ohio State University.

Konz, S. and Johnson, S. (2000). *Work Design: Industrial Ergonomics* (5th edition). Scottsdale, AZ: Holcomb Hathaway.

Konz, S. and Johnson, S. (2007). *Work Design: Occupational Ergonomics* (7th edition). Scottsdale AZ: Holcomb Hathaway.

Kramer, A.F. (1991). Physiological Metrics of Mental Workload: A Review of Recent Progress. In: D.L. Damos (ed.) *Multiple-Task Performance*. (pp. 279–328). London: Taylor and Francis.

Krulak, D.C. (2004). Human Factors in Maintenance: Impact on Aircraft Mishap Frequency and Severity. *Aviation, Space and Environmental Medicine*, 75, 429–32.

Kryter, K.D. (1970). *The Effects of Noise on Man*. New York, NY: Academic Press.

Kryter, K.D. (1985). *The Effects of Noise on Man* (2nd edition). New York, NY: Academic Press.

Kryter, K.D. and Pearsons, K.S. (1966). Some Effects of Spectrum Content and Duration on Perceived Noise Level. *Journal of the Acoustical Society of America*, 39, 451–64.

Landauer, T.K. and Nachbar, D.W. (1985). Selection from Alphabetic and Numeric Menu Trees Using a Touch Screen: Breadth, Depth and Width. In: *Proceedings of Human Factors in Computing Systems* (pp. 73–8). New York, NY: ACM SIGCHI.

Lauber, J.K. and Kayten, P.J. (1988). Sleepiness, Circadian Dysrhythmia, and Fatigue in Transportation System Accidents. *Sleep*, 11, 503–12.

Learmount, D. (1994). Unanswered questions. *Flight International*, 12–18 October, 40–1.

Lee, A.T. (2007). *Flight Simulation: Virtual Environments in Aviation*. Aldershot: Ashgate.

Lee, A.T. and Bussolari, S.R. (1989). Flight Simulator Platform Motion and Air Transport Pilot Training. *Aviation, Space and Environmental Medicine*, 60, 136–40.

Lee, J.D. and Moray, N. (1992). Trust, Control Strategies and Allocation of Function in Human Machine Systems. *Ergonomics*, 22, 671–91.

Lee, J.D. and Moray, N. (1994). Trust, Self-Confidence, and Operators' Adaptation to Automation. *International Journal of Human-Computer Studies*, 40, 153–84.

Lehto, M.R. (1997). Decision Making. In: G. Salvendy (ed.) *Handbook of Human Factors and Ergonomics* (2nd edition) (pp. 1201–48). New York, NY: John Wiley.

Lehto, M.R. and Nah, F. (2006). Decision-making models and decision support. In: G. Salvendy (ed.) *Handbook of Human Factors and Ergonomics* (3rd edition). New York, NY: John Wiley and Sons.

Leveson, N. (2002). *A New Approach to System Safety Engineering*. Cambridge, MA: MIT Press.

Leveson, N. (2004). A New Accident Model for Engineering Safer Systems. *Safety Science*, 42, 237–70.

Leveson, N.G. (2009). *Engineering a Safer World: System Safety for the 21st Century (or Systems Thinking Applied to Safety)*. Boston, MA: Massachusetts Institute of Technology. (Available at: http://sunnyday.mit.edu/book2.pdf.)

Leveson, N.G., Allen, P. and Storey, M.A. (2002). The Analysis of a Friendly Fire Accident using a Systems Model of Accidents. In: *Proceedings of the 20th International System Safety Conference, Denver, Colorado, 5–9 August 2002*. Denver, CO: International System Safety Society.

Levin, M. (1996). Technology Transfer in Organisational Development: An Investigation into the Relationship between Technology Transfer and Organisational Change. *International Journal of Technology Management*, 2, 297–308.

Li, W-C. and Harris, D. (2005a). HFACS Analysis of ROC Air Force Aviation Accidents: reliability analysis and cross-cultural comparison. *International Journal of Applied Aviation Studies*, 5, 65–81.

Li, W-C. and Harris, D. (2005b). Aeronautical Decision-Making Mnemonics: Instructor-Pilots Evaluation of Five Alternative Methods. *Aviation, Space and Environmental Medicine*, 76, 1156–61.

Li, W-C. and Harris, D. (2005c). Where Safety Culture Meets National Culture: The How and Why of the China Airlines CI-611 Accident. *Human Factors and Aerospace Safety*, 5, 345–53.

Li, W-C. and Harris, D. (2006a). Pilot Error and Its Relationship with Higher Organizational Levels: HFACS Analysis of 523 Accidents. *Aviation, Space, and Environmental Medicine*, 77, 1056–61.

Li, W-C. and Harris, D. (2006b). Breaking the Chain: An Empirical Analysis of Accident Causal Factors by Human Factors Analysis and Classification System (HFACS). In: *Proceedings of International Society of Air Safety Investigators Seminar 2006*. 11–14 September. Sterling, VA: International Society of Air Safety Investigators.

Li, W-C. and Harris, D. (2006c). The Evaluation of the Decision Making Processes Employed by Cadet Pilots Following a Short Aeronautical Decision-Making Training Program. *International Journal of Applied Aviation Studies*, 6, 315–33.

Li, W-C. and Harris, D. (2008). The Evaluation of the Efficiency of a Short Aeronautical Decision-Making Training Program for Military Pilots. *International Journal of Aviation Psychology*, 18, 135–52.

Li, W-C., Harris, D. and Chen, S-Y. (2007). Eastern Minds in Western Cockpits: Meta-Analysis of Human Factors in Mishaps from Three Nations. *Aviation, Space and Environmental Medicine, 78*, 420–25.

Li, W-C., Harris, D. and Yu, C.S. (2008). Routes to Failure: Analysis of 41 Civil Aviation Accidents from the Republic of China Using the Human Factors Analysis and Classification System. *Accident Analysis and Prevention, 40*, 426–34.

Li, W-C., Young, H-T., Wang, T. and Harris, D. (2007). International Cooperation and Challenges: Understanding Cross-cultural Issues in the Processes of Accident Investigation. *International Society of Air Safety Investigators Forum, 40*, 16–21.

Lindseth, P.D., Vacek, J.L. and Lindseth, G.N. (2001). Urinalysis Drug Testing Within a Civilian Pilot Training Program: Did Attitudes Change During The 1990s? *Aviation, Space and Environmental Medicine, 72*, 647–51.

Lintern, G. (2004). An Open Systems Approach to Safety Management (Paper No. 2004-01-01). In: M.S. Patankar (ed.) *Proceedings of the First Safety Across High-Consequence Industries Conference, 2004 (March 9–10, Saint Louis University USA)*. Saint Louis, MO: Saint Louis University.

Lintern, G., Roscoe, S.N. and Sivier, J.E. (1990). Display Principles, Control Dynamics, and Environmental Factors in Pilot Training and Transfer. *Human Factors, 32*, 299–317.

Lipshitz, R. (1993). Converging Themes in the Study of Decision Making in Realistic Settings. In: G.A. Klein, J. Orasanu, R. Calderwood and C. Zsambok (eds) *Decision Making in Action: Models and Methods* (pp. 103–37). Norwood, NJ: Ablex Publishing Corporation.

Lipshitz, R. (1997). Naturalistic Decision Making Perspectives on Decision Errors. In: C.E. Zsambok and G. Klein (eds) *Naturalistic Decision Making* (pp. 151–61). Mahwah, NJ: Lawrence Erlbaum.

Little, L.F., Gaffney, I.C., Rosen, K.H. and Bender, M.M. (1990). Corporate Instability is Related to Airline Pilots' Stress Symptoms. *Aviation, Space and Environmental Medicine, 61*, 977–82.

Liu, D., Blickensderfer, E.L., Macchiarella, N.D. and Vincenzi, D.A. (2009). Transfer of Training. In: D.A. Vincenzi, J.A. Wise, M. Mouloua and P.A. Hancock (eds) *Human Factors in Simulation and Training* (pp. 49–60). Boca Raton, FL: CRC Press.

Liu, D., Macchiarella, N.D. and Vincenzi, D.A. (2009). Simulation Fidelity. In: D.A. Vincenzi, J.A. Wise, M. Mouloua and P.A. Hancock (eds) *Human Factors in Simulation and Training* (pp. 61–73). Boca Raton, FL: CRC Press.

Loewenthal, K.M., Eysenck, M., Harris, D., Lubitsh, G., Gorton, T. and Bicknell, H. (2000). Stress, Distress and Air Traffic Incidents: Job Dysfunction and Distress in Airline Pilots in Relation to Contextually-Assessed Stress. *Stress Medicine, 16*, 179–83.

Lofaro, R. (1992). *Workshop on Integrated Crew Resource Management (CRM) DOT/FAAIRD-92/5* (p. A-5). Washington, DC: US Department of Transportation, Federal Aviation Administration.

Logie, R.H. (1995). *Visuo-Spatial Working Memory*. Hove: Lawrence Erlbaum Associates.

Longridge, T., Bürki-Cohen, J., Go, T.H. and Kendra, A.J. (2001). Simulator Fidelity Considerations for Training and Evaluation of Today's Airline Pilots. In: R.S. Jensen (ed.) *Proceedings of the 11th International Symposium on Aviation Psychology*. Columbus, OH: Ohio State University.

Lovell, S.A. (1998). *Technology Transfer: Testing a Theoretical Model of the Human, Machine, Mission, Management and Medium Components*. Unpublished MSc thesis. Cranfield: College of Aeronautics, Cranfield University.

Lovesey, E. (2004). Anthropometrics for Flight Deck Design. In: D. Harris (ed.) *Human Factors for Civil Flight Deck Design* (pp. 199–210). Aldershot: Ashgate.

Lower, M.C. and Bagshaw, M. (1996). Noise Levels and Communications on the Flight Decks of Civil Aircraft. In: *Proceedings of InterNoise '96, Liverpool, UK* (pp. 349–52). Indianapolis, IA: Institute of Noise Control Engineering International.

Luxhøj, J.T. and Hadjimichael, M. (2006). A Hybrid Fuzzy-Belief Network (HFBN) for Modelling Aviation Safety Risk Factors. *Human Factors and Aerospace Safety*, 6, 191–216.

Lysaght, R.J., Hill, S.G., Dick, A., Plamondon, B.D., Linto, P.M., Wierwille, W.W., Zaklad, A.L., Bittner, A.C. and Wherry, R.J. (1989). *Operator Workload: Comprehensive Review and Evaluation of Operator Workload Methodologies (Technical Report 851)*. Alexandra, VA: Army Research Institute for Behavioral and Social Sciences.

Mackworth, N.H. (1944). *Notes on the Clock Test – a New Approach to the Study of Prolonged Visual Perception to Find the Optimum Length of Watch for Radar Operators*. MRC Applied Psychology Unit, Cambridge University Report. Cambridge.

Mackworth, N.H. (1948). The Breakdown of Vigilance During Prolonged Visual Search. *Quarterly Journal of Experimental Psychology*, 1, 6–21.

Macmillan, A.J.F. (1988). The Pressure Cabin. In: J. Ernsting and P. King (eds) *Aviation Medicine* (2nd edition) (pp. 112–26). London: Butterworths.

Maier, M.W. (1998). Architecting Principles for System of Systems. *Systems Engineering*, 1, 267–84.

Marshall, A., Stanton, N., Young, M., Salmon, P., Harris, D., Demagalski, J., Waldmann, T. and Dekker, S. (2003). *Development of the Human Error Template – a New Methodology for Assessing Design Induced Errors on Aircraft Flight Decks. Final Report of the ERRORPRED Project E! 1970 (August 2003)*. London: Department of Trade and Industry.

Martin-Emerson, R. and Wickens, C.D. (1997). Superimposition, Symbology, Visual Attention and the Head-Up Display. *Human Factors*, 39, 581–601.

Martinussen, M. (1996). Psychological Predictors as Predictors of Pilot Performance: A Meta-Analysis. *International Journal of Aviation Psychology*, 6, 1–20.

Martinussen, M. and Hunter, D. (2010). *Aviation Psychology and Human Factors*. Boca Raton, FL: CRC Press.

Marx, D. (2001). *Patient Safety and the 'Just Culture': A Primer for Health Care Executives. Report for National Heart, Lung and Blood Institute*. New York, NY: Columbia University.

Maschke, P. (1987). *Temperament Structure Scales (TSS) (Tech. Report ESA-TT-1069)*. Oberpfaffenhofen: European Space Agency.

Matthews, G. and Desmond, P.A. (1997). Underload and Performance Impairment: Evidence from Studies of Stress and Simulated Driving. In D. Harris (ed.) *Engineering Psychology and Cognitive Ergonomics* (Volume 1) (pp. 355–61). Aldershot: Ashgate.

Maurino, D.E. (1999). Crew Resource Management: A Time for Reflection. In: D.J. Garland, J.A. Wise and V.D. Hopkin (eds) *Handbook of Aviation Human Factors* (pp. 215–34). Mahwah, NJ: Lawrence Erlbaum.

Maxwell, E. and Harris, D. (1999). Drinking and Flying: A Structural Model. *Aviation, Space and Environmental Medicine*, 70, 117–23.

McDonald, N. (2008). Modelling the Human Role in Operational Systems. In: S. Martorell, C. Guedes Soares and J. Barnett (eds) *Safety, Reliability and Risk Analysis: Theory, Methods and Applications*. London: CRC Press.

McDonald, N. and Ryan, F. (1992). Constraints on the Development of Safety Culture: A Preliminary Analysis. *Irish Journal of Psychology*, 13, 271–81.

McDowell, E.D. (1978). *The Development and Evaluation of Objective Frequency Domain Based Pilot Performance Measures in ASUPT*. Air Force Office of Scientific Research, Bollings AFB, DC.

McLucas, J.L., Drinkwater, F.J. and Leaf, H.W. (1981). *Report of the President's Task Force on Aircraft Crew Complement*. Washington, DC: US Government Printing Office.

McRuer, D.T. (1980). Human Dynamics in Man–Machine Systems. *Automatica*, 16, 237–53.

McRuer, D.T. (1982). *Pitfalls and Progress in Advanced Flight Control Systems (AGARD CP-321)*. Neulliy-sur-Seine: AGARD/NATO.

McRuer, D.T. and Jex, H.R. (1967). A Review of Quasi-Linear Pilot Models. *IEEE Transactions on Human Factors in Electronics, HFE-8*, 3, 231–49.

McRuer, D.T and Krendel, E.S. (1974). *Mathematical Models of Human Pilot Behavior (AGARDograph 188)*. Neuilly-sur-Seine: AGARD/NATO

Means, B., Salas, E., Crandall, B. and Jacobs, T.O. (1993). Training Decision Makers for the Real World. In: G. Klein, J. Orasanu, R. Calderwood and C.E. Zsambok (eds) *Decision Making in Action: Models and Methods*. Norwood, NJ: Ablex Publishing Corporation.

Merritt, A. and Helmreich, R.L. (1995). Creating and Sustaining a Safety Culture: Some Practical Suggestions. *Proceedings of the Third Australian Aviation Safety Symposium; 1995 November 20–24 Sydney Australia*. Australian Association of Aviation Psychologists.

Merritt, A. and Maurino, D. (2004). Cross-Cultural Factors in Aviation Safety. In: M. Kaplan (ed.) *Advances in Human Performance and Cognitive Engineering Research* (pp. 147–81). San Diego, CA: Elsevier Science.

Miller, C.O. (1988). System Safety. In: E.L. Wiener and D.C. Nagel (eds) *Human Factors in Aviation* (pp. 53–80). San Diego, CA: Academic Press.

Miller, G.A. (1956). The Magical Number Seven, Plus or Minus Two: Some Limits on Our Capacity for Processing Information. *Psychological Review, 63, 81–97*.

Ministry of Defence (1996). *Human Factors for Designers of Equipment: Defence Standard 00-25 (part 7 – Issue 2): Visual Displays*. London: Author.

Ministry of Defence (2001). *Human Factors Integration (HFI): Practical Guidance for IPTs*. London: Author.

Mjøs, K. (2001) Communication and Operational Failures in the Cockpit. *Human Factors and Aerospace Safety*, 1, 323–40.

Molineros, J., Behringer, R. and Tam, C. (2004). Vision-Based Augmented Reality for Pilot Guidance in Airport Runways and Taxiways. In: *Proceedings of Third IEEE and ACM International Symposium on Mixed and Augmented Reality (ISMAR 2004)*. 2–5 November 2004, Washington, DC. Washington, DC: IEEE Computer Society.

Montgomery, H. (1983). Decision rules and the search for a dominance structure: towards a process model of decision making. In: P.C. Humphreys, O. Svenson and A. Vari (eds) *Analyzing and Aiding Decision Processes* (pp. 343–69). Amsterdam, The Netherlands: North Holland.

Montgomery, H. (1989). From Cognition to Action: The Search for Dominance in Decision Making. In: H. Montgomery and O. Svenson (eds) *Process and Structure in Human Decision Making* (pp. 23–49). Chichester, UK: John Wiley and Sons.

Moray, N. (1979). *Mental Workload: Its Theory and Measurement (NATO Conference Series 3: Human Factors)*. New York: Plenum.

Moray, N. (1988). Mental Workload Since 1979. *International Reviews of Ergonomics*, 2, 123–50.

Morley, F.J. and Harris, D. (2006). Ripples in a Pond: An Open System Model of the Evolution of Safety Culture. *International Journal of Occupational Safety and Ergonomics*, 12, 3–15.

Moroney, W.F. and Moroney, B.W. (1999). Flight Simulation. In: D.J. Garland, J.A. Wise and V.D. Hopkin (eds) *Handbook of Aviation Human Factors* (pp. 355–88). Mahwah, NJ: Lawrence Erlbaum.

Moshansky, Hon. V.P. (1992). *Commission of Inquiry into the Air Ontario Crash at Dryden, Ontario: Final Report*. Ottawa, Canada: Minister of Supply and Services.

Moskowitz, H. and Fiorentino, D. (2000). *A Review of the Literature on the Effects of Low Doses of Alcohol on Driving Related Skills (DOT/HS/809/028)*. Washington, DC: US Department of Transportation.

Muir, B.M. (1994). Trust in Automation: Part I. Theoretical Issues in the Study of Trust and Human Intervention in Automated Systems. *Ergonomics*, 37, 1905–22.

Mulder, L.J.M. (1992). Measurement and Analysis Methods of Heart Rate and Respiration for Use in Applied Environments. *Biological Psychology*, 34, 205–36.

Mulder, L.J.M., de Waard, D. and Brookhuis, K. (2005). Estimating Mental Effort Using Heart Rate and Heart Rate Variability. In: N.A. Stanton, A. Hedge, K. Brookhuis, E. Salas and H. Hendrick (eds) *Handbook of Human Factors and Ergonomics Methods* (pp. 20-1–20-8). Boca Raton, FL: CRC Press.

Mulder, M. (2003a). An Information-Centered Analysis of the Tunnel-In-The Sky Display, Part One: Straight Tunnel Trajectories. *International Journal of Aviation Psychology*, 13, 49–72.

Mulder, M. (2003b). An Information-Centered Analysis of the Tunnel-in-the-Sky Display, Part Two: Curved Tunnel Trajectories. *International Journal of Aviation Psychology*, 13, 131–51.

Mulder, M., Pleijsant, J-M., der Vaart, H. and van Wieringen, P. (2000). The Effects of Pictorial Detail on the Timing of the Landing Flare: Results of a Visual Simulation Experiment. *International Journal of Aviation Psychology*, 10, 291–315.

Mumford, E. (2003). *Redesigning Human Systems*. London: Information Science Publishing.

Murphy, M. (1980). Review of Aircraft Incidents. In: G.E. Cooper, M.D. White and J.K. Lauber (eds) *Resource Management on the Flightdeck: Proceedings of a NASA/Industry Workshop (NASA CP-2120)*. Moffett Field, CA: NASA-Ames Research Center.

Murray, S.R. (1997). Deliberate Decision Making by Aircraft Pilots: A Simple Reminder to Avoid Decision Making Under Panic. *International Journal of Aviation Psychology*, 7, 83–100.

Naef, W. (1995). Practical Applications of CRM Concepts: Swissair's Human Aspects Development Program (HAD). In: R.S. Jensen (ed.) *Proceedings of the Eighth International Symposium on Aviation Safety* (pp. 597–602). Columbus, OH: Ohio State University.

National Institute of Occupational Safety and Health (1999). *STRESS … at Work. (NIOSH Publication No. 99-101)*. Cincinnati, OH: National Institute of Occupational Safety and Health.

National Transportation Safety Board (1973). *Eastern Airlines L-1011, Miami, Florida, December 29, 1972 (NTSB-AAR-73-14)*. Washington, DC: National Transportation Safety Board.

National Transportation Safety Board (1983). *Accident Investigation of Human Performance Factors*. Washington, DC: National Transportation Safety Board Human Performance Group.

National Transportation Safety Board (1988). *Northwest Airlines, Inc., McDonnell Douglas DC-9-82, N312RC, Detroit Metropolitan Wayne County Airport, Romulus Michigan, August 16, 1987 (NTSB-AAR-88-05)*. Washington, DC: National Transportation Safety Board.

National Transportation Safety Board (2000). *Aircraft Accident Brief: Sunjet Aviation Learjet Model 35 (N47BA) at Aberdeen, South Dakota, October 25, 1999*. Washington, DC: NTSB.

National Transportation Safety Board (2004). *In-Flight Separation of Vertical Stabilizer American Airlines Flight 587 Airbus Industrie A300-605R, N14053 Belle Harbor, New York November 12, 2001 (NTSB/AAR-04/04)*. Washington, DC: National Transportation Safety Board.

Neisser, U. (1976). *Cognition and Reality: Principles and Implications of Cognitive Psychology*. San Francisco, CA: W.H. Freeman.

Newman, R.L. (1995). *Head-Up Displays: Designing the Way Ahead*. Aldershot: Avebury Aviation.

Nisbett, R.E. (2003). *The Geography of Thought*. London: Nicholas Brealey Publishing.

Noble, D. (1993). A Model to Support Development of Situation Assessment Aids. In: G.A. Klein, J. Orasanu, R. Calderwood and C.E. Zsambok (eds) *Decision Making in Action: Models and Methods* (pp. 287–305). Norwood, NJ: Ablex.

Norman, D.A. (1988). *The Psychology of Everyday Things*. New York NY: Basic Books.

Norman, D.A. (1993). *Things That Make Us Smart*. New York NY: Perseus Books.

Norman, D.A. and Bobrow, D.J. (1975). On Data-Limited and Resource-Limited Processes. *Cognitive Psychology*, 7, 44–64.

Noyes, J. (2001). *Designing for Humans*. Hove: Psychology Press.

Noyes, J.M., Starr, A.F. and Kazem, M.L.N. (2004). Warning System Design in Civil Aircraft. In: D. Harris (ed.) *Human Factors for Civil Flight Deck Design* (pp. 141–55). Aldershot: Ashgate.

O'Connor, P., Flin, R. and Fletcher, G. (2002). Techniques Used to Evaluate Crew Resource Management Training: A Literature Review. *Human Factors and Aerospace Safety*, 2, 217–33.

O'Connor, P., Flin, R., Fletcher, G. and Hemsley, P. (2002). Methods used to Evaluate the Effectiveness of Flightcrew CRM Training in the UK Aviation Industry. *Human Factors and Aerospace Safety*, 2, 235–55.

O'Donnell, R.D. and Eggemeier, F.T. (1986). Workload Assessment Methodology. In: K.R. Boff, L. Kaufman and J.P. Thomas (eds) *Handbook of Perception and Human Performance*. New York, NY: John Wiley.

O'Hare, D. (1992). The Artful Decision Maker: A Framework Model for Aeronautical Decision Making. *International Journal of Aviation Psychology*, 2, 175–92.

O'Hare, D. (2006). Cognitive Functions and Performance Shaping Factors in Aviation Accidents and Incidents. *International Journal of Aviation Psychology*, 16, 145–56.

O'Leary, M. (1995). Too Bad We Have To Have To Have Confidential Reporting Programmes! Some Observations on Safety Culture. In: N. McDonald and N. Johnston and R. Fuller (eds) *Applications of Psychology to the Aviation System*. Aldershot: Avebury Aviation.

O'Leary, M. and Pidgeon, N. (1995). Too Bad We Have To Have To Have Confidential Reporting Programmes. *Flight Deck*, 16, 11–16.

Oborne, D. (1987). *Ergonomics and Work* (2nd edition). London: John Wiley.

Office of Statistical Control (1945). *Army Air Forces Statistical Digest – World War II*. (Available from the Air Force Historical Research Agency: http://afhra.maxwell.af.mil/)

Oliver, J.G. (1990). *Improving Situational Awareness Through the Use of Intuitive Pictorial Displays (Tech. Rep. No. 901829)*. Warrendale, PA: SAE International.

Onken, R. (1994). Human-Centred Cockpit Design through the Knowledge-Based Cockpit Assistant System CASSY. In: *NATO DRG Panel 8 Workshop on Improving Function Allocation for Integrated System Design*. Soesterberg, NL: TNO Institute for Perception.

Onken, R. (1997). *Functional Development and Field Test of CASSY – A Knowledge Based Cockpit Assistant System, Knowledge-Based Functions in Aerospace Systems. AGARD Lecture Series 200*. Neuilly-sur-Seine: AGARD/NATO.

Orasanu, J. (1993). Decision Making in the Cockpit. In: E.L. Wiener, B.G. Kanki and R.L. Helmreich (eds) *Cockpit Resource Management* (pp. 137–72). San Diego, CA: Academic Press.

Orasanu, J. and Connolly, T. (1993). The Reinvention of Decision Making. In: G.A. Klein, J. Orasanu, R. Calderwood and C.E. Zsambok (eds) *Decision Making in Action: Models and Methods* (pp. 3–20). Norwood, NJ: Ablex.

Orasau, J. and Fischer, U. (1992). Distributed Cognition in the Cockpit: Linguistic Control of Shared Problem Solving. In: *Proceedings of the 14th Annual Conference of the Cognitive Science Society* (pp. 189-194). Hillsdale, NJ: Lawrence Erlbaum Associates.

Orasanu, J. and Fischer, U. (1997). Finding Decisions in Natural Environments: The View from the Cockpit. In: C.E. Zsambok and G.A. Klein (eds) *Naturalistic Decision Making* (pp. 343–58). Mahwah, NJ: Lawrence Erlbaum.

Orasanu, J. and Martin, L. (1998). *Errors in Aviation Decision Making: A Factor in Accidents and Incidents*. (Downloaded on 30 June 2005 from http://www.dcs.gla.ac.uk/~johnson/papers/seattle_hessd/judithlynne.pdf.)

Orasanu, J., Martin, L. and Davison, J. (2001). Cognitive and Contextual Factors in Aviation Accidents: Decision Errors. In: E. Salas and G. Klein (eds) *Linking Expertise and Naturalistic Decision Making*. Mahwah, NJ: Lawrence Erlbaum Associates.

Osgood, C.E. (1949). The Similarity Paradox in Human Learning and Resolution. *Psychological Review*, 56, 132–43.

Owen, G. and Funk, K. (1997). *Flight Deck Automation Issues: Incident Report Analysis*. (Available at: http://www.flightdeckautomation.com/incidentstudy/incident-analysis.aspx.) Corvallis, OR: Oregon State University, Department of Industrial and Manufacturing Engineering.

Paradies, M., Unger, L., Haas, P. and Terranova, M. (1993). *Development of the NRC's Human Performance Investigation Process (HPIP), Investigators Manual, Division of Systems Research Office of Nuclear Regulatory Commission, NUREG/CR-5455, SI-92-101*. Washington, USA: System Improvements Inc.

Parasuraman, R., Bahri, T., Deaton, J.E., Morrison, J.G., and Barnes, M. (1992). *Theory and Design of Adaptive Automation in Aviation Systems (Technical Report No. NAWCADWAR-92033-60)*. Warminster, PA: Naval Air Warfare Center, Aircraft Division.

Parasuraman, R., Molloy, R. and Singh, I.L. (1993). Performance Consequences of Automation-Induced 'Complacency'. *International Journal of Aviation Psychology*, 3, 1–23.

Parasuraman, R., Mouloua, M., Molloy, R. and Hilburn, B. (1996). Monitoring of Automated Systems. In: R. Parasuraman and M. Mouloua (eds) *Automation and Human Performance: Theory and Applications*. (pp. 91–115). Mahwah, NJ: Lawrence Erlbaum Associates.

Parasuraman, R., Sheridan, T.B. and Wickens, C.D. (2000). A Model for Types and Levels of Human Interaction with Automation. *IEEE Transactions on Systems, Man, and Cybernetics—Part A: Systems and Humans*, 30, 286–97.

Parasuraman, R., Sheridan, T.B., and Wickens, C.D. (2008). Situation Awareness, Mental Workload, and Trust in Automation: Viable, Empirically Supported Cognitive Engineering Constructs. *Journal of Cognitive Engineering and Decision Making*, 2, 141–61.

Pariés, J. and Amalberti, R. (1995). *Recent Trends in Aviation Safety: From Individuals to Organisational Resources Management Training*. Risøe National Laboratory Systems Analysis Department Technical Report, Series 1 (pp. 216–28). Roskilde, Denmark: Risøe National Laboratory.

Parsons, K. (1990). Human Response to Thermal Environments. In: J.R. Wilson and E.N. Corlett (eds) *Evaluation of Human Work*. London: Taylor and Francis.

Pashler, H.E. (1998). *The Psychology of Attention*. Cambridge, MA: MIT Press.

Pastoor, S., Schwarz, E. and Beldie, I.P. (1983). The Relative Suitability of Four Dot Matrix Sizes for Text Presentation on Color Television Screens. *Human Factors*, 25, 265–72.

Patrick, J. (1993). *Training: Research and Practice*. London: Academic Press.

Patterson, R.D. (1990). Auditory Warning Sounds in the Work Environment. *Philosophical Transactions of the Royal Society London*, B 327, 485–92.

Patterson, R. and Milroy, R. (1979). *Existing and Recommended Levels for Auditory Warnings on Civil Aircraft. Civil Aviation Authority Report (Contract Number 7D/S/0142)*. Cambridge: MRC Applied Psychology Unit.

Payne, J.W., Bettman, J.R. and Johnson, E.J. (1988). Adaptive Strategy Selection in Decision-Making. *Journal of Experimental Psychology: Learning, Memory and Cognition*, 14, 534–52.

Perrow, C. (1984). *Normal Accidents*. New York NY: Basic Books.

Perry, N., Stevens, C., and Howell, C. (2006). Warning Signal Design: The Effect of Modality and Iconicity on Recognition Speed and Accuracy. In: *Proceedings of ess2006: Evolving System Safety: The 7th International Symposium of the Australian Aviation Psychology Association*. 9–12 November 2006, Manly Pacific Hotel, Sydney, Australia.

Peryer, G., Noyes, J., Pleydell-Pearce, K. and Lieven, N. (2005). Auditory Alert Characteristics: A Survey of Pilot Views. *International Journal of Aviation Psychology*, 15, 233–50.

Peterson, A.P.G. and Gross, E. (1972). *Handbook of Noise Measurement*. Concord, MA: GenRad Inc.

Petrie, K.J., Powell, D., and Broadbent, E. (2004). Fatigue Self-Management Strategies and Reported Fatigue in International Pilots. *Ergonomics*, 47, 461–8.

Pfeiffer, M.G., Horey, J.D. and Burrimas, S.K. (1991). Transfer of Simulated Instrument Training to Instrument and Contact Flight. *International Journal of Aviation Psychology*, 1, 219–29.

Pheasant, S. and Haslegrave, C.M. (2005). *Bodyspace: Anthropometry, Ergonomics and the Design of Work* (3rd edition). London: CRC Press.

Pierce, B.J., Geri, G.A. and Hitt, J.M. (1998). *Display Collimation and the Perceived Size of Flight Simulator Imagery (AFRL-HE-AZ-TR-1998-0058)*. Meza AZ: United States Airforce Research Laboratory: Human Effectiveness Directorate.

Poulton, E.C. (1976). Continuous Noise Interferes with Work by Masking Auditory Feedback and Inner Speech. *Applied Ergonomics*, 7, 79–44.

Poulton, E.C. (1977). Continuous Intense Noise Masks Auditory Feedback and Inner Speech. *Psychological Bulletin*, 84, 977–1001.

Prince, C. and Salas, E. (1993). Training and Research for Teamwork in the Military Aircrew. In: E.L. Wiener, B.G. Kanki and R.L. Helmreich (eds) *Cockpit Resource Management* (pp. 337–66). San Diego, CA: Academic Press.

Prince, C. and Salas, E. (1997). Situation Assessment for Routine Flight and Decision Making. *International Journal of Cognitive Ergonomics*, 1, 315–24.

Prince, C., Oser, R., Salas, E. and Woodruff, W. (1993). Increasing Hits and Reducing Misses in CRM/LOS Scenarios: Guidelines for Simulator Scenario Development. *International Journal of Aviation Psychology*, 3, 69–82.

Proctor, P. (1997, December 1). Economic, Safety Gains Ignite HUD Sales. *Aviation Week and Space Technology*, 54–7.

Rainford, D.J. and Gradwell, D.P. (2006). *Ernsting's Aviation Medicine* (4th edition). London: Hodder Arnold.

Ramsey, J., Burford, C., Beshir, M. and Jensen, R. (1983). Effects of Workplace Thermal Conditions on Safe Work Behavior. *Journal of Safety Research*, 14, 105–14.

Ramsey, J.D. (1985). Ergonomic Factors in Task Analysis for Consumer Product Safety. *Journal of Occupational Accidents*, 7, 113–23.

Rasmussen, J. (1974). *The Human Data Processor as a System Component: Bits and Pieces of a Model (Report No. Risø-M-1722)*. Roskilde, Denmark: Danish Atomic Energy Commission.

Rasmussen, J. (1983). Skill, Rules and Knowledge: Signals, Signs and Symbols, and Other Distinctions in Human Performance Models. *IEEE Transactions on Systems, Man and Cybernetics*, 13, 257–66.

Rasmussen, J. (1986). *Information Processing and Human–Machine Interaction*. Amsterdam: Elsevier.

Rasmussen, J. and Svedung, I. (2000). *Proactive Risk Management in a Dynamic Society*. Karlstad, Sweden: Swedish Rescue Services Agency.

Rausand, M. (2004). *System Reliability Theory*. Hoboken, NJ: Wiley-Interscience.

Rawlins, N. (2000). *Evaluation of a Selection Model Used in the Recruitment of Trainee Airline Pilots*. Unpublished MSc Thesis. Cranfield: Human Factors Group, College of Aeronautics, Cranfield University.

Reason, J.T. (1987). Generic Error-Modelling System (GEMS): A Cognitive Framework for Locating Human Error Forms. In: J. Rasmussen, K. Duncan and J. Leplat (eds) *New Technology and Human Error*. London: Wiley.

Reason, J.T. (1990). *Human Error*. Cambridge: Cambridge University Press.

Reason, J.T. (1997). *Managing the Risks of Organizational Accidents*. Aldershot: Ashgate.

Reason, J.T. and Brand, J.J. (1975). *Motion Sickness*. London: Academic Press.

Ree, T.R. and Carretta, M.J. (1996). Central Role of 'g' in Military Pilot Selection. *International Journal of Aviation Psychology*, 6, 111–23.

Reid, G.B. and Nygren, T.E. (1988). The Subjective Workload Assessment Technique: A Scaling Procedure for Measuring Mental Workload. In P.A. Hancock and N. Meshkati (eds) *Human Mental Workload* (pp. 185–214). Amsterdam: North-Holland.

Reid, L.D. and Nahon, M.A. (1988). Response of Airline Pilots to Variations in Flight Simulator Motion Algorithms. *Journal of Aircraft*, 25, 639–46.

Reigeluth, C.M. (1999). The Elaboration Theory: Guidance for Scope and Sequence Decisions. In: C.M. Reigeluth (ed.) *Instructional-Design Theories and Models: A New Paradigm of Instructional Theory* (Volume II). Hillsdale, NJ: Lawrence Erlbaum Associates.

Reinach, S. and Viale, A. (2006). Application of a Human Error Framework to Conduct Train Accident/Incident Investigations. *Aviation, Space and Environmental Medicine*, 30, 396–406.

Reising, J.M., Ligget, K.K. and Munns, R.C. (1998). Controls, Displays and Workplace Design. In: D.J. Garland, J.A. Wise and V.D. Hopkin (eds) *Handbook of Aviation Human Factors* (pp. 327–54). Mahwah, NJ: Lawrence Erlbaum Associates.

Rigby, L.V. (1970). The Nature of Human Error. In: *Annual Technical Conference Transactions of the ASQC* (pp. 457–66). Milwaukee, WI: American Society for Quality Control.

Rignér, J. and Dekker, S.W.A. (2000). Sharing the Burden of Flight Deck Automation Training. *International Journal of Aviation Psychology*, 10, 317–26.

Robertson, I.T. and Smith, M. (2001). Personnel Selection. *Journal of Occupational and Organizational Psychology*, 74, 441–72.

Rogers, Y. (1997). *A Brief Introduction to Distributed Cognition*. (Retrieved from: http://mcs.open.ac.uk/yr258/papers/dcog/dcog-brief-intro.pdf 14 December 2009.)

Rolfe, J.M. and Staples, K.J. (1986). *Flight Simulation*. Cambridge: Cambridge University Press.

Romahn, S. and Schaefer, D. (1997). Error Analysis as a Means for User Interface Evaluation: A Comparison of Graphically Interactive and Traditional FMS User Interfaces. In: D. Harris (ed.) *Engineering Psychology and Cognitive Ergonomics* (Volume 1) (pp. 101–9). Aldershot: Ashgate.

Roscoe, A.H. (1984). Assessing Pilot Workload in Flight. Flight Test Techniques. In: *Proceedings of the Advisory Group for Aerospace Research and Development (Conference Proceedings No. 373): Flight Test Techniques*. Neuilly-sur-Seine: AGARD/NATO.

Roscoe, A.H. (1992). Assessing Pilot Workload. Why Measure Heart Rate, HRV and Respiration? *Biological Psychology*, 34, 259–87.

Roscoe, A.H. (1993). Heart Rate as Psychophysiological Measure for In-Flight Workload Assessment. *Ergonomics*, 36, 1055–62.

Roscoe, S.N. (1968). Airborne Displays for Flight and Navigation. *Human Factors*, 10, 321–32.

Roscoe, S.N. (1971). Incremental Transfer Effectiveness. *Human Factors*, 13, 561–7.

Roscoe, S.N. (1997). *The Adolescence of Engineering Psychology*. Santa Monica, CA: Human Factors and Ergonomics Society

Roscoe, S.N. and Williges, B.H. (1980) Measurement of Transfer of Training. In: S.N. Roscoe (ed.) *Aviation Psychology*. Ames, IA: Iowa State University Press.

Rosekind, M.R., Gander, P.H., Miller, D.L., Gregory, K.B., Smith, R.M., Weldon, K.J., Co, E.L., McNally, K.L. and Lebacqz, J.V. (1994). Fatigue in Operational Settings: Examples from the Aviation Environment. *Human Factors*, 36, 327–38.

Rosenman, R.H., Freidman, M. and Straus, R. (1964). A Predictive Study of CHD. *Journal of the Medical Association*, 189, 15–22.

Ross, L.R. and Ross, S.M. (1990). Pilots' Knowledge of Blood Alcohol Levels and The 0.04% Blood Alcohol Concentration Rule. *Aviation, Space and Environmental Medicine*, 62, 412–17.

Ross, L.R. and Ross, S.M. (1992). Professional Pilots' Evaluation of the Extent Causes and Reduction of Alcohol Use in Aviation. *Aviation, Space and Environmental Medicine*, 63, 805–8.

Ross, S.M. and Ross, L.R. (1995). Professional Pilots. Views of Alcohol Use in Aviation and the Effectiveness of Employee Assistance Programs. *International Journal of Aviation Psychology, 5,* 199–213.

Rotter, J.B. (1954). *Social Learning and Clinical Psychology.* New York: Prentice-Hall.

Rotter, J.B. (1966). Generalized Expectancies of Internal Versus External Control of Reinforcements. *Psychological Monographs, 80,* all.

Rowlands, G.F. (1977). *The Transmission of Vertical Vibration to the Heads and Shoulders of Seated Men.* Royal Aircraft Establishment Technical Report 77088. London: Ministry of Defence.

Ruffell Smith, H.P. (1979). *A Simulator Study of the Interaction of Pilot Workload with Errors, Vigilance, and Decisions. NASA Technical Memorandum 78482.* Moffett Field, CA: NASA Ames Research Center.

Russel, J.C. and Davis, A.W. (1995). *Alcohol Rehabilitation of Airline Pilots (FAA/DOT Report no. DOT/FAA-AM-85-12).* Washington, DC: Federal Aviation Administration.

RyanAir (2009). Full Year Results, 2009. (Available from: http://www.ryanair.com/doc/investor/2009/q4_2009_doc.pdf Accessed 17 August 2010.)

Salas, E., Wilson, K.A. and Edens, E. (2009). *Crew Resource Management: Critical Essays.* Aldershot: Ashgate.

Salazar, G.J. (2007). *Fatigue in Aviation: Medical Facts for Pilots (Publication no. OK-07-193).* Oklahoma City, OK: FAA Civil Aerospace Medical Institute

Salvendy, G. and Karwowski, W. (eds) (2010). *Advances in Physical Ergonomics.* Boca Raton, FL: CRC Press.

Sanders, A.F. (1983). Towards a Model of Stress and Human Performance. *Acta Psychologica, 53,* 67-97.

Sanders, M.S. and McCormick, E.J. (1987). *Human Factors in Engineering and Design* (6th edition). New York, NY: McGraw Hill.

Sanders, M.S. and McCormick, E.J. (1993). *Human Factors in Engineering and Design* (7th edition). New York, NY: McGraw Hill.

Sarter, N.B. and Woods, D.D. (1992). Pilot Interaction with Cockpit Automation I: Operational Experience with the Flight Management System. *International Journal of Aviation Psychology, 2,* 303–21.

Sarter, N.B. and Woods, D.D. (1994). Pilot Interaction with Cockpit Automation: An Experimental Study of Pilot's Mental Model and Awareness of the Flight Management System. *International Journal of Aviation Psychology, 4,* 1–28.

Sauter, S.L., Murphy, L.R. and Hurrell, J.J. (1990). Prevention of Work-Related Psychological Disorders. *American Psychologist, 45,* 1146–53.

Savage, L.J. (1954). *The Foundations of Statistics.* New York NY: Wiley.

Scarborough, A. and Pounds, J. (2001). Retrospective Human Factors Analysis of ATC Operational Errors. In: R.S. Jensen (ed.) *Proceedings of 11th International Symposium on Aviation Psychology.* Columbus, OH: Ohio State University.

Scerbo, M.W. (2001). Adaptive Automation. In: W. Karwowski (ed.) *International Encyclopedia of Ergonomics and Human Factors* (pp. 1077–9). London: Taylor and Francis.

Scerbo, M.W., Freeman, F.G., Mikulka, P.J., Parasuraman, R., Di Nocero, F. and Prinzel, L.J. (2001). *The Efficacy of Psychophysiological Measures for Implementing Adaptive Technology (NASA TP-2001-211018).* Hampton, VA: NASA Langley Research Center.

Schein, E.H. (1992). *Organizational Culture and Leadership* (2nd edition). San Francisco, CA: Jossey-Bass.

Schiewe, A. (1995). CRM Training and Transfer: the 'Behavioral Business Card' as an Example for the Transition of Plans into Actual Behavior. In: N. Johnston, R. Fuller, and N. McDonald (eds) *Aviation Psychology: Training and Selection* (pp. 38–44). Aldershot: Avebury Aviation.

Seamster, T.L., Redding, R.E. and Kaempf, G.L. (1998). *Applied Task Analysis in Aviation*. Aldershot: Ashgate.

Shackel, B. (1986). Ergonomics in Design for Usability. In: M.D. Harrison and A.F. Monk (eds) *People and Computers: Designing for Usability. Proceedings of HCI 86* (pp. 44–64). Cambridge: Cambridge University Press.

Shackel, B. (1991). Usability – Context, Framework, Design and Evaluation. In: B. Shackel and S. Richardson (eds) *Human Factors for Informatics Usability* (pp. 21–38). Cambridge: Cambridge University Press.

Shappell, S.A., Detwiler, C., Holcomb, K., Hackworth, C., Boquet, A. and Wiegmann, D.A. (2007). Human Error and Commercial Aviation Accidents: An Analysis Using the Human Factors Analysis and Classification System. *Human Factors*, 49, 227–42.

Shappell, S.A. and Wiegmann, D.A. (2001). Applying Reason: the Human Factors Analysis and Classification System (HFACS). *Human Factors and Aerospace Safety*, 1, 59–86.

Shappell, S.A. and Wiegmann, D.A. (2004). HFACS Analysis of Military and Civilian Aviation Accidents: A North American Comparison. In: *Proceedings of International Society of Air Safety Investigators conference* (Australia, Queensland, 2–8 November 2004). Sterling, VA: International Society of Air Safety Investigators.

Sharit, J. and Salvendy, G. (1982). Occupational Stress: Review and Appraisal. *Human Factors*, 24, 129–62.

Sharp, G.R. and Anton, D.J. (1988). Toxic Gases and Vapours in Flight. In: J. Ernsting and P. King (eds) *Aviation Medicine* (2nd edition) (pp. 127–35). London; Butterworths.

Sheridan, T.B. (1987). Supervisory Control. In: G. Salvendy (ed.) *Handbook of Human Factors* (1st edition) (pp. 1243–68). New York: John Wiley and Sons.

Sheridan, T.B. (1992). *Telerobotics, Automation, and Human Supervisory Control*. Cambridge, MA: MIT Press.

Sheridan, T.B. (1999). Automation. In: P.A Hancock (ed.) *The Human Face of Technology: Selected Presidential Addresses of the Human Factors and Ergonomics Society*. San Diego, CA: Human Factors and Ergonomics Society.

Sheridan, T.B. and Verplank, W.L. (1978). *Human and Computer Control of Undersea Teleoperators (Technical Report, Engineering Psychology Program)*. Cambridge, MA: Department of Mechanical Engineering, MIT.

Shiffrin, R.M. and Schneider, W. (1977). Controlled and Automatic Human Information Processing II: Perceptual Learning, Automatic Attending, and A General Theory. *Psychological Review*, 84, 127–90.

Shneiderman, B. (1992). *Designing the User Interface* (2nd edition). Reading, MA: Addison-Wesley.

Shneiderman, B. and Plaisant, C. (2009). *Designing the User Interface* (5th edition). Reading MA. Addison-Wesley.

Simard, M. and Marchand, A. (1997). Workgroups' Propensity to Comply with Safety Rules: The Influence of Micro-Macro Organizational Factors. *Ergonomics*, 40, 172–88.

Simon, H.A. (1955). A Behavioral Model of Rational Choice. *Quarterly Journal of Economics*, 69, 87–103.

Singer, G. (1999). Filling the Gaps in the Human Factors Certification Net. In: S. Dekker and E. Hollnagel (eds) *Coping With Computers in the Cockpit* (pp. 87–108). Aldershot: Ashgate.

Sirevaag, E., Kramer, A.F., De Jong, R. and Mecklinger, A. (1988). A Psychophysiological Analysis of Multi-Task Processing Demands. *Psychophysiology*, 25, 482.

Skinner, J. and Ree, M.J. (1987). *Air Force Officer Qualifying Test (AFOQT): Item and Factor Analysis of Form O (Technical Report No. AFHRL-TR-86-68)*. Brooks AFB, TX: Air Force Human Resources Laboratory.

Sklar, A.E. and Sarter, N.B. (1999). Good Vibrations: Tactile Feedback in Support of Attention Allocation and Human-Automation Coordination in Event-Driven Domains. *Human Factors*, 41, 543–52.

Sloane, H.R. and Cooper, C.L. (1984). Health-Related Lifestyle Habits in Commercial Airline Pilots. *British Journal of Aviation Medicine*, 2, 32–41.

Sloane, S.J. and Cooper, C.L. (1986). *Pilots Under Stress*. New York, NY: Routledge.

Smith, F.J. and Harris, D. (1994). The Effects of Low Blood Alcohol Levels on Pilots' Prioritisation of Tasks During a Radio Navigation Task. *International Journal of Aviation Psychology*, 4, 349–58.

Smith, K. and Hancock, P.A. (1995). Situation Awareness is Adaptive, Externally Directed Consciousness. *Human Factors*, 37, 137–48.

Soeters, J.L. and Boer, P.C. (2000). Culture and Flight Safety in Military Aviation. *International Journal of Aviation Psychology*, 10, 111–33.

Sperry, W. (1978). Aircraft and Airport Noise Control. In: D. Lipscomb and A. Taylor (eds) *Noise Control: Handbook of Principles and Practices*. New York, NY: Van Nostrand Reinhold.

Stanney, K.M., Maxey, J. and Salvendy, G. (1997). Socially Centred Design. In: G. Salvendy (ed.) *Handbook of Human Factors and Ergonomics* (2nd edition) (pp. 637–56). New York, NY: John Wiley and Sons.

Stanton, N.A. and Edworthy, J. (1999). *Human Factors in Auditory Warnings*. Aldershot: Ashgate.

Stanton, N.A., Baber, C. and Harris, D. (2008). *Event Analysis of Systemic Teamwork: Analysis of Command and Control Activities*. Aldershot: Ashgate.

Stanton, N.A., Baber, C., Walker, G.H., Salmon, P. and Green, D. (2004). Toward a Theory of Agent-Based Systemic Situational Awareness. In: D.A. Vincenzi, M. Mouloua and P.A. Hancock (eds) *Proceedings of the Second Human Performance, Situation Awareness and Automation Conference (HPSAAII)*, Daytona Beach, FL, 22–25 March.

Stanton, N.A., Harris, D., Salmon, P., Demagalski, J.M., Marshall, A., Young, M.S., Dekker, S.W.A. and Waldmann, T. (2006). Predicting Design Induced Pilot Error using HET (Human Error Template) – A New Formal Human Error Identification Method for Flight Decks. *The Aeronautical Journal*, 110, 107–15.

Stanton, N.A., Salmon, P.M., Walker, G.H., Baber, C. and Jenkins, D.P. (2006). *Human Factors Methods: A Practical Guide for Engineering and Design*. Aldershot: Ashgate.

Stanton, N.A., Stewart, R.J., Baber, C., Harris, D., Houghton, R.J., Mcmaster, R., Salmon, P., Hoyle, G., Walker, G., Young, M.S., Linsell, M. and Dymott, R. (2006). Distributed Situational Awareness in Dynamic Systems: Theoretical Development and Application of an Ergonomics Methodology. *Ergonomics*, 49, 1288–311.

Stead, G. (1995). Qantas Pilot Selection Procedures; Past to Present. In: N. Johnston, R. Fuller and N. McDonald (eds) *Aviation Psychology: Selection and Training* (pp. 176–81). Aldershot: Avebury.

Stephens, D. (1979). Developments in Ride Quality Criteria. *Noise Control Engineering*, 12, 6–14.

Steurs, M., Mulder, M. and van Paassen, M.M. (2004). A Cybernetic Approach to Assess Flight Simulator Fidelity. In: *Proceedings of the AIAA Modelling and Simulation Technologies Conference and Exhibition (Providence RI, 16–19 August 2004). Paper AIAA-2004-5442*. Reston, VA: American Institute for Aeronautics and Astronautics.

Stewart, R.J., Stanton, N.A., Harris, D., Baber, C., Salmon, P., Mock, M., Tatlock, K., Wells, L. and Kay, A. (2007). Distributed Situational Awareness in an Airborne Warning and Control Aircraft: Application of a Novel Ergonomics Methodology. *Cognition, Technology and Work*, 10, 221–9.

Stewart, S. (2005). Crew Utilisation Project: Aircrew Fatigue, Scheduling and Performance. *Association of Aerospace Medical Examiners Conference.* 9 April 2005. Stafford: UK.

Stokes, A.F. and Kite, K. (1997). *Flight Stress.* Aldershot: Ashgate.

Stolzer, A.J., Halford, C.D. and Goglia, J.J. (2009). *Safety Management Systems in Aviation.* Aldershot, UK: Ashgate.

Stott, J.R.R. (1980). Mechanical Resonance of the Eyeball. In: *Proceedings of the Human Response to Vibration Seminar.* University of Wales, Swansea. October, 1980.

Stott, J.R.R. (1988). Vibration. In: J. Ernsting and P. King (eds) *Aviation Medicine* (2nd edition) (pp. 185–99). London: Butterworths.

Strauch, B. (2004). *Investigating Human Error: Incidents, Accidents, and Complex Systems.* Aldershot: Ashgate.

Stuart, G.W., McAnally, K.I. and Meehan, J.W. (2001). Head-Up Displays and Visual Attention: Integrating Data and Theory. *Human Factors and Aerospace Safety,* 1, 103–25.

Suirez, J., Barborek, S., Nikore, V. and Hunter, D.R. (1994). *Current Trends in Pilot Hiring and Selection (AAM-240-94-1).* Washington, DC: Federal Aviation Administration.

Swain, A.D. and Guttmann, H.E. (1983). *Handbook of Human Reliability Analysis with Emphasis on Nuclear Powerplant Operations.* Sandia National Laboratories, NUREG/CR-1278). Washington, DC: US Nuclear Regulatory Commission.

Szasz, S., Gauci, J., Zammit-Mangion, D., Zammit, B., Sammut, A. and Harris, D. (2008). Design of Experiment for the Pilot Evaluation of an Airborne Runway Incursion Alerting System. *International Council of the Aeronautical Sciences Conference, 2008.* 14–19 September 2008, Anchorage, Alaska.

Tan, H.Z., Barbagli, F., Salisbury, K., Ho, C. and Spence, C. (2006). Force-Direction Discrimination is not Influenced by Reference Force Direction. *Haptics-e,* 4, 1–6. http://www.haptics-e.org/

Taylor, R.M. (1990). Situational Awareness Rating Technique (SART): The Development of a Tool for Aircrew System Design. In: *Situational Awareness in Aerospace Operations (AGARD-CP-478).* Neuilly-sur-Seine: NATO/AGARD.

Taylor, R.M. and Selcon, S.J. (1991). Subjective Measurement of Situational Awareness. In: *Proceedings of the 11th Congress of the International Ergonomics Association.* London: Taylor and Francis.

Tenney, Y.J., Rogers, W.H. and Pew, R.W. (1998). Pilot Opinions on Cockpit Automation Issues. *International Journal of Aviation Psychology,* 8, 103–20.

Terry, D., Callan, V.J. and Sartori, G. (1996). Employee Adjustment to an Organizational Merger: Stress, Coping and Intergroup Differences. *Stress Medicine,* 12, 105–22.

The Myers and Briggs Foundation (n.d.). *MBTI Basics.* (Retrieved from: http://www.myersbriggs.org/my-mbti-personality-type/mbti-basics 27 April 2011.)

Thomas, M.J.W. (2003). Improving Organisational Safety through the Integrated Evaluation of Operational and Training Performance: An Adaptation on the Line Operations Safety Audit (LOSA) Methodology. *Human Factors and Aerospace Safety,* 3, 25–45.

Tiller, A.R. (2002). *The Measure of Man and Woman: Human Factors in Design.* New York NY: John Wiley.

Transport Canada (2004). *Safety Management Systems for Small Aviation Operations - A Practical Guide to Implementation (TP 14135).* Ottawa, ON: Transport Canada.

Treisman, A. (1960). Contextual Cues in Selective Listening. *Quarterly Journal of Experimental Psychology,* 12, 242–8.

Turner, G.M. and Birch, C. (1996). *UK Military Aircrew Anthropometric Growth Trends. DERA Report PLSD/CHS/CR96071/1.* Farnborough: DERA.

Tvarnyas, A.P., Thompson, W.T. and Constable, S.H. (2006). Human Factors in Remotely Piloted Aircraft Operations: HFACS Analysis of 221 Mishaps Over 10 Years. *Aviation, Space, and Environmental Medicine, 77*, 724–32.

Tversky, A. (1972). Elimination by Aspects: A Theory of Choice. *Psychological Review, 79*, 281–99.

Tversky, A. and Kahneman, D. (1974). Judgement under Uncertainty: Heuristics and Biases. *Science,* 185, 1124–31.

Ulfvengren, P. (2003). Associability: A Comparison of Sounds in a Cognitive Approach to Auditory Alert Design. *Human Factors and Aerospace Safety, 3*, 313–31.

Ulleberg, T., Sten, T., Rosness, R., Ingstadt, O., Hudson, P., Harris, D. and Elwell, R.S. (1990). *Helicopter Safety Study (Main Report).* Trondheim, Norway: SINTEF (Stiftelsen For Industriell og Forskining Ved Norges Teknisk) Division of Safety and Reliability.

US Department of Defense (n.d.). *Department of Defense Human Factors Analysis and Classification System: A mishap investigation and data analysis tool.* (Available from: http://www.uscg.mil/safety/docs/ergo_hfacs/hfacs.pdf Accessed 20 August 2010.)

US Department of Defense (1993). *Military Standard: System Safety Program Requirements (MIL-STD-882C).* Washington, DC: US Department of Defense.

US Department of Transportation (1999). Aviation Rulemaking Advisory Committee; Transport Airplane and Engine: Notice of new task assignment for the Aviation Rulemaking Advisory Committee (ARAC). *Federal Register* 22 July 1999, 64, 140.

US Department of Transportation (1974). *Federal Aviation Regulations (Part 25 – Airworthiness Standards).* Revised 1 January 2009. US Department Of Transportation, Washington, DC,

van Avermaete, J.A.G. (1998). *NOTECHS: Non-Technical Skill Evaluation in JAR-FCL. NLR-TP-98518.* Amsterdam: National Aerospace Laboratory (NLR).

van Veen, H.A.H.C. and van Erp, J.B.F. (2001). Tactile Information Presentation in the Cockpit. In: S. Brewster and R. Murray-Smith (eds) *Haptic Human–Computer Interaction (LNCS 2058)* (pp. 174–81). Berlin: Springer-Verlag.

Veillette, P. (1995). Differences in Aircrew Manual Skills in Automated and Conventional Flight Decks. *Transport Research Record,* 1480, 43–50.

Veltman, J.A. and Gaillard, A.W.K. (1993). Indices of Mental Workload in a Complex Task Environment. *Neuropsychobiology,* 28, 72–5.

Ververs, P.M. and Wickens, C.D. (1998). Head-Up Displays: Effects of Clutter, Display Intensity and Display Location on Pilot Performance. *International Journal of Aviation Psychology, 8*, 377–403.

Vidulich, M.A. (1988). The Cognitive Psychology of Subjective Mental Workload. In: P.A. Hancock and N. Meshkati (eds) *Human Mental Workload.* Amsterdam: North-Holland.

Vidulich, M.A. and Wickens, C.D. (1986). Causes and Dissociation Between Subjective Workload Measures and Performance. *Applied Ergonomics, 17*, 291–6.

Vidulich, M.A., Crabtree, M.S. and McCoy, A.L. (1993). Developing Subjective and Objective Metrics of Situation Awareness. In: R.S. Jensen and D. Neumeister (eds) *Proceedings of the 7th International Symposium on Aviation Psychology.* Columbus, OH: Ohio State University.

Vincente, K.J. (1999). *Cognitive Work Analysis.* Mahwah, NJ: Lawrence Erlbaum Associates.

Vincenzi, D.A., Wise, J.A. Mouloua, M. and Hancock, P.A. (2008). *Human Factors in Simulation and Training.* Boca Raton, FL: CRC Press.

Vingilis, E.R. and Salutin, L. (1980). A Prevention Programme for Drinking and Driving. *Accident Analysis and Prevention, 12*, 267–74.

von Berthalanfry L. (1956). General Systems Theory: General Systems. *Yearbook of the Society of General Systems Theory,* 1, 1–10.

Waag, W.L. (1981). *Training Effectiveness of Visual and Motion Simulation (Report AFHRL-TR-79-72).* Brooks Air Force Base, Texas: Airforce Human Resource Laboratory.

Waag, W.L. and Houck, M.R. (1994). Tools for Assessing Situation Awareness in an Operational Fighter Environment. *Aviation, Space and Environmental Medicine*, 65, A13–A19.

Wainwright, W. (1987). Flight Test Evaluation of Crew Workload. In: A.H. Roscoe (ed.) *The Practical Assessment of Pilot Workload. AGARDograph No. 282* (pp. 60–68). Neuilly-sur-Seine: AGARD/NATO.

Warr, P. and Wall, T. (1975). *Work and Well-Being*. Harmondsworth: Penguin.

Weintraub, D.J. and Ensing, M.J. (1992). *Human Factors Issues in Head-Up Display Design: The Book of HUD (SOAR CSERIAC State of the Art Report 92-2)*. Dayton, OH: Crew System Ergonomics Information Analysis Center, Wright Patterson AFB.

Weintraub, D.J., Haines, R.F. and Randle, R.J. (1984). The Utility of Head-Up Displays: Eye-Focus vs. Decision Times. In: *Proceedings of the 28th Annual Meeting of the Human Factors Society* (pp. 529–33). Santa Monica, CA: Human Factors Society.

Weiser, M. (1991). The Computer for the Twenty-First Century. *Scientific American* (September), 94–110.

Wernimont, P.F. and Campbell, J.P. (1968). Signs, Samples and Criteria. *Journal of Applied Psychology*, 52, 372–6.

Wertheim, A.H. (1998). Working in a Moving Environment. *Ergonomics*, 41, 1845–58.

Westrum, R. and Adamski, A.J. (1999). Organizational Factors Associated with Safety and Mission Success in Aviation Environments. In: D.J. Garland, J.A. Wise and V.D. Hopkin (eds) *Handbook of Aviation Human Factors* (pp. 67–104). Mahwah, NJ: Lawrence Erlbaum.

Wheale, J. (1983). Crew Coordination on the Flight Deck of Commercial Transport Aircraft. In: *Proceedings of the Flight Operations Symposium* (pp. 19–20). Dublin: Irish Airline Pilots Association/Aer Lingus.

Wichman, H. and Ball, J. (1983). Locus of Control, Self-Serving Biases and Attitudes Towards Safety in General Aviation Pilots. *Aviation, Space and Environmental Medicine*, 53, 6507–10.

Wickens, C.D. (1984). Processing Resources in Attention. In: R. Parasuraman and R. Davies (eds) *Varieties of Attention* (pp. 63–101). New York, NY: Academic Press.

Wickens, C.D. (2002). Multiple Resources and Performance Prediction. *Theoretical Issues in Ergonomics Science*, 2, 150–77.

Wickens, C.D. (2005). Multiple Resource Time Sharing Models. In: N. Stanton, A. Hedge, K. Brookhuis, E. Salas and H. Hendrick (eds) *Handbook of Human Factors and Ergonomics Methods* (pp. 40-1–40-6). Boca Raton, FL: CRC Press.

Wickens, C.D. (2008). Multiple Resources and Mental Workload. *Human Factors*, 50, 449–55.

Wickens, C.D. and Carswell, C.M. (1995). The Proximity Compatibility Principle: Its Psychological Foundation and Its Relevance to Display Design. *Human Factors*, 37, 473–94.

Wickens, C.D. and Flach, J.M. (1988). Information Processing. In: E.L. Wiener and D.C. Nagel (eds) *Human Factors in Aviation* (pp. 111–55). San Diego, CA: Academic Press.

Wickens, C.D. and Hollands, J.G. (1999). *Engineering Psychology and Human Performance* (3rd edition). Upper Saddle River, NJ: Prentice Hall.

Wickens, C.D. and Long, J. (1995). Object vs Space-Based Models of Visual Attention: Implications for the Design of Head-Up Displays. *Journal of Experimental Psychology: Applied*, 1, 179–93.

Wickens, C.D., Lee, J.D., Liu, Y. and Gordon Becker, S.E. (2004). *An Introduction to Human Factors Engineering* (2nd edition). Upper Saddle River, NJ: Pearson Education.

Wickens, C.D., Ververs, P.M. and Fadden, S. (2004). Head-Up Displays. In: D. Harris (ed.) *Human Factors for Civil Flight Deck Design* (pp. 103–40). Aldershot: Ashgate.

Widders, R. and Harris, D. (1997). Pilots' Knowledge of the Relationship between Alcohol Consumption and Levels of Blood Alcohol Concentration. *Aviation, Space and Environmental Medicine*, 68, 531–7.

Wiegmann, D.A. and Shappell, S.A. (1997). Human Factors Analysis of Postaccident Data: Applying Theoretical Taxonomies of Human Error. *International Journal of Aviation Psychology*, 7, 67–81.

Wiegmann, D.A. and Shappell, S.A. (2001a). Human Error Analysis of Commercial Aviation Accidents: Application of the Human Factors Analysis and Classification System. *Aviation, Space and Environmental Medicine*, 72, 1006–16.

Wiegmann, D.A. and Shappell, S.A. (2001b). Applying the Human Factors Analysis and Classification System to the Analysis of Commercial Aviation Accident Data. In: R.S. Jensen (ed.) *Proceedings of 11th International Symposium on Aviation Psychology*. Columbus, OH: Ohio State University.

Wiegmann, D.A. and Shappell, S.A. (2003). *A Human Error Approach to Aviation Accident Analysis: The Human Factors Analysis and Classification System*. Aldershot: Ashgate.

Wiegmann, D.A., Zhang, H., von Thaden, T., Sharma, G. and Mitchell, A. (2002). *Safety Culture: A Review (Technical Report ARL-02-3/FAA-02-2)*. University of Illinois at Urbana-Champaign. Aviation Research Lab, Institute of Aviation.

Wiener, E.L. (1981). Complacency: Is the term useful for air safety? In: *Proceedings of the 26th Corporate Aviation Safety Seminar* (pp. 116–25). Denver: Flight Safety Foundation, Inc.

Wiener, E.L. (1989). *Human Factors of Advanced Technology ('Glass Cockpit') Transport Aircraft. NASA Contractor Report 177528*. Moffett Field, CA: NASA Ames Research Center.

Wiener, E.L. and Curry, R.E. (1980). Flight Deck Automation: Promises and Problems. *Ergonomics*, 23, 995–1011.

Wierwille, W.W. (1988). Important Remaining Issues in Workload Estimation. In: P.A. Hancock and N. Meshkati (eds) *Human Mental Workload*. Amsterdam: North Holland.

Wierwille, W.W. and Casali, J.G. (1983). A Validated Rating Scale for Global Mental Workload Measurement Application. In: *Proceedings of the Human Factors Society 27th Annual Meeting* (pp. 129–33). Santa Monica, CA: Human Factors Society.

Wierwille, W.W. and Connor, S.A. (1983). Evaluation of 20 Workload Measures Using a Psychomotor Task in a Using Base Aircraft Simulator. *Human Factors*, 25, 1–16.

Wightman, D.C. and Lintern, G. (1985). Part-Task Training for Tracking and Manual Control. *Human Factors*, 27, 267–83.

Wightman, D.C. and Sistrunk, F. (1987). Part-Task Training Strategies in Simulated Carrier Landing Final-Approach Training. *Human Factors*, 29, 245–54.

Williges, R.C., Williges, B.H. and Fainter, R.G. (1988). Software Interfaces for Aviation Systems. In: E.L. Wiener and D.C. Nagel (eds) *Human Factors in Aviation*. San Diego, CA: Academic Press.

Wilson, G.F. (1992). Applied Use of Cardiac and Respiration Measures: Practical Considerations and Precautions. *Biological Psychology*, 34, 163–78.

Wilson, G.F. and Eggemeier, F.T. (1991). Psychophysiological Assessment of Workload in Multi-Task Environments. In: D.L. Damos (ed.) *Multiple-Task Performance* (pp. 329–60). London: Taylor and Francis.

Winget, C.M., DeRoshia, C.W., Markley, C.L. and Holley, D.C. (1984). A Review of Human Physiological and Performance Changes Associated with Desynchronosis of Biological Rhythms. *Aviation, Space and Environmental Medicine*, 55, 1085–96.

Wohl, J.G. (1981). Force Management Decision Requirements for Air Force Tactical Command and Control. *IEEE Transactions on Systems, Mans, and Cybernetics*, SMC-11, 618–39.

Wolf, J.D. (1978). *Crew Workload Assessment: Development of a Measure of Operator Workload (AFFDL-TR-78-165)*. Wright-Patterson Air Force Base, OH: Air Force Flight Dynamics Laboratory.

Wood, S.J. (2004). *Flight Crew Reliance on Automation. (CAA Paper No 2004/10)*. London, UK: Civil Aviation Authority.

Wood, S.J. (2009). *A Study to Develop a New Methodology for Automation Training for a Modern Highly Automated Transport Aircraft. Report to the Civil Aviation Authority.* Flight Operations Research Centre of Excellence: Cranfield University.

Wood, S.J. and Huddlestone, J. (2006). Requirements for a Revised Syllabus to Train Pilots in the Use of Advanced Flight Deck Automation. *Human Factors and Aerospace Safety*, 6, 359–70.

Woods, D.D., Johannesen, L.J., Cook, R.I. and Sarter, N.B. (1994). Behind Human Error: Cognitive Systems, Computers and Hindsight. *CSERIAC Gateway*, 5. Columbus, OH: CSERIAC.

Woodworth, R.S. (1899). The Accuracy of Voluntary Movement. *Psychological Review* (Monograph Supplement), 3, 1–119.

Yeh, Y-Y. and Wickens, C.D. (1988). The Dissociation of Subjective Measures of Mental Workload and Performance. *Human Factors*, 30, 111–20.

Yerkes, R.M. and Dodson, J.D. (1908). The Relation of Strength of Stimulus to Rapidity of Habit-Formation. *Journal of Comparative Neurology and Psychology*, 18, 459–82.

Yesavage, J.A. and Leirer, Von O. (1986). Hangover Effects on Aircraft Pilots 14 Hours after Alcohol Ingestion: A Preliminary Report. *American Journal of Psychiatry*, 143, 1546–50.

Young, M.S. and Stanton, N.A. (2001). Size Matters: The Role of Attentional Capacity in Explaining the Effects of Mental Underload on Performance. In: D. Harris (ed.) *Engineering Psychology and Cognitive Ergonomics* (Volume 5) (pp. 357–64). Aldershot, UK: Ashgate.

Young, M.S. and Stanton, N.A. (2002). Malleable Attentional Resources Theory: A New Explanation for the Effects of Mental Underload on Performance. *Human Factors*, 44, 365–75.

Zakay, D. and Tsal, J. (1993). The Impact of Using Forced Decision-Making Strategies on Post-Decisional Confidence, *Journal of Behavioral Decision Making*, 6, 53–68.

Zambarbieri, D., Schmid, R., Magenes, G. and Prablanc, C. (1982). Saccadic Responses Evoked by Presentation of Visual and Auditory Targets. *Experimental Brain Research*, 47, 417–27.

Zeier, H. (1979). Concurrent Physiological Activity of Driver and Passenger When Driving With and Without Automatic Transmission in Heavy City Traffic. *Ergonomics*, 22, 799–810.

Zeitlin, L.R. (1995). Estimates of Driver Mental Workload: A Long-Term Field Trial of Two Subsidiary Tasks. *Human Factors*, 37, 611–21.

Zsambok, C.E. (1997). Naturalistic Decision Making: Where Are We Now? In: C.E. Zsambok and G.A. Klein (eds) *Naturalistic Decision Making*. Mahwah, NJ: Lawrence Erlbaum Associates.

Zsambok, C.E. and Klein, G.A. (1997). *Naturalistic Decision Making*. Mahwah, NJ: Lawrence Erlbaum Associates.

Index